Lecture Notes in Mathematics

Edited by A. Dold and B. Eckmann

1323

Douglas R. Anderson
Hans J. Munkholm

Boundedly Controlled Topology

Foundations of Algebraic Topology and Simple Homotopy Theory

Springer-Verlag
Berlin Heidelberg New York London Paris Tokyo

Authors

Douglas R. Anderson
Department of Mathematics, Syracuse University
Syracuse, N.Y. 13210, USA
(Email: Anderson at SUVM.Bitnet)

Hans J. Munkholm
Institut for Matematik og Datalogi, Odense Universitet
DK-5230 Odense M, Denmark
(Email: HJM at DKDOU01.Bitnet)

Mathematics Subject Classification (1980): 57Q10, 57Q05, 57R80, 55N99, 55P99, 18F25, 18D99, 18B40, 18A25

ISBN 3-540-19397-9 Springer-Verlag Berlin Heidelberg New York
ISBN 0-387-19397-9 Springer-Verlag New York Berlin Heidelberg

Library of Congress Cataloging-in-Publication Data. Anderson, Douglas R. (Douglas Ross), 1940-. Boundedly controlled topology foundations of algebraic topology and simple homotopy theory / Douglas R. Anderson, Hans J. Munkholm. p. cm–(Lecture notes in mathematics; 1323) Bibliography: p. Includes index. ISBN 0-387-19397-9 (U.S.) 1. Piecewise linear topology. 2. Homptopy theory. 3. Complexes. 4. Categories (Mathematics) I. Munkholm, Hans J. (Hans Jørgen), 1940-. II. Title. III. Series: Lecture notes in mathematics (Springer-Verlag); 1323. QA3.L28 no. 1323 [QA613.4] 510 s–dc 19 [514'.22] 88-16054

Printing and binding: Druckhaus Beltz, Hemsbach/Bergstr.
2146/3140-543210

INTRODUCTION

This book gives an exposition of the foundations of "boundedly controlled topology" and, in particular, of the algebraic topology and simple homotopy theory of spaces boundedly controlled over a space Z. The theory presented here is one aspect of the general theory of the topology of spaces "parametrized over a space Z".

The first named author's interest in this area arose around 1973 in conjunction with the early stages of his joint work with Hsiang [AH1], [AH2], and [AH3]. The geometric essence of that work was an investigation of whether a naturally arising non-compact h-cobordism W with lower boundary of the form $M\times\mathbb{R}^k$, where M was compact, had a product structure. At the time, the only tools available for studying such h-cobordisms were the work of Siebenmann [Si] on proper h-cobordisms and the "torus trick" of Kirby [EK], which converted the non-compact problem to a compact one. The latter approach was chosen since it made some needed compactification arguments possible.

Anderson, however, observed that there was a map $p: W \longrightarrow \mathbb{R}^k$ with $p|M\times\mathbb{R}^k = \text{proj}$ and that the needed compactification arguments could still be made if there were a "bounded homeomorphism" $h: M\times\mathbb{R}^k\times I \longrightarrow W$ with $h|M\times\mathbb{R}^k\times 0 = \text{id}$. In this formulation, $M\times\mathbb{R}^k\times I$ was parametrized over $\mathbb{R}^k$ via the projection and h was bounded in the sense of Kirby [Ki], i.e. it was required that there be a number $N>0$ such that $|ph(x)-\text{proj}(x)| < N$ for each $x \in M\times\mathbb{R}^k\times I$, where absolute value denotes the box metric. This observation quickly leads to the construction of the category of the category $\mathcal{T}op^c/\mathbb{R}^k$, whose objects are pairs (X,p) with $p: X \longrightarrow \mathbb{R}^k$ and whose morphisms $f: (X,p) \longrightarrow (Y,q)$ are the maps $f: X \longrightarrow Y$ bounded in the above sense, and to the problem of classifying h-cobordisms in $\mathcal{T}op^c/\mathbb{R}^k$.

More generally, if Z is a metric space with metric ρ, let $\mathcal{T}op^c/Z$ be the category whose objects are pairs (X,p) where p: X ⟶ Z and whose morphisms f: (X,p) ⟶ (Y,q) are the continuous maps f: X ⟶ Y for which the function $\rho(qf(\ ),p(\ ))\colon X \longrightarrow \mathbb{R}$ is bounded. We call $\mathcal{T}op^c/Z$ the *category of boundedly controlled spaces over* Z *and boundedly controlled maps* and one of the objectives of this book is to investigate the following problem:

Problem: Under what circumstances does a boundedly controlled h-cobordism have a boundedly controlled product structure?

The reader should note that we have chosen the term "boundedly controlled" to describe our work to contrast it with the study of topology with ε-control described briefly below. In the sequel to simplify terminology, we shall use "bc" as an abbreviation for "boundedly controlled".

For several years, there was no attempt made to attack this problem, even in the case when $Z = \mathbb{R}^k$ with the box metric. However, the work of Pedersen [Pe1] and [Pe2] revived interest in this problem. Although the first of these papers, stimulated while Anderson was visiting Munkholm and Pedersen at Odense University during the 1981-82 academic year, gave a new and purely algebraic description of the lower K-theoretic groups $K_{-k}(R)$ of a ring R, it was conceived from the viewpoint of bc topology. The second of these papers used the first in the expected way to define, via the algebra of chain complexes, lower K-theoretic valued invariants of a "finitely dominated" bc CW complex over $\mathbb{R}^k$ and of a bc homotopy equivalence between "finite" bc CW complexes. Since it was quite clear that these algebraically defined invariants were not complete, except in some very special cases, no attempt was made in [Pe2] to supply the geometric arguments needed to study the completeness problem.

The motivation for the work presented in this book was the desire to provide the geometric analysis of bc CW complexes that was missing in [Pe2]. The earliest stages of this work, shaped by Pedersen's work and described in [An], concentrated on simple homotopy theory with the emphasis on some very benign special cases. They revealed, however, that a full understanding of boundedly controlled simple homotopy theory required the development of an algebraic milieu, richer than Pedersen's, in which analogues of such standard algebraic invariants of spaces as homotopy groups, homology, and chain groups could be defined and such basic results as Hurewicz and Whitehead Theorems could be established.

This phase of our work was completed, for the most part, in 1984, was described in [AM1], and is contained in chapters I and II of this book. In addition, the algebraic study of simple homotopy theory requires an adequate algebra of "free modules". Since Pedersen's algebra is not sufficient, the development of the needed algebra occupies Chapter IV.

On the geometric side, we wanted simple homotopy theory to develop along classical lines beginning with a some notion of elementary expansions and collapses. This theory, of course, had to be sharp enough for torsions to be combinatorial invariants and to provide the setting in which a boundedly controlled s-cobordism theorem could be formulated and proven. In fact, it took us some time to find definitions of elementary expansions and collapses that met these conditions without being redundant. Our earlier attempts at doing this are described in [AM2] and [AM3]. This material has been reworked and simplified for this book and is presented in Chapter III; while Chapter V contains the proof of the combinatorial invariance of torsions and Chapter VI contains the boundedly controlled s-cobordism theorem. The first third of Chapter VII is devoted to a careful comparison of our results with those of Pedersen.

We should remark that Pedersen [Pe3] also investigated and proved an s-cobordism theorem in a case in which his algebra is sufficient. The results of Chapter VII show that this result is a special case of ours.

No summary of developments in the area of topology parametrized over a space Z is complete if it fails to mention the early work by Connell and Hollingsworth, [CH], Ferry's basic results on ε-maps, [Fe], and the vigorous development of topology with ε-control initiated by Quinn [Qn1] and [Qn2] and Chapman [Ch]. In Chapman's and Quinn's approach one starts with a map $p: M \longrightarrow Z$ from a manifold, say, to a metric space Z, fixes a number $\varepsilon>0$, and tries to answer some question about M to within the given ε. For example, if $\partial M = \partial_- M \amalg \partial_+ M$ and $\delta>0$, then Chapman [Ch] and Quinn [Qn1] define what it means to say that $(M,\partial_- M)$ is a (δ,h)-cobordism and solve the following problem:

Problem: Let $\varepsilon>0$. Under what circumstances does a (δ,h)-cobordism have an ε product structure?

In the last few years, Farrell and Jones [FJ1] and [FJ2] and Steinberger and West [SW1] and [SW2] have found significant refinements and extensions of the Chapman-Quinn theory.

Since boundedly controlled topology and topology with ε-control present two aspects of topology parametrized over a space Z, it is natural to ask for a precise comparison between them. Although there is strong evidence suggesting that such a comparison is possible, this book does not explore that question. The authors expect to return to it in a future paper.

We refer the reader who would like a more detailed summary of the results contained in this book to our paper [AM2] with the caveat that the notion of an elementary expansion used there has been simplified in this account with a resulting simplification in the description of the algebraically defined Whitehead group. The reader should see section 2 of Chapter III and section 3 of Chapter IV for these simplifications.

We now give a more detailed summary of the contents of each chapter of this book.

Chapter I opens with the definition of the concept that the category $\mathfrak{B}$ is a "category with endomorphism". This is one of the central algebraic ideas in this book. Section 1 also contains the most important examples of categories with endomorphism; while section 2 considers properties of a category of fractions $\mathcal{B}(\Sigma^{-1})$ that arises naturally from a category with endomorphism. All the algebraic invariants of bc spaces considered in this book lie in such categories. Many of the remaining results in chapter I are largely technical in nature and are collected there to provide the background and references needed in the rest of the book.

After introducing the idea of a "boundedness control structure on Z" and the categories $\mathcal{T}op^c/Z$, $\mathcal{CW}^c/Z$, and $\mathcal{CW}^c_f/Z$, generalizing the category $\mathcal{T}op^c/Z$ described above and its analogues for CW complexes and finite CW complexes, respectively, Chapter II describes the elementary algebraic topology of boundedly controlled spaces. This is done by first introducing the homology and homotopy of such spaces and then establishing such useful results as the Absolute Hurewicz Theorem (Theorem 6.1), the Relative Hurewicz Theorem (Theorem 7.1), sharpened versions of the Hurewicz theorems in an appropriate "simply connected" setting (Theorems 8.1 and 8.2), two versions of the Whitehead Theorem (Corollaries 10.4 and 10.5) and a Cellular Approximation Theorem (Corollary 10.3).

Let R be a reasonable ring and $\mathcal{B}$ be a category with endomorphism. Chapter IV is devoted to the study of "free R$\mathcal{B}$ modules". Section 1 defines free R$\mathcal{B}$ modules and their "bases" and establishes the basic mapping properties of these modules. In section 2, $\mathcal{B}$ is specialized to have the form $\mathcal{P}$G that arises most naturally in geometric applications. The notions, motivated by geometry, of "boundedly finitely generated free R$\mathcal{P}$G modules" and "boundedly finitely generated basis" are introduced and compared. In particular, it is shown that the isomorphism classification of boundedly finitely generated free R$\mathcal{P}$G modules is the same as the isomorphism classification of boundedly finitely generated bases. The groups $K_1(R\mathcal{P}G)$ and Wh(R$\mathcal{P}$G), that are the boundedly controlled analogues of the usual K_1 and Whitehead groups, are explored in sections 3 and 4. The algebraically defined Whitehead group of a bc space (X,p), $Wh\mathbb{Z}\mathcal{G}_1(X,p)$, is included in this discussion as a special case. Finally, in section 5, Stiefel-Whitney classes are defined and shown to give rise to involutions on $K_1(R\mathcal{G})$ and $Wh(R\mathcal{G})$.

The main geometric developments in this book are contained in Chapters III, V, and VI. Let (X,p) be in $\mathcal{CW}^c_{\mathcal{L}}/Z$. The main thrust of Chapter III is to define the group of boundedly controlled simple homotopy types on (X,p) and to establish its main properties. This group, denoted $Wh^c(X,p)$, is defined geometrically using appropriate notions of expansions and collapses and is shown to be a homotopy functor.

Chapter V has two main objectives: to show that there is a natural isomorphism τ: $Wh^c(X,p) \longrightarrow Wh\mathbb{Z}\mathcal{G}_1(X,p)$ and to prove that bc Whitehead torsion is a combinatorial invariant. The first two sections contain the material needed to define the function τ; while section 3 contains the proof that τ is an isomorphism. The combinatorial invariance of bc torsions is proven in section 4.

Chapter VI is devoted to developing the results needed to state and prove a Boundedly Controlled s-Cobordism Theorem (Theorem 6.1) and a companion Realization Theorem (Theorem 7.1). It also contains a Duality Theorem (Theorem 7.3) for the torsion of a bc PL h-cobordism. Along the way, Chapter VI investigates bc collars (section 1), bc handles and handle decompositions (sections 2 and 3), techniques for modifying handle decompositions (section 4), and the geometric connectivity of manifolds (section 5).

Chapter VII opens with the comparison of our results with those of Pedersen [Pe1], [Pe2], and [Pe3] and their extensions by Pedersen and Weibel [PW1] and [PW2]. Sections 2 and 3 of that chapter are based on the observation that there is a forgetful homomorphism $\square$: $Wh^c(X,p) \longrightarrow \mathcal{S}(X)$ where $\mathcal{S}(X)$ is Siebenmann's group of proper simple homotopy types on X [Si] and investigates that homomorphism in the cases when (X,p) is $(M\times\mathbb{R}^k,\text{proj})$ for M a closed PL manifold and $k=1,2$. The last section calculates $Wh^c(X,p)$ in a special, but quite interesting, case.

We have used the chapter-section-number system for referring to the results in this book. Thus IV.2.6 refers to the sixth result in section 2 of Chapter IV. The chapter number is omitted from references to results within the same chapter.

In addition to the people acknowledged above, we would like to thank A.A. Ranicki with whom we had a very stimulating conversation during the formative stages of this work; L.G. Lewis with whom we had many discussions about category theory that helped clarify the material in Chapter I; Ib Madsen whose questions pushed us to get the definitions of the geometric and algebraic Whitehead groups "right" and who raised the question of the comparison of our work with Siebenmann's; and E.P. Martin from whose love of words the terms "fragment" and "fragmentation" were drawn.

We would also like to thank all of the agencies that have provided financial support for this research. Among them are the Scientific Affairs Division of NATO whose grant number 670/84 supported both authors, the National Science Foundation whose grants numbered MCS-8201776 and DMS-8504320 supported the first named author, and Statens Naturvidenskabelige Forskningsråd (Denmark) which supported the second named author under grant number 81 4056.

Several institutions have extended their kind hospitality to one or both of the authors during the period this book was in preparation. In particular, Anderson would like to thank the Mathematics Department at the University of Connecticut for providing a refuge in which the first phases of this work were conceived, Odense University's Department of Mathematics and Computer Science for its welcoming atmosphere during several visits, and the Institute for Advanced Study.

Munkholm extends his thanks to the Department of Mathematics of the

University of Maryland where much of his share in the work was done during the year of 1983/84, to the Department of Mathematics at Syracuse University for hospitality on several occasions, and to the Institute for Advanced Study.

Both authors would like to extend special thanks to Lisbeth Larsen for her patient, efficient, and beautiful typing of this manuscript.

Finally Professor Anderson would like to thank his wife and family for the sacrifices they have made and their great forbearance during the several long absences from home his work on this project necessitated.

Similar sacrifices made by the Munkholm family were, at least partially, offset by the fact that the project took them all to the US for a total of 15 months which they all thoroughly enjoyed.

TABLE OF CONTENTS

V. THE ISOMORPHISM BETWEEN THE GEOMETRIC AND ALGEBRAIC WHITEHEAD GROUPS 176
1. Whitehead torsion for based chain complexes 176
2. Based $\mathbb{Z}\mathcal{G}$G modules in bc algebraic topology 183
3. The isomorphism τ: $Wh^c(X,p) \longrightarrow Wh\mathbb{Z}\mathcal{G}_1(X,p)$ 194
4. Combinatorial invariance of Whitehead torsion 205
5. Bumpy homotopy equivalences 210

VI. BOUNDEDLY CONTROLLED MANIFOLDS AND THE s-COBORDISM THEOREM 212
1. Boundedly controlled collars 214
2. Attaching boundedly controlled handles and disks 217
3. Handle decompositions exist 223
4. Modifying handle presentations 228
5. Geometric connectivity of manifolds 237
6. The boundedly controlled s-cobordism theorem 251
7. The realization and duality theorems 256

VII. TOWARD COMPUTATIONS 260
1. A comparison of $Wh^c(Y\times\mathbb{R}^k,proj)$ with $\tilde{K}_{1-k}\mathbb{Z}\pi_1(Y)$ 261
2. A comparison of $Wh^c(N\times\mathbb{R}^k,proj)$ with $\mathcal{S}(N\times\mathbb{R}^k)$ 269
3. The algebra of germs at infinity 275
4. An example over $\mathbb{R}$ 286

BIBLIOGRAPHY 301

INDEX 305

CHAPTER I
CATEGORY THEORETIC FOUNDATIONS

The main objective of this chapter is to develop the category theoretic tools needed for the study of spaces in the category $\mathcal{Top}^c/Z$. In particular, we construct the categories in which the main algebraic topological invariants of spaces in $\mathcal{Top}^c/Z$ lie and explore their properties. These categories are all categories of fractions $\mathfrak{B}(\Sigma^{-1})$ where $\mathfrak{B}$ is some category that is naturally associated with (X,p) and Σ is a distinguished family of morphisms in $\mathfrak{B}$. It turns out that the most basic structural features of $\mathfrak{B}$ (in fact, those that determine Σ) are inherited from similar features of the category $\mathcal{P}$ obtained from the boundedness control structure (P,C) on Z by regarding the poset P as a category $\mathcal{P}$ and $C\colon \mathcal{P} \longrightarrow \mathcal{P}$ as a functor. Thus these features are abstracted into the notion of *category with endomorphism* which is introduced in section 1. In that section, we also observe that the endomorphism structure on a category with endomorphism $\mathfrak{B}$ determines the family of morphisms Σ, that Σ is left calculable, and we give a description of the morphisms in $\mathfrak{B}(\Sigma^{-1})$ that is used repeatedly in this book.

In section 2, we study the properties of the category $\mathfrak{B}(\Sigma^{-1})$. One of the most useful results is Corollary 2.5 which describes circumstances under which $\mathfrak{B}(\Sigma^{-1})$ is abelian. This result is subsequently used to show that many of the categories in which our algebraic invariants lie are abelian. This in turn makes it possible, for example, to use most standard exact sequence arguments without change.

In section 3, we examine functors between categories with endomorphism. The point is that different algebraic invariants of the same space or similar algebraic invariants of different spaces may lie in different categories and it is often necessary to compare such invariants by introducing appropriate functors between the categories in which they lie. For example, if X and Y are spaces and $f: X \longrightarrow Y$ is a map, then the higher homotopy groups of X (respectively, Y) are in the category of $\mathbb{Z}\pi_1(X)$-modules (respectively, of $\mathbb{Z}\pi_1(Y)$-modules) and these are compared by means of a functor $f^*: \mathbb{Z}\pi_1(Y)\text{-modules} \longrightarrow \mathbb{Z}\pi_1(X)\text{-modules}$ induced by the homomorphism $f_*: \pi_1(X) \longrightarrow \pi_1(Y)$.

In section 4, we study functor categories as categories with endomorphism. This study arises since it is functor categories that arise in constructing algebraic invariants for spaces in $\mathcal{T}op^c/\mathrm{Z}$. In particular, in Corollary 4.3 we decide when the category of fractions $\mathcal{C}^{\mathcal{B}}(\Sigma^{-1})$ obtained from the functor category $\mathcal{C}^{\mathcal{B}}$ is abelian and in Lemmas 4.5 to 4.7 we collect some results needed to make exact sequence arguments in $\mathcal{C}^{\mathcal{B}}(\Sigma^{-1})$ in some of the cases when this category is not abelian.

In section 5, we study functors between functor categories regarded as categories with endomorphism. This study continues the study begun in section 3 by applying those results and specializing them.

Finally in section 6, we give some conditions which are sufficient to assure that a functor $F: \mathcal{B}_1 \longrightarrow \mathcal{B}_2$ induces an equivalence between the categories $\mathcal{C}^{\mathcal{B}_1}(\Sigma^{-1}) \longrightarrow \mathcal{C}^{\mathcal{B}_2}(\Sigma^{-1})$.

It would probably help the reader understand the motivation behind the ideas developed in this chapter to proceed directly to Chapter II, read the first couple of sections of it, and then return to this chapter. In fact, it would probably be advisable to read Chapters I and II in conjunction with each other, focusing on Chapter II but referring to this chapter as it becomes necessary.

1. Categories with endomorphism.

This section describes the categories that appear repeatedly in this book and establishes their basic properties. We begin by introducing some conventions and notations.

Following [Sc, p. 17], we shall let U be a universe containing the set of natural numbers. By the word *category* we shall mean a U-category; that is, the objects of $\mathscr{C}$ form a U-*class* (which we denote by $|\mathscr{C}|$), and for any pair of objects, C_0 and C_1, the morphisms from C_0 to C_1 form a U-*set*. In the sequel, we will often suppress mention of the universe U and will write simply *set* instead of the more precise U-*set*.

A *category with endomorphism* is a triple $(\mathscr{B},C,\tau)$ where $\mathscr{B}$ is a category, $C: \mathscr{B} \longrightarrow \mathscr{B}$ is a functor, and $\tau: I_{\mathscr{B}} \longrightarrow C$ is a natural transformation from the identity functor on $\mathscr{B}$ to C such that $C\tau = \tau C$ (that is, for all $B \in |\mathscr{B}|$, $C(\tau_B) = \tau_{CB}$ where τ_D denotes τ evaluated at D).

Example 1.1 Let (Z,ρ) be a metric space where the metric $\rho: Z\times Z \longrightarrow \mathbb{R}_+$ is *proper* in the sense that for every $z \in Z$, $\rho(z, \ ): Z \longrightarrow \mathbb{R}_+$ is a proper map and suppose

(A) If $B(z,r) \subseteq B(u,s)$, then $B(z,r+1) \subseteq B(u,s+1)$.

Let $P = \{B(z,n) \mid z\in Z, n\in\mathbb{Z}_+\}$ be the family of closed balls in Z. Define $C(B(z,n)) = B(z,n+1)$. If we regard P as a category $\mathscr{P}$ with objects the balls $B(z,n)$ and morphisms inclusions, it is easy to see that $C: \mathscr{P} \longrightarrow \mathscr{P}$ is a functor. Furthermore, the inclusions $B(z,n) \subseteq B(z,n+1)$ define a natural transformation $\tau: I_{\mathscr{P}} \longrightarrow C$ such that $C\tau = \tau C$. Thus $(\mathscr{P},C,\tau)$ is a category with endomorphism.

More generally, we have the following example:

Example 1.2 Let (P,C) be a pair consisting of a partially ordered set P and an order preserving function $C: P \longrightarrow P$ such that for all $a \in P$, $a \leq C(a)$. Then we can regard (P,C) as a category with endomorphism $(\mathcal{P},C,\tau)$ by first regarding P as a category $\mathcal{P}$ with objects the elements of P and with $\mathcal{P}(a,b)$ a single morphism if $a \leq b$ and empty otherwise. Then $C: \mathcal{P} \longrightarrow \mathcal{P}$ becomes a functor and the unique morphisms $\tau_a: a \longrightarrow C(a)$ $(a\in|\mathcal{P}|)$ define the needed natural transformation $\tau: I_{\mathcal{P}} \longrightarrow C$.

The examples described in 1.2 will reappear in II.1 in the definition of a boundedness control structure on a space Z.

Example 1.3 Let $(\mathcal{B},C,\tau)$ be a category with endomorphism (e.g. $(\mathcal{P},C,\tau)$ as in 1.2) and $G: \mathcal{B} \longrightarrow \mathcal{G}\mathit{poid}$ be a functor into the category $\mathcal{G}\mathit{poid}$ of U-small groupoids. (Recall that a groupoid is a category in which every morphism is invertible.) We define a new category $\mathcal{B}G$ as follows: $|\mathcal{B}G| = \{(x,A) \mid A\in|\mathcal{B}| \text{ and } x\in|G(A)|\}$. A morphism is a pair $(\omega,i): (x,A) \longrightarrow (y,B)$ where $i \in \mathcal{B}(A,B)$ and induces the functor $G(i): G(A) \longrightarrow G(B)$ and where $\omega: G(i)(x) \longrightarrow y$ is a morphism in G(B). Composition in $\mathcal{B}G$ is defined by setting

$$(\omega_2,i_2)(\omega_1,i_1) = (\omega_2 \circ G(i_2)(\omega_1),i_2 i_1).$$

It is easily verified that $\mathcal{B}G$ is a category. We insist that our functors G take values which are U-*small* categories because in that way $\mathcal{B}G$ is U-small whenever $\mathcal{B}$ itself is U-small. This is important because we want to be able to apply 5.4 below to categories of the form $\mathcal{B}G$.

If we now define $\bar{C}: \mathcal{B}G \longrightarrow \mathcal{B}G$ and $\bar{\tau}: I_{\mathcal{B}G} \longrightarrow \bar{C}$ by setting $\bar{C}(x,A) = (G(\tau_A)(x),CA)$, $\bar{C}(\omega,i) = (G(\tau_B(\omega)),C(i))$, and $\bar{\tau}_{(x,A)} = (1,\tau_A)$, then $(\mathcal{B}G,\bar{C},\bar{\tau})$ is a category with endomorphism. We call $(\bar{C},\bar{\tau})$ the *endomorphism structure induced* by $(\mathcal{B},C,\tau)$.

We note that there is a forgetful functor $\rho: \mathcal{B}G \longrightarrow \mathcal{B}$ that sends (x,A) to A and (ω,i) to i. This functor has the pleasant property that

$\rho\bar{C} = C\rho$ and $\rho(\bar{\tau}) = \tau$. Furthermore, for any $A \in |\mathcal{B}|$, the "inverse image" of A under ρ is the groupoid G(A).

Examples of this form play a fundamental role in our theory. The first topological use of such examples occurs in II.2 where the fundamental groupoid is introduced. We shall occasionally write $\mathcal{B}(G)$ rather than $\mathcal{B}G$.

Let $(\mathcal{B},C,\tau)$ be a category with endomorphism, $B \in |\mathcal{B}|$, and $n \in \mathbb{Z}_+ = \{0,1,2,\ldots\}$. Define τ^n_B by $\tau^0_B = 1_B$ and $\tau^n_B = \tau_{C^{n-1}B}\tau^{n-1}_B$ for $n \geq 1$ where $C^0(B) = B$. Let $\Sigma = \{\tau^n_B \mid n\in\mathbb{Z}_+, B\in|\mathcal{B}|\} \subseteq \mathrm{Mor}\,\mathcal{B}$. In the sequel, we shall write simply τ^n rather than τ^n_B when B is clear from the context.

Lemma 1.4 *The family* Σ *admits a calculus of left fractions.*

Proof: Clearly all identities are in Σ and Σ is closed under composition. Hence, Σ has the first two of the four properties given by Schubert [Sc, p. 258] to define when a class of morphisms admits a calculus of left fractions. Since τ is a natural transformation, so is τ^n. Hence, any diagram of the form $C^nA \xleftarrow{\tau^n} A \xrightarrow{f} B$ can be embedded in the commutative square

$$\begin{array}{ccc} A & \xrightarrow{\quad f \quad} & B \\ {\scriptstyle \tau^n}\big\downarrow & & \big\downarrow{\scriptstyle \tau^n} \\ C^nA & \xrightarrow{\quad C^nf \quad} & C^nB \end{array}$$

and Σ has Schubert's third property since $\tau^n \in \Sigma$. Suppose now that $f,g \in \mathcal{B}(C^nA,B)$ are such that the two composites

$$A \xrightarrow{\quad \tau^n \quad} C^nA \underset{g}{\overset{f}{\rightrightarrows}} B$$

are equal and consider the commutative diagram

$$\begin{array}{ccc} C^nA & \overset{f}{\underset{g}{\rightrightarrows}} & B \\ {\scriptstyle \tau^n}\downarrow & & \downarrow{\scriptstyle \tau^n} \\ C^{2n}A & \overset{C^nf}{\underset{C^ng}{\rightrightarrows}} & C^nB \end{array}$$

Since $C\tau = \tau C$, $C^n\tau^n = \tau^n C^n$ and $\tau^n f = C^n f \circ \tau = C^n f \circ C^n \tau = C^n(f\tau^n)$. Similarly, $\tau^n g = C^n(g\tau^n)$ and since $f\tau^n = g\tau^n$, $\tau^n f = \tau^n g$. Thus Σ also has Schubert's fourth property and admits a calculus of left fractions.

Let $\mathcal{B}(\Sigma^{-1})$ be the category of fractions obtained from $\mathcal{B}$ by inverting the elements of Σ and let $Q\colon \mathcal{B} \longrightarrow \mathcal{B}(\Sigma^{-1})$ be the obvious functor. We recall that the objects of $\mathcal{B}(\Sigma^{-1})$ are the same as the objects of $\mathcal{B}$ (i.e. $Q(A) = A$ for $A \in |\mathcal{B}|$), and that every morphism in $\mathcal{B}(\Sigma^{-1})(A,B)$ can be written in the form $Q(\tau^n_B)^{-1}Q(f)$ for some $f \in \mathcal{B}(A,C^nB)$. Furthermore, $Q(\tau^{n_1})^{-1}Q(f_1) = Q(\tau^{n_2})^{-1}Q(f_2)$ if and only if there exist $m_1, m_2 \in \mathbb{Z}_+$ such that $\tau^{m_1}f_1 = \tau^{m_2}f_2$ where τ^{m_i} is evaluated on $C^{n_i}B$ $(i=1,2)$.

The remarks above may be used to give a more intuitive description of the morphisms in $\mathcal{B}(\Sigma^{-1})$ that will often be used. Let $A,B \in |\mathcal{B}| = |\mathcal{B}(\Sigma^{-1})|$. An *eventual morphism* from A to B is any element $f \in \mathcal{B}(A,C^nB)$ for $n \in \mathbb{Z}_+$. Two eventual morphisms $f_i \in \mathcal{B}(A,C^{n_i}B)$ $(i=1,2)$ are *eventually equivalent*, denoted $f_1 \sim f_2$, if there exist $m_1, m_2 \in \mathbb{Z}_+$ such that $\tau^{m_1}f_1 = \tau^{m_2}f_2$ where τ^{m_i} is evaluated on $C^{n_i}B$ $(i=1,2)$. It is easily seen that $\sim$ is an equivalence relation. Let $E(A,B)$ denote the eventual equivalence classes of eventual morphisms from A to B.

Lemma 1.5 *There is a natural bijection* $\alpha\colon E(A,B) \longrightarrow \mathcal{B}(\Sigma^{-1})(A,B)$.

Proof: Define $\tau_*\colon \mathcal{B}(A,C^nB) \longrightarrow \mathcal{B}(A,C^{n+1}B)$ to be composition with τ_{C^nB}. Then clearly $E(A,B)$ is the colimit of the diagram

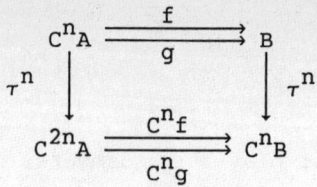

$$\mathcal{B}(A,B) \xrightarrow{\tau_*} \mathcal{B}(A,CB) \longrightarrow \cdots \longrightarrow \mathcal{B}(A,C^nB) \xrightarrow{\tau_*} \mathcal{B}(A,C^{n+1}B) \longrightarrow \cdots$$

Define α_n: $\mathcal{B}(A,C^nB) \longrightarrow \mathcal{B}(\Sigma^{-1})(A,B)$ by $\alpha_n(f) = Q(\tau_B^n)^{-1}Q(f)$ and note that $\alpha_n = \alpha_{n+1}\tau_*$. Hence $\{\alpha_n \mid n\in\mathbb{Z}_+\}$ induces a function α: $E(A,B) \longrightarrow \mathcal{B}(\Sigma^{-1})(A,B)$ which is easily seen to be bijective.

In the sequel, we shall say that $f_n \in \mathcal{B}(A,C^nB)$ *represents* $f \in \mathcal{B}(\Sigma^{-1})(A,B)$ if $f = \alpha(f_n) = Q(\tau_B^n)^{-1}Q(f_n)$. We also note that if $f_n \in \mathcal{B}(A,C^nB)$ represents $f \in \mathcal{B}(\Sigma^{-1})(A,B)$ and $g_m \in \mathcal{B}(B,C^mD)$ represents $g \in \mathcal{B}(\Sigma^{-1})(B,D)$, then $C^ng_m \circ f_n \in \mathcal{B}(A,C^{n+m}D)$ represents $gf \in \mathcal{B}(\Sigma^{-1})(A,D)$.

2. Properties of the category $\mathcal{B}(\Sigma^{-1})$.

In this section we investigate the properties of the category $\mathcal{B}(\Sigma^{-1})$. In particular, we discuss which properties of $\mathcal{B}$ are preserved by Q: $\mathcal{B} \longrightarrow \mathcal{B}(\Sigma^{-1})$. The main results are the following:

Proposition 2.1 *Let $(\mathcal{B},C,\tau)$ be a category with endomorphism. Then*

(i) *The functor Q: $\mathcal{B} \longrightarrow \mathcal{B}(\Sigma^{-1})$ preserves finite colimits, initial objects, and terminal objects (those that exist).*

(ii) *If $\mathcal{B}$ is preadditive, then so is $\mathcal{B}(\Sigma^{-1})$ and Q is an additive functor.*

Corollary 2.2 (i) *If $\mathcal{B}$ has any of the following properties so does $\mathcal{B}(\Sigma^{-1})$: finite coproducts, coequalizers, finite colimits, a zero object, or cokernels.*

(ii) *If $f \in \mathcal{B}(A,B)$ is an epimorphism, then $Q(f) \in \mathcal{B}(\Sigma^{-1})(Q(A),Q(B))$ is an epimorphism.*

Proofs of 2.1 and 2.2: All of these results except part (ii) of 2.2 are standard consequences of the fact that Σ is left calculable (cf. for example [Sc, Thm. 19.2.8] and [Po, pp. 154-157]). Part (ii) of 2.2 follows from the observation that f is an epimorphism if and only if the diagram

$$\begin{array}{ccc} A & \xrightarrow{f} & B \\ {\scriptstyle f}\downarrow & & \downarrow{\scriptstyle 1} \\ B & \xrightarrow{1} & B \end{array}$$

is a pushout and from the fact that Q preserves colimits.

Proposition 2.3 *Let* $(\mathcal{B},C,\tau)$ *be a category with endomorphism. If* $D\colon \mathcal{J} \longrightarrow \mathcal{B}$ *is a finite diagram in* $\mathcal{B}$ *such that for all* $n \in \mathbb{Z}_+$, $\lim_i C^n(D_i)$ *exists and equals* $C^n(\lim_i D_i)$, *then* $\lim_i Q(D_i)$ *exists and equals* $Q(\lim_i D_i)$.

To say the functor $D\colon \mathcal{J} \longrightarrow \mathcal{B}$ is a finite diagram simply means $\mathcal{J}$ is a finite category; that is, the set of all morphisms in $\mathcal{J}$ is finite. We write D_i instead of $D(i)$.

The proofs of 2.3 and the following two corollaries are temporarily deferred.

Corollary 2.4 *Let* $(\mathcal{B},C,\tau)$ *be a category with endomorphism.*

(i) *If* $\mathcal{B}$ *has finite products which are preserved by* C, *then* $\mathcal{B}(\Sigma^{-1})$ *has finite products and* Q *preserves finite products.*

(ii) *If* $\mathcal{B}$ *has equalizers which are preserved by* C, *then* $\mathcal{B}(\Sigma^{-1})$ *has equalizers and* Q *preserves equalizers.*

(iii) *If* $\mathcal{B}$ *has finite limits which are preserved by* C, *then* $\mathcal{B}(\Sigma^{-1})$ *has finite limits and* Q *preserves finite limits.*

(iv) *Suppose* $\mathcal{B}$ *has kernels which are preserved by* C. *Let* $f \in \mathcal{B}(\Sigma^{-1})(Q(A),Q(B))$ *be represented by* $f_n \in \mathcal{B}(A,C^nB)$ *and let* (K_n,i_n) *be the kernel of* f_n. *Then* f *has a kernel given by*

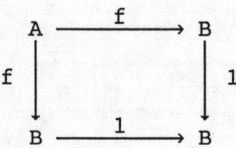

$(Q(K_n), Q(i_n))$. *Furthermore, if* $A_0 \xrightarrow{f} A_1 \xrightarrow{g} A_2$ *is exact in* $\mathcal{B}$, *then* $Q(A_0) \xrightarrow{Q(f)} Q(A_1) \xrightarrow{Q(g)} Q(A_2)$ *is exact in* $\mathcal{B}(\Sigma^{-1})$.

(v) *If* $f \in \mathcal{B}(A,B)$ *is such that* $C^n(f)$ *is monic for all* $n \in \mathbb{Z}_+$, *then* $Q(f)$ *is monic.*

The following result is an important corollary of the results above:

Corollary 2.5 *Let* $(\mathcal{B}, C, \tau)$ *be a category with endomorphism such that* $\mathcal{B}$ *is abelian and* C *preserves finite products and kernels. Then* $\mathcal{B}(\Sigma^{-1})$ *is abelian.*

We turn now to the proofs of these results.

Proof of 2.3: Let $f_i: A \longrightarrow Q(D_i)$ $(i \in |\mathcal{J}|)$ be a family of morphisms in $\mathcal{B}(\Sigma^{-1})$ such that for all $g \in \mathcal{J}(i,j)$, the following diagram commutes

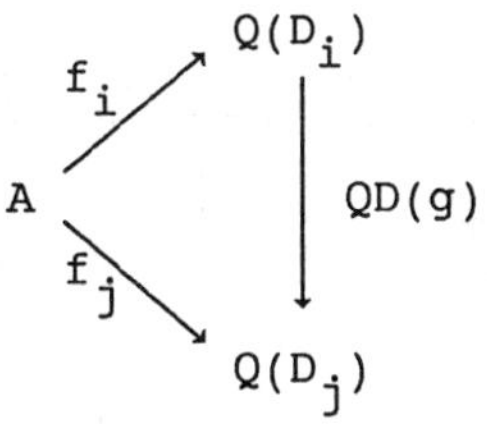

For each i, represent f_i by a morphism $f_i^{n_i}: A \longrightarrow C^{n_i}D_i$. Since $D: \mathcal{J} \longrightarrow \mathcal{B}$ is a finite diagram, without loss of generality, we may assume $n_i = n_j = n$ for all i,j.

Now consider the following diagram in $\mathcal{B}$

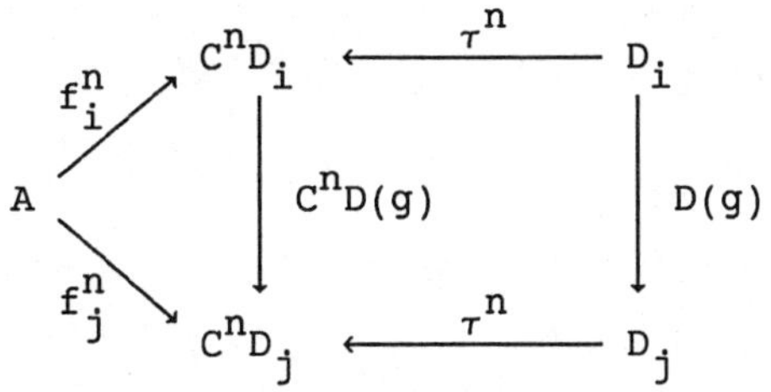

in which the square commutes since τ^n is a natural transformation from $I_{\mathcal{B}}$ to C^n. Since $f_k = Q(\tau^n)^{-1}Q(f_k^n)$ $(k=i,j)$,

$$Q(\tau^n)^{-1}Q(C^nD(g))Q(f_i^n) = QD(g)Q(\tau^n)^{-1}Q(f_i^n) = QD(g)f_i$$
$$= f_j = Q(\tau^n)^{-1}Q(f_j^n).$$

Hence $Q(C^nD(g)f_i^n) = Q(f_j^n)$. This implies that there exists a power $p = p(g)$ such that $\tau^p f_j^n = \tau^p C^n D(g) f_i^n$ where τ^p is evaluated at $C^n D_j$. Since the diagram

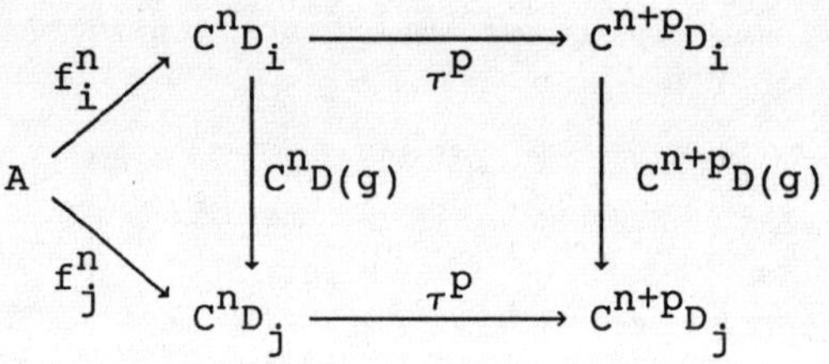

then commutes, by setting $m = n+p$ and $f_k^m = \tau^p f_k^n$ for all $k \in |\mathcal{I}|$, we may assume that for the given $g \in \mathcal{I}(i,j)$, $C^m D(g) f_i^m = f_j^m$.

A simple induction argument now shows that we may assume that n has been chosen so that for all $g \in \text{Mor}\,\mathcal{I}$, the following diagram in $\mathcal{B}$ commutes

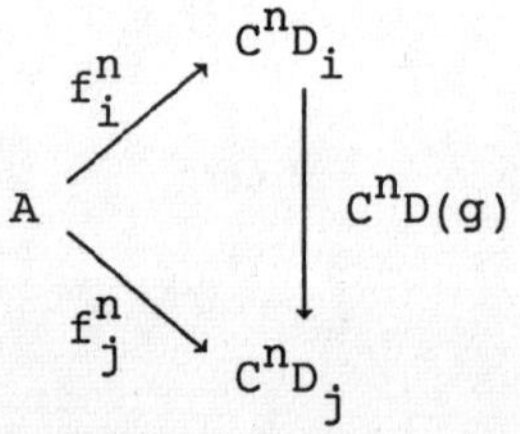

But then there exists a unique morphism $f_n\colon A \longrightarrow \lim_i C^n(D_i) = C^n(\lim_i D_i)$ such that for all i, $f_i^n = C^n(\ell_i)f_n$ where $\ell_i\colon \lim_i D_i \longrightarrow D_i$ is the structure map for the limit. Let $f\colon A \longrightarrow Q(\lim_i D_i)$ be the morphism represented by f_n. An argument, similar to the one given but simpler, shows that f is the unique map such that $Q(\ell_i)Q(f) = Q(f_i)$ for all i. The lemma follows.

Proof of 2.4: The proofs of parts (i), (ii), and the first part of (iv) all follow the same strategy. It is to observe that the property involved can be expressed as a limit of an appropriate diagram $\mathcal{B}(\Sigma^{-1})$, that the diagram can essentially be lifted back to $\mathcal{B}$, that the hypothesis guarantees that the lifted diagram satisfies the hypothesis of 2.3, and then to apply 2.3. We give the proof of (ii) as a sample and leave proofs of (i) and the first part of (iv) to the reader.

Let $f,g \in \mathcal{B}(\Sigma^{-1})(A,B)$ be represented by $f_n, g_n \in \mathcal{B}(A, C^n B)$. Then f_n and g_n have an equalizer (D,h) in $\mathcal{B}$ which is the limit of the diagram

$$\begin{array}{ccc} A & \xrightarrow{1} & A \\ {\scriptstyle 1}\downarrow & & \downarrow{\scriptstyle f_n} \\ A & \xrightarrow{g_n} & C^n B \end{array}$$

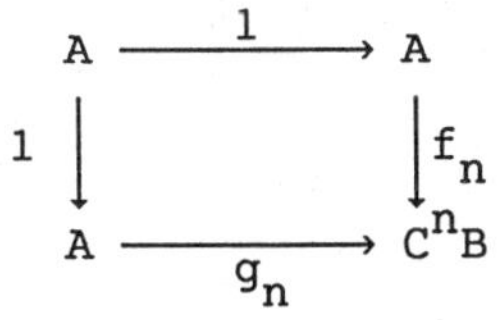

Since C preserves equalizers, 2.3 applies to this diagram and $(Q(D), Q(h))$ is the limit of the outer square in the commutative diagram

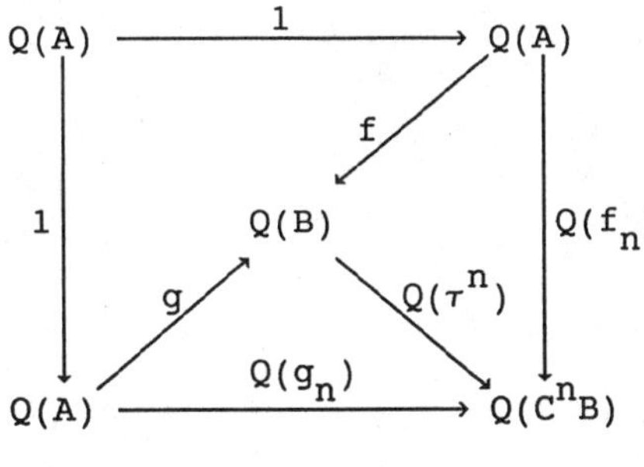

Since $Q(\tau^n)$ is an isomorphism, it is easy to see that $(Q(D), Q(h))$ is also the limit of the "triangular quadrilateral" in the upper left hand corner. Hence $Q(h)$ is an equalizer for f and g.

To show (iii), it suffices to recall that a category has finite limits if and only if it has finite products and equalizers [Sc, Thm. 7.4.2]. Thus (iii) follows immediately from (i) and (ii).

To show the last part of (iv), recall that $A \xrightarrow{f} B \xrightarrow{g} D$ is exact if and only if $gf = 0$ and if, in the factorization $f = if'$ of f, f' is an epimorphism

$$\begin{array}{ccccc} A & \xrightarrow{f} & B & \xrightarrow{g} & D \\ & {\scriptstyle f'}\searrow & \uparrow{\scriptstyle i} & & \\ & & \operatorname{Ker} g & & \end{array}$$

But by (iii), $Q(\ker g) = \ker Q(g)$ and by (ii) of 2.2, $Q(f')$ is an epimorphism. Hence, in the factorization $Q(i)Q(f')$ of $Q(f)$ through $\ker Q(g)$, $Q(f')$ is an epimorphism as required.

Finally (v) is immediate from 2.3 since $f\colon A \longrightarrow B$ is monic if and only if the angle

$$\begin{array}{ccc} A & \xrightarrow{1} & \\ {\scriptstyle 1}\downarrow & & \\ & & \end{array}$$

is the limit of the diagram

$$\begin{array}{ccc} & & A \\ & & \downarrow{\scriptstyle f} \\ A & \xrightarrow{f} & B \end{array}$$

and the hypothesis of (v) guarantees that C^n preserves limits of such diagrams for all $n \in \mathbb{Z}_+$.

Proof of 2.5: We verify that $\mathfrak{B}(\Sigma^{-1})$ satisfies the definition of an abelian category given in [Po, p.27]. $\mathfrak{B}(\Sigma^{-1})$ has finite products by 2.4, is additive by 1.5, and has cokernels by 2.2, and kernels by 2.4. Furthermore, if $f \in \mathfrak{B}(\Sigma^{-1})(A,B)$ is represented by $f_n\colon A \longrightarrow C^nB$ and the canonical factorization of f_n in $\mathfrak{B}$ is given by the diagram

$$\begin{array}{ccc} \operatorname{coker} j_n = \operatorname{coim} f_n & \xrightarrow{\bar{f}_n} & \operatorname{im} f_n = \ker p_n \\ {\scriptstyle q_n}\uparrow & & \downarrow{\scriptstyle i_n} \\ A & \xrightarrow{f_n} & C^nB \\ {\scriptstyle j_n}\uparrow & & \downarrow{\scriptstyle p_n} \\ \ker f_n & & \operatorname{coker} f_n \end{array}$$

then the image of this diagram in $\mathfrak{B}(\Sigma^{-1})$ gives the canonical factorization of $Q(f_n)$. Hence, since $\mathcal{B}$ is abelian and $\bar{f}_n$ is an isomorphism, so is $Q(\bar{f}_n)$. Since the diagram

$$\begin{array}{ccc}
\text{coim } f & \xrightarrow{\ \bar{f}\ } & \text{im } f \\
q \uparrow & & \downarrow i \\
A & \xrightarrow{\ f\ } & B \\
j \uparrow & & \downarrow p \\
\ker f & & \text{coker } f
\end{array}$$

giving the canonical factorization of f in $\mathcal{B}(\Sigma^{-1})$ is isomorphic to the image of the above diagram, it follows that $\bar{f}$ is an isomorphism and that $\mathcal{B}(\Sigma^{-1})$ is abelian.

3. Functors between categories with endomorphism.

In this section we introduce a class of functors between categories with endomorphism that induce functors on the corresponding categories of fractions. We also obtain some of the basic properties of these functors and of natural transformations between them.

Let $(\mathcal{B},C_i,\tau_i)$ $(i=1,2)$ be categories with endomorphism. A functor $F\colon \mathcal{B}_1 \longrightarrow \mathcal{B}_2$ is *almost endomorphism preserving* if for some $n \geq 0$, there exists a natural transformation $\alpha\colon FC_1 \longrightarrow C_2^nF$ such that the diagram of natural transformations

$$\begin{array}{ccc}
F & \xrightarrow{\ F\tau_1\ } & FC_1 \\
& \searrow^{\tau_2^nF} & \downarrow \alpha \\
& & C_2^nF
\end{array}$$

commutes where for $B \in |\mathcal{B}_1|$, $(F\tau_1)(B) = F(\tau_{1B})$ and $(\tau_2^nF)(B) = \tau_{2FB}^n$.

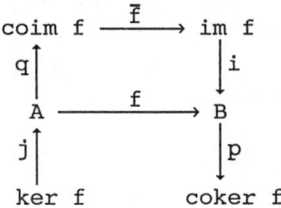

Example 3.1 If $f\colon \mathcal{B}_1 \longrightarrow \mathcal{B}_2$ commutes with C and τ (i.e. $FC_1 = C_2F$ and $F\tau_1 = \tau_2F$), then by taking $n = 1$ and $\alpha = 1$, it is seen that F is

almost endomorphism preserving. Although such functors appear most natural from an algebraic point of view, we emphasize the almost endomorphism preserving functors because they arise more easily in topological applications.

Proposition 3.2 *Let* $(\mathcal{B}_i, C_i, \tau_i)$ $(i=1,2)$ *be categories with endomorphism and* $F\colon \mathcal{B}_1 \longrightarrow \mathcal{B}_2$ *be almost endomorphism preserving. Then* F *induces a unique functor* $F_!\colon \mathcal{B}_1(\Sigma_1^{-1}) \longrightarrow \mathcal{B}_2(\Sigma_2^{-1})$ *such that the diagram*

$$\begin{array}{ccc} \mathcal{B}_1 & \xrightarrow{\ F\ } & \mathcal{B}_2 \\ {\scriptstyle Q_1}\downarrow & & \downarrow{\scriptstyle Q_2} \\ \mathcal{B}_1(\Sigma_1^{-1}) & \xrightarrow{\ F_!\ } & \mathcal{B}_2(\Sigma_2^{-1}) \end{array}$$

commutes. Furthermore, $(1_{\mathcal{B}})_! = 1_{\mathcal{B}(\Sigma^{-1})}$ *and* $(GF)_! = G_!F_!$.

Proof: By the universal mapping properties of $\mathcal{B}_1(\Sigma_1^{-1})$, it suffices to prove that for all $B \in |\mathcal{B}_1|$ and all $n \in \mathbb{Z}_+$, $Q_2F(\tau_{1B}^n)$ is an isomorphism. To do this, it suffices to show $Q_2F(\tau_{1B})$ is an isomorphism for all $B \in |\mathcal{B}_1|$.

Now consider the following diagram

$$\begin{array}{ccccc} FB & \xrightarrow{\ F(\tau_1)\ } & FC_1B & \xrightarrow{\ FC_1(\tau_1)\ } & FC_1^2B \\ & {\scriptstyle \tau_2^n}\searrow & \downarrow{\scriptstyle \alpha} & {\scriptstyle \tau_2^n}\searrow & \downarrow{\scriptstyle \alpha} \\ & & C_2^nFB & \xrightarrow{\ (C_2^nF)(\tau_1)\ } & C_2^nFC_1B \end{array}$$

where we have written $F(\tau_1)$ instead of $F(\tau_{1B})$, etc. The triangle on the left commutes since F is almost endomorphism preserving, while the square on the right commutes since α is a natural transformation. On the other hand, since $C_1\tau_1 = \tau_1C_1$, the upper triangle on the right is just the commutative diagram

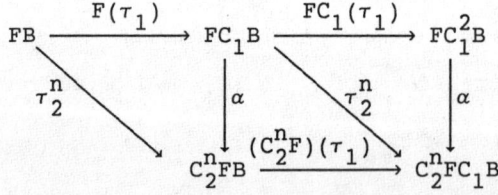

$$\begin{array}{ccc} F(C_1B) & \xrightarrow{F(\tau_1)} & FC_1(C_1B) \\ & {\scriptstyle \tau_2^n} \searrow & \downarrow {\scriptstyle \alpha} \\ & & C_2^nF(C_1B) \end{array}$$

It follows easily that the lower triangle also commutes as does the diagram

$$\begin{array}{ccccc} Q_2FB & \xrightarrow{Q_2F(\tau_1)} & Q_2FC_1B & & \\ & {\scriptstyle Q_2(\tau_2^n)} \searrow & \downarrow {\scriptstyle Q_2(\alpha)} & \searrow {\scriptstyle Q_2(\tau_2^n)} & \\ & & Q_2C_2^nFB & \xrightarrow{Q_2C_2^nF(\tau_1)} & Q_2C_2^nFC_1B \end{array}$$

in $\mathcal{B}_2(\Sigma_2^{-1})$. Since the diagonal maps are isomorphisms in $\mathcal{B}_2(\Sigma_2^{-1})$, the other maps in diagram are also isomorphisms. In particular, $Q_2F(\tau_1) = Q_2F(\tau_{1B})$ is an isomorphism and the first part of the proposition follows.

The last sentence follows immediately from the uniqueness of $F_!$.

Let $F_1, F_2 \colon \mathcal{B}_1 \longrightarrow \mathcal{B}_2$ be almost endomorphism preserving functors between categories with endomorphism. If $\nu \colon F_1 \longrightarrow F_2$ is a natural transformation, so is $\nu_! \colon F_{1!} \longrightarrow F_{2!}$ given by $\nu_{!Q_1(A)} = Q_2(\nu_A)$ for any object $Q_1(A)$ in $\mathcal{B}_1(\Sigma_1^{-1})$. The following proposition, whose proof is left to the reader, summarizes the properties of these natural transformations:

Proposition 3.3 *Let* $F_i \colon (\mathcal{B}_1, C_1, \tau_1) \longrightarrow (\mathcal{B}_2, C_2, \tau_2)$ $(i=1,2,3)$ *be almost endomorphism preserving functors and* $\nu \colon F_1 \longrightarrow F_2$ *and* $\mu \colon F_2 \longrightarrow F_3$ *be natural transformations.*

(i) *If* $F_2 = F_1 = F$ *and* $\nu = 1_F$, *then* $\nu_! = 1_{F_!}$.

(ii) $(\mu\nu)_! = \mu_!\nu_!$.

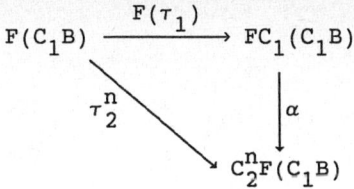

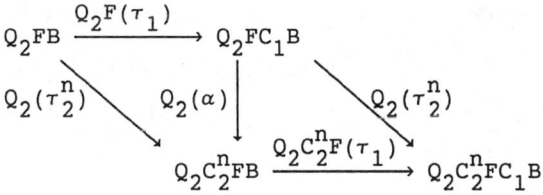

If also $F'_j \colon (\mathcal{B}_2, C_2, \tau_2) \longrightarrow (\mathcal{B}_3, C_3, \tau_3)$ $(j=1,2)$ *are almost endomorphism*

preserving and $\nu' : F'_1 \longrightarrow F'_2$ *is a natural transformation, then*

(iii) $(F'_1\nu)_! = F'_{1!}\nu_! : F'_{1!}F_{1!} \longrightarrow F'_{1!}F_{2!}$;

(iv) $(\nu' F_1)_! = \nu'_! F_{1!} : F'_{1!}F_{1!} \longrightarrow F'_{2!}F_{1!}$.

Example 3.4 Let $(\mathcal{B},C,\tau)$ be a category with endomorphism. Then $C: \mathcal{B} \longrightarrow \mathcal{B}$ commutes with C trivially and with τ by definition. Thus C is almost endomorphism preserving and induces $C_! : \mathcal{B}(\Sigma^{-1}) \longrightarrow \mathcal{B}(\Sigma^{-1})$; while $\tau: I_{\mathcal{B}} \longrightarrow C$ induces $\tau_! : I_{\mathcal{B}(\Sigma^{-1})} \longrightarrow C_!$. Since it is easy to check that $C_!\tau_! = \tau_!C_!$, $(\mathcal{B}(\Sigma^{-1}),C_!,\tau_!)$ is a category with endomorphism. Furthermore, since $\tau_{!Q(A)} = Q(\tau_A)$ is an isomorphism for all Q(A) in $\mathcal{B}(\Sigma^{-1})$, $\tau_!$ is a natural equivalence of functors.

Let $F,G: (\mathcal{B}_1,C_1,\tau_1) \longrightarrow (\mathcal{B}_2,C_2,\tau_2)$ be almost endomorphism preserving functors. We say that F and G are *eventually equal* if there exist integers m, n, p, and q such that $C_2^mFC_1^p = C_2^nGC_1^q$.

Corollary 3.5 *If* F *is eventually equal to* G, *then there is a natural equivalence* $\sigma: F_! \longrightarrow G_!$.

Proof: Suppose $C_2^mFC_1^p = C_2^nGC_1^q$. Since $\tau_{2!}^mF_!\tau_{1!}^p : F_! \longrightarrow C_{2!}^mF_!C_{1!}^p$ and $\tau_{2!}^nG_!\tau_{1!}^q : G_! \longrightarrow C_{2!}^nG_!C_{1!}^q$ are easily seen to be natural equivalences by 3.4, $\sigma = (\tau_{2!}^nG_!\tau_{1!}^q)^{-1}(\tau_{2!}^mF_!\tau_{1!}^p)$ is the desired natural equivalence.

This natural equivalence will be called the *canonical natural equivalence* since it is constructed essentially from the equivalences $\tau_{1!}$ and $\tau_{2!}$.

Since adjoint functors will be needed in the sequel, the following proposition is useful:

Proposition 3.6 *Let* $(\mathcal{B}_i,C_i,\tau_i)$ (i=1,2) *be categories with endomorphism and* $F: \mathcal{B}_1 \longrightarrow \mathcal{B}_2$ *be an almost endomorphism preserving functor. Let* $G: \mathcal{B}_2 \longrightarrow \mathcal{B}_1$ *be a left adjoint for* F. *Then the following are* *equivalent*

(i) G *is almost endomorphism preserving; and*

(ii) *For some integer* $m \geq 0$, *there exists a natural transformation* $\beta\colon C_2F \longrightarrow FC_1^m$ *such that the following diagram commutes*

$$\begin{array}{ccc} F & \xrightarrow{\tau_2 F} & C_2F \\ & \searrow^{F\tau_1^m} & \downarrow{\scriptstyle\beta} \\ & & FC_1^m \end{array}$$

If either (i) *or* (ii) *holds, then the induced functor* $G_!$ *is left adjoint for* $F_!$.

Proof: Suppose (ii) holds. Let $\eta\colon 1_{\mathcal{B}_2} \longrightarrow FG$ and $\varepsilon\colon GF \longrightarrow 1_{\mathcal{B}_1}$ be the unit and counit of the adjunction. We define natural transformations $\gamma\colon GC_2 \longrightarrow C_1^mG$ and $\delta\colon G \longrightarrow C_1^mG$ to be the composites of the rows in the diagram

$$\begin{array}{ccccccc} G & \xrightarrow{G(\eta)} & G(FG)=G(F)G & \xrightarrow{G(F\tau_1^m)G} & G(FC_1^mG)=(GF)C_1^mG & \xrightarrow{\varepsilon C_1^mG} & C_1^mG \\ \downarrow{\scriptstyle G(\tau_2)} & & \downarrow{\scriptstyle G(\tau_2F)G} & & \| & & \| \\ GC_2 & \xrightarrow[GC_2(\eta)]{} & GC_2(FG)=G(C_2F)G & \xrightarrow[G(\beta)G]{} & G(FC_1^mG)=(GF)C_1^mG & \xrightarrow[\varepsilon C_1^mG]{} & C_1^mG \end{array}$$

Since $\tau_2\colon I_{\mathcal{B}_2} \longrightarrow C_2$ is a natural transformation, the diagram

$$\begin{array}{ccc} I_{\mathcal{B}_2} & \xrightarrow{\eta} & FG \\ \downarrow{\scriptstyle\tau_2} & & \downarrow{\scriptstyle(\tau_2F)G} \\ C_2 & \xrightarrow{C_2(\eta)} & C_2FG \end{array}$$

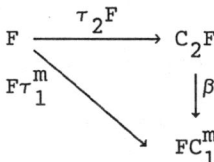

commutes. The left hand square in the top diagram is obtained by applying G to the diagram. Hence it commutes. The middle square is obtained by applying G( )G to the diagram in (ii). Hence, it commutes also, as does the triangle

$$
\begin{array}{ccc}
G & & \\
{\scriptstyle G\tau_2}\downarrow & \searrow^{\delta} & \\
 & & C_1^m G \\
 & \nearrow_{\gamma} & \\
GC_2 & &
\end{array}
$$

We claim that $\delta = \tau_1^m G$. To see this note that since $\varepsilon\colon GF \longrightarrow 1_{\mathfrak{B}_1}$ is a natural transformation, the square in the diagram

$$
\begin{array}{ccccc}
G & \xrightarrow{G(\eta)} & GFG & \xrightarrow{GF(\tau_1^m G)} & GF(C_1^m G) \\
 & {\scriptstyle 1}\searrow & \downarrow{\scriptstyle \varepsilon G} & & \downarrow{\scriptstyle \varepsilon C_1^m G} \\
 & & G & \xrightarrow{\tau_1^m G} & C_1^m G
\end{array}
$$

commutes. But it is standard from the adjointness of F and G, that the triangle commutes. Hence $\delta = \tau_1^m G$ as claimed and (i) of the proposition follows.

The proof that (i) implies (ii) is similar and is left to the reader. The fact that $G_!$ is a left adjoint for $F_!$ follows easily from [Sc, Prop. 16.5.7] using 3.2 and 3.3.

We recall for the reader's benefit some standard facts about adjoint functors which we will want to use, perhaps implicity, later in this book. In particular, the left adjoint of a functor $F\colon \mathcal{C} \longrightarrow \mathfrak{D}$ is not really well defined, but given any two left adjoints G_1, G_2 for F there is a canonical natural equivalence $\alpha\colon G_1 \longrightarrow G_2$. Thus, when F has a left adjoint, we shall occasionally use the notation F^ℓ to denote a choice of left adjoint for F. The following result is well known:

Lemma 3.7 (i) *One may choose* $(1_{\mathfrak{B}})^\ell = 1_{\mathfrak{B}}$.

(ii) *There is a canonical natural equivalence* $\alpha(F_2, F_1)\colon (F_2F_1)^\ell \longrightarrow F_1^\ell F_2^\ell$ *such that the following diagram commutes*

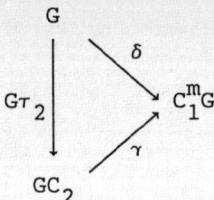

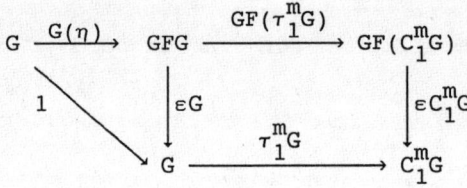

$$\begin{array}{ccc}
(F_3F_2F_1)^{\ell} & \xrightarrow{\alpha(F_3,F_2F_1)} & (F_2F_1)^{\ell}F_3^{\ell} \\
\Big\downarrow{\scriptstyle \alpha(F_3F_2,F_1)} & & \Big\downarrow{\scriptstyle \alpha(F_2,F_1)F_3^{\ell}} \\
F_1^{\ell}(F_3F_2)^{\ell} & \xrightarrow{F_1^{\ell}\alpha(F_3,F_2)} & F_1^{\ell}F_2^{\ell}F_3^{\ell}
\end{array}$$

Now if $F,G\colon \mathcal{C} \longrightarrow \mathcal{D}$ both have left adjoints and $\upsilon\colon F \longrightarrow G$ is a natural transformation, then there is a unique *conjugate* natural transformation $\upsilon^{\ell}\colon G^{\ell} \longrightarrow F^{\ell}$ (cf. [Ma, p. 98]) whose properities are summarized in the following lemma:

Lemma 3.8 (i) *If* $\upsilon=1\colon F \longrightarrow F$, *then* $\upsilon^{\ell}=1\colon F^{\ell} \longrightarrow F^{\ell}$.

(ii) *If* $\upsilon\colon F_1 \longrightarrow F_2$ *and* $\mu\colon F_2 \longrightarrow F_3$, *then* $(\mu\upsilon)^{\ell} = \upsilon^{\ell}\mu^{\ell}\colon F_3^{\ell} \longrightarrow F_1^{\ell}$.

(iii) *If* $F\colon \mathcal{C} \longrightarrow \mathcal{D}$ *and* $G_1,G_2\colon \mathcal{D} \longrightarrow \mathcal{E}$ *are functors with left adjoints and* $\upsilon\colon G_1 \longrightarrow G_2$ *is a natural transformation, then the following diagram commutes*

$$\begin{array}{ccc}
(G_2F)^{\ell} & \xrightarrow{(\upsilon F)^{\ell}} & (G_1F)^{\ell} \\
\Big\downarrow{\scriptstyle \alpha(G_2,F)} & & \Big\downarrow{\scriptstyle \alpha(G_1,F)} \\
F^{\ell}G_2^{\ell} & \xrightarrow[F^{\ell}\upsilon^{\ell}]{} & F^{\ell}G_1^{\ell}
\end{array}$$

We leave the proofs of 3.7 and 3.8 to the reader. We also leave it to the reader to formulate the analogue of (iii) of 3.8 in the case when $F_1,F_2\colon \mathcal{C} \longrightarrow \mathcal{D}$, $G\colon \mathcal{D} \longrightarrow \mathcal{E}$, and $\upsilon\colon F_1 \longrightarrow F_2$ is a natural transformation.

4. Functor categories as categories with endomorphism.

In this section we show how categories of functors (or, simply, functor categories) become categories with endomorphism and give the main properties of such categories.

Let $\mathscr{B}$ and $\mathscr{C}$ be categories and $\mathscr{C}^{\mathscr{B}}$ be the functor category whose objects are functors $F\colon \mathscr{B} \longrightarrow \mathscr{C}$ and whose morphisms are natural transformations. If $(\mathscr{B},C,\tau)$ is a category with endomorphism, define a functor $\bar{C}\colon \mathscr{C}^{\mathscr{B}} \longrightarrow \mathscr{C}^{\mathscr{B}}$ and a natural transformation $\bar{\tau}\colon I \longrightarrow \bar{C}$ (where I is the identity functor on $\mathscr{C}^{\mathscr{B}}$) by setting $\bar{C}(F) = FC$, $\bar{C}(\sigma) = \sigma C\colon FC \longrightarrow GC$ when $\sigma\colon F \longrightarrow G$ is a natural transformation, and $\bar{\tau}(F) = F(\tau)$. Then clearly $\bar{C}\bar{\tau} = \bar{\tau}\bar{C}$, so $(\mathscr{C}^{\mathscr{B}},\bar{C},\bar{\tau})$ is a category with endomorphism. We shall call $(\bar{C},\bar{\tau})$ the *endomorphism structure induced by* (C,τ). In the sequel, we will assume $\mathscr{C}^{\mathscr{B}}$ is equipped with this structure unless we state otherwise. We also note that any morphism in $\mathscr{C}^{\mathscr{B}}(\Sigma^{-1})$ can be represented by an eventual natural transformation $\sigma_n\colon F \longrightarrow GC^n = \bar{C}^n(G)$. When referring to $\mathscr{C}^{\mathscr{B}}$ we shall often write C or τ instead of $\bar{C}$ or $\bar{\tau}$.

The following proposition and its corollaries give the most important properties of the categories $\mathscr{C}^{\mathscr{B}}$ viewed as categories with endomorphism:

Proposition 4.1 *Let* $(\mathscr{B},C,\tau)$ *be a category with endomorphism and* $\mathscr{C}$ *be finite-complete. Then* $\mathscr{C}^{\mathscr{B}}$ *is finite-complete and* $\bar{C}$ *preserves finite limits.*

Corollary 4.2 *Let* $(\mathscr{B},C,\tau)$ *be a category with endomorphism and* $\mathscr{C}$ *be a category with zero object, kernels, and pushouts. Then* $\mathscr{C}^{\mathscr{B}}$ *has a zero object, kernels, and pushouts, and the following statements are equivalent.*

(i) *The sequence* $F_0 \xrightarrow{\sigma} F_1 \xrightarrow{\rho} F_2$ *is exact in* $\mathscr{C}^{\mathscr{B}}$.

(ii) *For all* $A \in |\mathscr{B}|$, $F_0(A) \xrightarrow{\sigma_A} F_1(A) \xrightarrow{\rho_A} F_2(A)$ *is exact in* $\mathscr{C}$.

If $\mathscr{C}$ *is also finite-complete, either* (i) *or* (ii) *implies that* $Q(F_0) \longrightarrow Q(F_1) \longrightarrow Q(F_2)$ *is exact in* $\mathscr{C}^{\mathscr{B}}(\Sigma^{-1})$.

Corollary 4.3 *Let* $(\mathcal{B}, C, \tau)$ *be a category with endomorphism and* $\mathcal{C}$ *be an abelian category. Then* $\mathcal{C}^{\mathcal{B}}(\Sigma^{-1})$ *is abelian.*

Proof of 4.1: Let $D: \mathcal{J} \longrightarrow \mathcal{C}^{\mathcal{B}}$ be a finite diagram. Abusing notation slightly, we denote this diagram more simply by $\{D_i \mid i \in \mathcal{J}\}$. Then, for any object $A \in |\mathcal{B}|$, evaluation at A gives a finite diagram $\{D_i(A) \mid i \in \mathcal{J}\}$ in $\mathcal{C}$. Since $\mathcal{C}$ is finite-complete, $\lim_i (D_i(A))$ exists in $\mathcal{C}$. Now define $\lim_i D_i: \mathcal{B} \longrightarrow \mathcal{C}$ by $(\lim_i D_i)(A) = \lim_i (D_i(A))$. It is easily verified that $\lim_i D_i$ is a functor and is the limit of $\{D_i \mid i \in \mathcal{J}\}$ in $\mathcal{C}^{\mathcal{B}}$.

With this description of $\lim_i D_i$, it is easy to see that $\bar{C}$ preserves limits for $[\bar{C}(\lim_i D_i)](A) = (\lim_i D_i)(CA) = \lim_i [D_i(CA)] = \lim_i [\bar{C}(D_i)(A)] = [\lim_i \bar{C}(D_i)](A)$.

Proof of 4.2: The zero object of $\mathcal{C}^{\mathcal{B}}$ is the zero functor $F(A) = 0$ for all $A \in |\mathcal{B}|$. If $\sigma: F \longrightarrow G$ is a morphism in $\mathcal{C}^{\mathcal{B}}$, it is easily verified that setting $(\ker \sigma)(A) = \ker\{\sigma_A: F(A) \longrightarrow G(A)\}$ defines a functor $\ker \sigma: \mathcal{B} \longrightarrow \mathcal{C}$ and that $\ker \sigma$ is the kernel of σ. That is, kernels in $\mathcal{C}^{\mathcal{B}}$ are given "pointwise". Similarly, pushouts are given pointwise.

If (i) holds, then σ factors through an epic morphism σ' as indicated in the following diagram

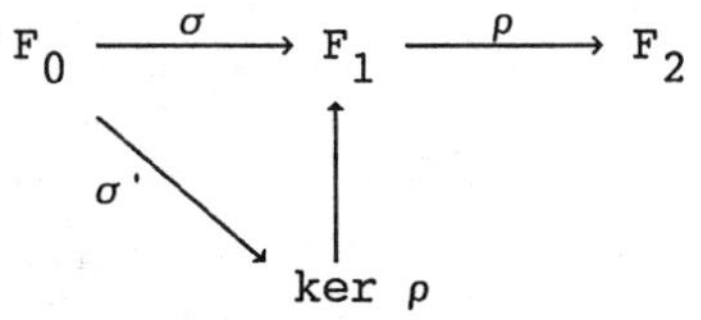

Since kernels in $\mathcal{C}^{\mathcal{B}}$ are given pointwise, for any $A \in |\mathcal{B}|$, we have $(\ker \rho)(A) = \ker \rho_A$ in the diagram

$$\begin{array}{ccccc}
F_0(A) & \xrightarrow{\sigma_A} & F_1(A) & \xrightarrow{\rho_A} & F_2(A) \\
& {}_{\sigma'_A}\searrow & \uparrow & & \\
& & \ker\rho(A) & & \\
& & \| & & \\
& & \ker\rho_A & &
\end{array}$$

Thus, it suffices to prove σ'_A is epic in $\mathscr{C}$. But in any category $\mathscr{D}$ a morphism f is epic if and only if the diagram

$$\begin{array}{ccc}
C & \xrightarrow{f} & D \\
{\scriptstyle f}\downarrow & & \downarrow{\scriptstyle 1} \\
D & \xrightarrow{1} & D
\end{array}$$

is a pushout. By applying this principle to σ' and using the fact that pushouts in $\mathscr{C}^{\mathscr{B}}$ are given pointwise, one sees easily that σ_A is epic. Hence, (ii) follows.

The routine proof that (ii) implies (i) is omitted.

The last sentence is an immediate consequence of 4.1 and 2.4 (iv).

Proof of 4.3 If $\mathscr{C}$ is abelian, it is well known that $\mathscr{C}$ is finite-complete and that $\mathscr{C}^{\mathscr{B}}$ is abelian (cf. [Sc, Prop. 12.1.6]). This corollary now follows directly from 4.1 and 2.5.

Example 4.4 Let R be a ring with unit and R-*mod* be the category of left R modules. Let $(\mathscr{B},C,\tau)$ be a category with endomorphism. Then R-$mod^{\mathscr{B}}(\Sigma^{-1})$ is called the *category of left* R$\mathscr{B}$ *modules* and will be denoted by R$\mathscr{B}$-*mod*.

Many of the algebraic topological invariants we will consider lie in one of the categories R$\mathscr{B}$-*mod*. Since these categories are abelian by 4.3, the standard exact sequence arguments hold without modification. Some of our algebraic topological invariants, however, lie in the categories $\mathscr{C}^{\mathscr{B}}(\Sigma^{-1})$ where $\mathscr{C}$ is either $\mathscr{G}p$, the category of groups, or $\mathscr{S}et_*$, the category of pointed sets, which are not abelian. Thus, we collect for future reference some results concerning these categories that will be useful in making exact sequence arguments.

Lemma 4.5 *Let* $(\mathcal{B}, C, \tau)$ *be a category with endomorphism and* $\mathcal{C}$ *be a category with zero object. Let* $M \in |\mathcal{C}^{\mathcal{B}}(\Sigma^{-1})|$. *Then* $M = 0$ *if and only if there exists a* $p \geq 0$ *such that for all* $B \in |\mathcal{B}|$, $M(\tau_B^p)\colon M_B \longrightarrow M_{C^pB}$ *is the zero morphism.*

The reader should note the specializations of 4.5 to the cases when $\mathcal{C}$ is R-*mod*, $\mathcal{G}p$, or $\mathcal{S}et_*$.

Proof of 4.5: We note that the zero object of $\mathcal{C}^{\mathcal{B}}(\Sigma^{-1})$ is $Q(F)$ where $F\colon \mathcal{B} \longrightarrow \mathcal{C}$ is the zero functor $F(B) = 0$ for all $B \in |\mathcal{B}|$ and that the (unique) morphisms $\rho\colon M \longrightarrow 0$ and $\sigma\colon 0 \longrightarrow M$ are represented by the natural transformations ρ_0 and σ_0 consisting of the unique morphisms $\rho_B\colon M_B \longrightarrow 0$ and $\sigma_B\colon 0 \longrightarrow M_B$ in $\mathcal{C}$ where $B \in |\mathcal{B}|$. But $M = 0$ if and only if $0 = \sigma\rho$ is the identity in $\mathcal{C}^{\mathcal{B}}(\Sigma^{-1})$. By 1.5 this is equivalent to saying that $0 = \sigma\rho = \tau_M^p$ for some p. The lemma follows.

Let $\sigma\colon M \longrightarrow N$ be a morphism in $\mathcal{C}^{\mathcal{B}}$ where $\mathcal{C}$ is R-*mod*, $\mathcal{G}p$, or $\mathcal{S}et_*$. We say that σ is *eventually monomorphic* (respectively, *eventually epimorphic*) (respectively, *eventually zero*) if there exists an integer $p \geqq 0$ such that for all $B \in |\mathcal{B}|$, if $x_1, x_2 \in M_B$ are such that $\sigma_B(x_1) = \sigma_B(x_2)$, then $\tau_B^p(x_1) = \tau_B^p(x_2)$ (respectively, for all $x \in N_B$, $\tau_B^p(x) \in \mathrm{Im}\{\sigma_{C^pB}\colon M_{C^pB} \longrightarrow N_{C^pB}\}$) (respectively, $N(\tau_B^p)\sigma = 0$).

If $\sigma\colon M \longrightarrow N$ is a morphism in $\mathcal{C}^{\mathcal{B}}(\Sigma^{-1})$, we say σ is *eventually monomorphic*, *eventually epimorphic*, or *eventually zero*, respectively, if it has a representative $\sigma_n\colon M \longrightarrow C^nN$ with the corresponding property in $\mathcal{C}^{\mathcal{B}}$.

Lemma 4.6 *Let* $\sigma: M \longrightarrow N$ *be a morphism in* $\mathscr{C}^{\mathscr{B}}(\Sigma^{-1})$ *where* $\mathscr{C}$ *is any of the categories* R-*mod*, $\mathscr{G}p$, *or* $\mathscr{S}et_*$. *Then each of the following statements implies the next*

(i) σ *is eventually monomorphic*

(ii) σ *is monic*

(iii) $\ker \sigma = 0$.

If $\mathscr{C}$ *is either* R-*mod or* $\mathscr{G}p$, *then* (iii) *implies* (i). *In addition, each of the following statements implies the next*

(iv) σ *is eventually epimorphic*

(v) σ *is epic*

(vi) $\operatorname{coker} \sigma = 0$.

If $\mathscr{C}$ *is* R-*mod, then* (vi) *implies* (iv).

Proof of 4.6: Let $\sigma: M \longrightarrow N$ be eventually monomorphic. We wish to show σ is monic. Thus, let $\mu, \upsilon: P \longrightarrow M$ be such that $\sigma\mu = \sigma\upsilon$ and choose representatives $\mu_m, \upsilon_m: P \longrightarrow C^m M = MC^m$ and $\sigma_n: M \longrightarrow NC^n$ for μ, υ, and σ. Then $\sigma_n C^m \mu_m$ and $\sigma_n C^m \upsilon_m$ represent $\sigma\mu$ and $\sigma\upsilon$ respectively and we may assume m,n chosen so that $\sigma_n C^m \mu_m = \sigma_n C^m \upsilon_m$. Let $B \in |\mathscr{B}|$ and $x \in P(B)$. Then $\sigma_n C^m \mu_m(x) = \sigma_n C^m \upsilon_m(x)$ and since σ is eventually monomorphic, there exists a $p \geq 0$ such that $M(\tau^p_{C^m B})\mu_m(x) = M(\tau^p_{C^m B})\upsilon_m(x)$. It follows that $\mu = \upsilon$ and σ is monic. The very similar proof that (iv) implies (v) is left to the reader.

Suppose (ii) holds. Since $\mathscr{C}$ has kernels, so does $\mathscr{C}^{\mathscr{B}}$ and by 4.1 they are preserved by C. Hence by 2.4 (iv), $\mathscr{C}^{\mathscr{B}}(\Sigma^{-1})$ has kernels. Since the kernel of a monic morphism in a category with kernels is the zero object, (iii) follows. The proof that (v) implies (vi) is similar; one need only replace the words "kernel" and "monic" with "cokernel" and "epic" respectively.

Now suppose (iii) holds and $\mathscr{C}$ is either R-*mod* or $\mathscr{G}p$. Let σ be represented by $\sigma_n: M \longrightarrow NC^n$. Then $\ker \sigma$ is represented by the functor $B \longrightarrow \ker\{\sigma_{nB}: M(B) \longrightarrow N(C^n B)\}$. Since $\ker \sigma = 0$, it follows from 4.5

that there exists a $p \geq 0$ such that $M(\tau_B^p)|$: $\ker \sigma_{nB} \longrightarrow \ker \sigma_{n_{C^pB}}$ is the zero map. But this is equivalent to the definition that σ be eventually monomorphic.

Finally suppose (vi) holds and $\mathscr{C}$ = R-*mod*. Let σ be represented σ_n: $M \longrightarrow NC^n$. Since R-*mod* has cokernels, so does R-$mod^{\mathscr{B}}$ and they are given pointwise. It now follows from 2.1 that coker σ is represented by the functor $B \longrightarrow \mathrm{coker}\{\sigma_{nB}: M(B) \longrightarrow NC^n(B)\}$. Since coker σ = 0, there exists an integer $p \geq 0$ such that $(\mathrm{coker}\ \sigma)(\tau_B^p) = 0$ for all $B \in |\mathscr{B}|$. A simple chase of the following commutative diagram with exact rows now shows that σ is eventually epimorphic

$$\begin{array}{ccccc} M(B) & \xrightarrow{\sigma_{n_B}} & N(C^nB) & \longrightarrow & (\mathrm{coker}\ \sigma)(B) \\ & & \Big\downarrow N(\tau_C^p n_B) & & \Big\downarrow (\mathrm{coker})(\tau_B^p) \\ M(C^pB) & \xrightarrow[\sigma_{n_{C^pB}}]{} & N(C^{n+p}B) & \longrightarrow & (\mathrm{coker}\ \sigma)(C^pB) \end{array}$$

This completes the proof of 4.6.

We note that the proof that (vi) implies (iv) given above fails in $\mathscr{G}p^{\mathscr{B}}(\Sigma^{-1})$ because the rows in the above diagram need not be exact in $\mathscr{G}p$. In general, if σ: $G \longrightarrow H$ is a group homomorphism, coker σ = H/N where N is the smallest normal subgroup of H containing Im σ.

Lemma 4.7 *Let* $\mathscr{C}$ *be any of the categories* R-*mod*, $\mathscr{G}p$, *or* $\mathscr{S}et_*$ *and let* $F_1 \xrightarrow{\sigma_1} F_2 \xrightarrow{\sigma_2} F_3$ *be exact in* $\mathscr{C}^{\mathscr{B}}$. *Then*

(i) *If* σ_2 *is eventually monomorphic,* σ_1 *is eventually zero. If* $\mathscr{C}$ *is either* R-*mod or* $\mathscr{G}p$, *the converse holds.*

(ii) σ_1 *is eventually epimorphic if and only if* σ_2 *is eventually zero.*

(iii) $Q(\sigma_1)$ *is an isomorphism if and only if* σ_1 *is eventually monomorphic and eventually epimorphic.*

(iv) σ_1 *and* σ_2 *are eventually zero if and only if* $Q(F_2) = 0$ *in* $\mathscr{C}^{\mathscr{B}}(\Sigma^{-1})$.

Proof of 4.7: The proof is a straightforward exercise and is left to the reader.

5. Functors between functor categories.

In this section we derive some basic results concerning functors between functor categories regarded as categories with endomorphism as in Section 4. These results will be used repeatedly in this book.

Lemma 5.1 *Let* $(\mathscr{B},C,\tau)$ *be a category with endomorphism and* $F\colon \mathscr{C} \longrightarrow \mathscr{D}$ *be a functor. Then*

(i) *Composition on the left with* F *induces a functor* $F_*\colon \mathscr{C}^{\mathscr{B}} \longrightarrow \mathscr{D}^{\mathscr{B}}$ *that commutes with* C *and* τ. *Hence,* F_* *induces a unique functor* $F_!$ *such that the following diagram commutes*

$$\begin{array}{ccc} \mathscr{C}^{\mathscr{B}} & \xrightarrow{F_*} & \mathscr{D}^{\mathscr{B}} \\ {\scriptstyle Q}\downarrow & & \downarrow{\scriptstyle Q} \\ \mathscr{C}^{\mathscr{B}}(\Sigma^{-1}) & \xrightarrow{F_!} & \mathscr{D}^{\mathscr{B}}(\Sigma^{-1}) \end{array}$$

(ii) *If* $G\colon \mathscr{D} \longrightarrow \mathscr{C}$ *is a left adjoint for* F, *then* $G_!\colon \mathscr{D}^{\mathscr{B}}(\Sigma^{-1}) \longrightarrow \mathscr{C}^{\mathscr{B}}(\Sigma^{-1})$ *is a left adjoint for* $F_!$.

Proof: That F_* commutes with C and τ follows from an easy calculation. The rest of part (i) now follows from 3.1 and 3.2.

To prove (ii), let η: I $\longrightarrow$ FG and ε: GF $\longrightarrow$ I be the unit and counit (cf. [Ma, p. 81]) of the adjunction. It is easily seen that η and ε induce natural transformations $\eta_*: I \longrightarrow F_*G_*$ and $\varepsilon_*: G_*F_* \longrightarrow I$, where the two functors I are the identities of $\mathcal{D}^{\mathcal{B}}$ and $\mathcal{C}^{\mathcal{B}}$ respectively, that form the unit and counit of an adjunction between F_* and G_*. Part (ii) now follows from 3.6.

To illustrate the uses of 5.1, let the usual forgetful functor U: R-*mod* $\longrightarrow$ $\mathcal{S}ets$ play the role of F and the free module functor F: $\mathcal{S}ets$ $\longrightarrow$ R-*mod* that of G. Then by 5.1, U and F induce adjoint functors $R\mathcal{B}\text{-}mod \underset{F_!}{\overset{U_!}{\rightleftarrows}} \mathcal{S}ets^{\mathcal{B}}(\Sigma^{-1})$.

Let $(\mathcal{B}_i, C_i, \tau_i)$ (i=1,2) be categories with endomorphism. An almost endomorphism preserving functor F: $\mathcal{B}_1 \longrightarrow \mathcal{B}_2$ is said to be *endomorphism preserving* if there exists an integer $m \geq 0$ and a natural transformation $\beta: C_2F \longrightarrow FC_1^m$ such that the diagram

$$\begin{array}{ccc} F & \xrightarrow{\tau_2 F} & C_2F \\ & {\scriptstyle F\tau_1^m}\searrow & \downarrow{\scriptstyle\beta} \\ & & FC_1^m \end{array}$$

commutes. The reader will note that this is just condition (ii) of 3.6.

Lemma 5.2 *Let* $\mathcal{C}$ *be a category,* $(\mathcal{B}_i, C_i, \tau_i)$ (i=1,2) *be categories with endomorphism, and* F: $\mathcal{B}_1 \longrightarrow \mathcal{B}_2$ *be an endomorphism preserving functor. Then composition on the right induces an endomorphism preserving functor* $F^*: \mathcal{C}^{\mathcal{B}_2} \longrightarrow \mathcal{C}^{\mathcal{B}_1}$ *and, hence, a functor* $F^!$ *such that the* *following diagram commutes*

$$\begin{array}{ccc} \mathscr{C}^{\mathscr{B}_2} & \xrightarrow{\;F^*\;} & \mathscr{C}^{\mathscr{B}_1} \\ {\scriptstyle Q_2}\Big\downarrow & & \Big\downarrow{\scriptstyle Q_1} \\ \mathscr{C}^{\mathscr{B}_2}(\Sigma^{-1}) & \xrightarrow{\;F^!\;} & \mathscr{C}^{\mathscr{B}_1}(\Sigma^{-1}) \end{array}$$

If $\mathscr{C}$ *is a category with zero object and kernels, then* $F^!$ *preserves exact sequences that come from* $\mathscr{C}^{\mathscr{B}_2}$ *in the sense that if* $G_0 \longrightarrow G_1 \longrightarrow G_2$ *is exact in* $\mathscr{C}^{\mathscr{B}_2}$, *then* $F^!Q_2G_0 \longrightarrow F^!Q_2G_1 \longrightarrow F^!Q_2G_2$ *is exact in* $\mathscr{C}^{\mathscr{B}_1}(\Sigma^{-1})$.

Proof: It is well known that F^* is a functor. To see that F^* is endomorphism preserving, let $G \in |\mathscr{C}^{\mathscr{B}_2}|$ and apply G to the diagrams

$$\begin{array}{ccc} F & \xrightarrow{\;\tau_2 F\;} & C_2F \\ & {\scriptstyle F\tau_1^m}\searrow & \Big\downarrow{\scriptstyle \beta} \\ & & FC_1^m \end{array} \qquad\qquad \begin{array}{ccc} F & \xrightarrow{\;F\tau_1\;} & FC_1 \\ & {\scriptstyle \tau_2^n F}\searrow & \Big\downarrow{\scriptstyle \alpha} \\ & & C_2^nF \end{array}$$

respectively, to define $\bar{\alpha}$ and $\bar{\beta}$, respectively, on G and to show $\bar{\alpha}$ and $\bar{\beta}$ fit into commutative diagrams

$$\begin{array}{ccc} F^* & \xrightarrow{\;F^*\bar{\tau}_2\;} & F^*\bar{C}_2 \\ & {\scriptstyle \bar{\tau}_1^n F^*}\searrow & \Big\downarrow{\scriptstyle \bar{\alpha}} \\ & & \bar{C}_1^m F^* \end{array} \qquad\qquad \begin{array}{ccc} F^* & \xrightarrow{\;\bar{\tau}_1 F\;} & \bar{C}_1F \\ & {\scriptstyle F^*\bar{\tau}_2^n}\searrow & \Big\downarrow{\scriptstyle \bar{\beta}} \\ & & F^*\bar{C}_2^n \end{array}$$

The rest of the first part of the lemma follows from 3.2.

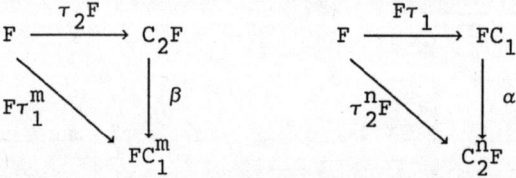

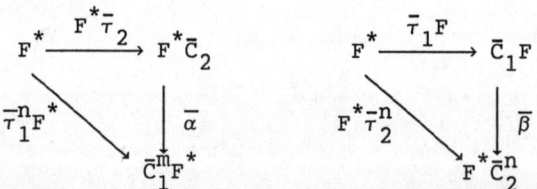

To prove the second part of 5.2, it now suffices to prove that if $G_0 \longrightarrow G_1 \longrightarrow G_2$ is exact in $\mathscr{C}^{\mathscr{B}_2}$, then $F^*G_0 \longrightarrow F^*G_1 \longrightarrow F^*G_2$ is exact in $\mathscr{C}^{\mathscr{B}_1}$. Let $B_1 \in |\mathscr{B}_1|$. Then by 4.2, $G_0(F(B_1)) \longrightarrow G_1(F(B_1)) \longrightarrow G_2(F(B_1))$ is exact in $\mathscr{C}$. Hence $F^*(G_0)B_1 \longrightarrow F^*(G_1)B_1 \longrightarrow F^*(G_2)B_1$ is exact in $\mathscr{C}$ by the definition of F^*. Since $B_1 \in |\mathcal{B}_1|$ is arbitrary, $F^*(G_0) \longrightarrow F^*(G_1) \longrightarrow F^*(G_2)$ is exact in $\mathcal{C}^{\mathcal{B}_1}$ by 4.2 and the proof of 5.2 is completed.

The following result is well known. A sketch of its proof is included here for later use.

Lemma 5.3 *Let $\mathcal{B}_i$ be U-small categories and $\mathcal{C}$ be U-small cocomplete. Let $F: \mathcal{B}_1 \longrightarrow \mathcal{B}_2$. Then the functor $F^*: \mathcal{C}^{\mathcal{B}_2} \longrightarrow \mathcal{C}^{\mathcal{B}_1}$ has a left adjoint $F_*: \mathcal{C}^{\mathcal{B}_1} \longrightarrow \mathcal{C}^{\mathcal{B}_2}$.*

Proof: To construct a left adjoint F_*, it suffices to find for each $G_1 \in |\mathcal{C}^{\mathcal{B}_1}|$, an object $F_*G_1 \in |\mathcal{C}^{\mathcal{B}_2}|$ and a morphism $\eta(G_1): G_1 \longrightarrow F^*F_*G_1$ with the following universal property: For any $G_2 \in |\mathcal{C}^{\mathcal{B}_2}|$ and any $\nu_1 \in \mathcal{C}^{\mathcal{B}_1}(G_1, F^*G_2)$, there is a unique $\nu_2 \in \mathcal{C}^{\mathcal{B}_2}(F_*G_1, G_2)$ such that $\nu_1 = F^*(\nu_2)\eta(G_1)$ (cf. [Ma, p.81] for example).

Thus let $G_1 \in |\mathcal{C}^{\mathcal{B}_1}|$. For any $B_2 \in |\mathcal{B}_2|$, let $F\backslash B_2$ be the category whose objects are pairs (B_1, f) where $B_1 \in |\mathcal{B}_1|$ and $f: F(B_1) \longrightarrow B_2$ is a morphism in $\mathcal{B}_2$. A morphism $g: (B_1, f) \longrightarrow (B_1', f')$ is a morphism $g: B_1 \longrightarrow B_1'$ in $\mathcal{B}_1$ such that $f = f'F(g)$.

Let $G_{1,B_2}: F\backslash B_2 \longrightarrow \mathcal{C}$ be the functor given by $G_{1,B_2}(B_1, f) = G_1(B_1)$ and $G_{1,B_2}(g) = G_1(g)$. Since $\mathcal{B}_1$ and $\mathcal{B}_2$ are U-small, so is $F\backslash B_2$. Since $\mathcal{C}$ is U-small cocomplete, the functor G_{1,B_2} has a colimit in $\mathcal{C}$. Let the value of $F_*(G_1)$ on B_2 be the colimit.

We leave it to the reader to construct the arrow $\eta(G_1): G_1 \longrightarrow F^*F_*G_1$ satisfying the mapping property above.

The following result is now an immediate corollary of 5.3 and 3.6.

Corollary 5.4 *Let $(\mathscr{B}_i, C_i, \tau_i)$ be U-small categories with endomorphism and $\mathscr{C}$ be U-small cocomplete. Let $F: \mathscr{B}_1 \longrightarrow \mathscr{B}_2$ be endomorphism preserving. Then the functor $F^!$ has a left adjoint $F_!$ such that the following diagram commutes*

$$\begin{array}{ccc} \mathscr{C}^{\mathscr{B}_1} & \xrightarrow{\ F_*\ } & \mathscr{C}^{\mathscr{B}_2} \\ {\scriptstyle Q_1}\big\downarrow & & \big\downarrow{\scriptstyle Q_2} \\ \mathscr{C}^{\mathscr{B}_1}(\Sigma^{-1}) & \xrightarrow{\ F_!\ } & \mathscr{C}^{\mathscr{B}_2}(\Sigma^{-1}) \end{array}$$

We illustrate the behavior of $F_!$ by considering the following example. It will be used several times in this book.

Example 5.5 Let $(\mathscr{B}, C, \tau)$, $G: \mathscr{B} \longrightarrow \mathscr{G}poid$ and $\mathscr{B}G$ be as in 1.3 and let $\rho: \mathscr{B}G \longrightarrow \mathscr{B}$ be the forgetful functor. Then ρ is endomorphism preserving and for any U-small cocomplete category $\mathscr{C}$, $\rho^!$ has a left adjoint $\rho_!: \mathscr{C}^{\mathscr{B}G}(\Sigma^{-1}) \longrightarrow \mathscr{C}^{\mathscr{B}}(\Sigma^{-1})$.

For any $B \in |\mathscr{B}|$, let $J_B: G(B) \longrightarrow \rho\backslash B$ be the functor given by $J_B(x) = ((x,B),1)$, $J_B(g) = (g,1)$ for $x \in |G(B)|$ and $g \in G(B)(x,y)$. If $f: B_1 \longrightarrow B_2$, then f induces a functor $f_*: \rho\backslash B_1 \longrightarrow \rho\backslash B_2$ and a natural transformation $\eta(f): f_*J_{B_1} \longrightarrow J_{B_2}G(f)$ by setting $f_*((x',B_1'),h) = ((x',B_1'),fh)$, $f_*(g,h) = (g,h)$, and $\eta(f)(x) = (1,f)$.

Let $F: \mathscr{B}G \longrightarrow \mathscr{C}$ and consider the following diagram of categories and functors

$$\begin{array}{ccccc} G(B_1) & \xrightarrow{\ J_{B_1}\ } & \rho\backslash B_1 & & \\ {\scriptstyle G(f)}\big\downarrow & & \big\downarrow{\scriptstyle f_*} & \searrow^{F_{B_1}} & \\ & & & & \mathscr{C} \\ & & & \nearrow_{F_{B_2}} & \\ G(B_2) & \xrightarrow{\ J_{B_2}\ } & \rho\backslash B_2 & & \end{array}$$

and pass to colimits. It is readily verified that $\eta(f)$ induces a morphism such that the following diagram commutes

$$\begin{array}{ccc} \operatorname{colim}\,(F_{B_1}J_{B_1}) & \longrightarrow & \operatorname{colim} F_{B_1} \\ {\scriptstyle \operatorname{colim}\eta(f)}\downarrow & & \downarrow \\ \operatorname{colim}\,(F_{B_2}J_{B_2}) & \longrightarrow & \operatorname{colim} F_{B_2} \end{array}$$

where the other morphisms are induced by the identity natural transformations. Furthermore, $H = H(F)\colon \mathcal{B} \longrightarrow \mathcal{C}$ defined by setting

(i) $H(B) = \operatorname{colim}(F_B J_B)$, for $B \in |\mathcal{B}|$; and

(ii) $H(f) = \operatorname{colim}\eta(f)$, for $f \in \mathcal{B}(B_1,B_2)$

is a functor.

Lemma 5.6 *For any* $F \in |\mathcal{C}^{\mathcal{B}G}(\Sigma^{-1})|$, $\rho_!(F) = Q(H)$ *where* H *is the functor described above. In particular, for any* $B \in |\mathcal{B}|$,

$$H(B) = \coprod F(x_\lambda,B)/\mathrm{Aut}_1(x_\lambda,B)$$

where $\{x_\lambda\}$ *is a set of representatives for the components of the category* $G(B)$ *and* $\mathrm{Aut}_1(x_\lambda,B) = \{(f,1) \in \mathcal{B}G((x_\lambda,B),(x_\lambda,B))\}$.

In this lemma $F(x_\lambda,B)/\mathrm{Aut}_1(x_\lambda,B)$ is a (suggestive) notation for the colimit of the diagram with the single object $F(x_\lambda,B)$ and all morphisms of the form $F(f,1)$ for $(f,1) \in \mathrm{Aut}_1(x_\lambda,B)$.

Proof: The proof of 5.3 shows that $\rho_!(F)$ is represented by the functor whose value on $f\colon B_1 \longrightarrow B_2$ is the right hand vertical arrow in the above diagram; while $H(f)$ is given by the left hand vertical arrow. Since it is easy to see that $G(B)$ is cofinal in $\rho\backslash B$, the horizontal arrows in that diagram are isomorphisms. The lemma now follows easily.

On rare occasions in the sequel, we will need to know how F_* and $F_!$ behave with respect to a change of base functor. So suppose that $\mathcal{C}$ and $\mathcal{D}$ are both U-small cocomplete categories and let $G\colon \mathcal{C} \longrightarrow \mathcal{D}$ be a functor. Then for any diagram $A\colon \mathcal{J} \longrightarrow \mathcal{C}$ the universal property of colimits defines a morphism in $\mathcal{D}$, $\nu\colon \operatorname{colim} GA \longrightarrow G(\operatorname{colim} A)$ where both colimits are taken over $i \in |\mathscr{I}|$. We say that G *preserves colimits*, if υ is an isomorphism for all diagrams in $\mathscr{C}$.

Proposition 5.7 *Let* $F\colon \mathscr{B}_1 \longrightarrow \mathscr{B}_2$ *be a functor,* $\mathscr{C}$ *and* $\mathscr{D}$ *be* U-*small cocomplete categories, and* $G\colon \mathscr{C} \longrightarrow \mathscr{D}$ *preserve colimits. Then* υ *induces a natural equivalence* $\upsilon_*\colon F_*^{\mathscr{D}} G_* \longrightarrow G_* F_*^{\mathscr{C}}$; *i.e. the diagram*

$$\begin{array}{ccc}
\mathscr{C}^{\mathscr{B}_1} & \xrightarrow{\;F_*^{\mathscr{C}}\;} & \mathscr{C}^{\mathscr{B}_2} \\
{\scriptstyle G_*}\downarrow & & \downarrow{\scriptstyle G_*} \\
\mathscr{D}^{\mathscr{B}_1} & \xrightarrow{\;F_*^{\mathscr{D}}\;} & \mathscr{D}^{\mathscr{B}_2}
\end{array}$$

commutes up to a natural equivalence. Furthermore, if $\eta\colon I \longrightarrow F^* F_*$ *is the unit of the adjunction of* 5.3, *then the following diagram of natural transformations commutes*

$$\begin{array}{ccc}
G_* & \xrightarrow{\;G_*(\eta_{\mathscr{C}})\;} & G_* F_{\mathscr{C}}^* F_*^{\mathscr{C}} \\
 & & \| \\
{\scriptstyle 1}\downarrow & & F_{\mathscr{D}}^* G_* F_*^{\mathscr{C}} \\
 & & \uparrow{\scriptstyle F_{\mathscr{D}}^*(\upsilon_*)} \\
G_* & \xrightarrow{\;\eta_{\mathscr{D}}\;} & F_{\mathscr{D}}^* F_*^{\mathscr{D}} G_*
\end{array}$$

We note that a simple computation shows that $G_* F_{\mathscr{C}}^* = F_{\mathscr{D}}^* G_*$. The proof of the proposition, which is a tedious exercise using the constructions of F_* and η in the proof of 5.3, is left to the reader.

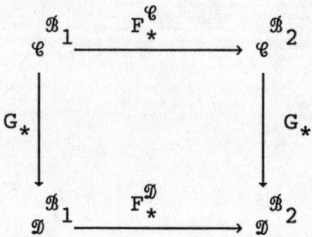

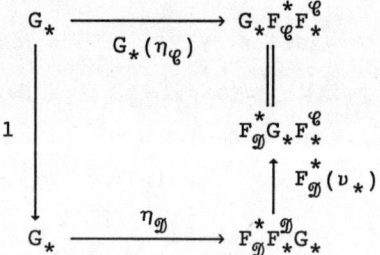

Corollary 5.8 *The following diagram commutes*

$$
\begin{array}{ccc}
\mathscr{C}^{\mathscr{B}_2}(F_*A,\ B) & \xrightarrow{\ \varphi_{\mathscr{C}}\ } & \mathscr{C}^{\mathscr{B}_1}(A,\ F^*B) \\
\Big\downarrow G_\# & & \Big\downarrow G_\# \\
\mathscr{D}^{\mathscr{B}_2}(G_*F_*A,\ B) & & \mathscr{D}^{\mathscr{B}_1}(G_*A,\ G_*F^*B) \\
\Big\uparrow \upsilon^\# & & \Big\| \\
\mathscr{D}^{\mathscr{B}_2}(F_*G_*A,\ B) & \xrightarrow{\ \varphi_{\mathscr{D}}\ } & \mathscr{D}^{\mathscr{B}_1}(G_*A,\ F^*G_*B)
\end{array}
$$

In this diagram φ *is the isomorphism defining the adjunction between* F^* *and* F_*, $C_\#$ *is obtained by applying the functor* G, *and* $\upsilon^\#$ *is obtained by composition on the right with* υ.

Proof: This follows immediately from 5.7 and [Ma, Thm 1(i), p.80].

Remark 5.9 The results described in 5.7 and 5.8 carry over to $F_!$ and $F^!$ with only the obvious changes of notation.

6. Equivalences of Categories.

Let $(\mathscr{B}_i, C_i, \tau_i)$ (i=1,2) be a category with endomorphism, $F\colon \mathscr{B}_1 \longrightarrow \mathscr{B}_2$ be an endomorphism preserving functor, and $\mathscr{C}$ be a cocomplete category. In this section we give some conditions on F sufficient to guarantee that the functors $F_!\colon \mathscr{B}_1(\Sigma^{-1}) \longrightarrow \mathscr{B}_2(\Sigma^{-1})$ and $F_!\colon \mathscr{C}^{\mathscr{B}_1}(\Sigma^{-1}) \longrightarrow \mathscr{C}^{\mathscr{B}_2}(\Sigma^{-1})$ of 3.2 and 5.4 are equivalences of categories. A specialization of the main result of this section, Theorem 6.2, will be given in II.5.1 and will be used repeatedly in this book.

Throughout this section we will assume that F commutes with C and τ (i.e. that $FC_1 = C_2F$ and $F(\tau_1) = \tau_{2F}$). Although this is a somewhat restrictive hypothesis, it will greatly simplify both the proofs and the exposition in this section.

We recall that a functor $F: \mathcal{C} \longrightarrow \mathcal{D}$ is called an *equivalence of categories* if there exist a functor $G: \mathcal{D} \longrightarrow \mathcal{C}$, called a *pseudo-inverse* for F, and natural equivalences $\eta: GF \longrightarrow I_{\mathcal{C}}$ and $\upsilon: FG \longrightarrow I_{\mathcal{D}}$.

Let $F: \mathcal{B}_1 \longrightarrow \mathcal{B}_2$ be a functor between categories with endomorphism that commutes with C and τ. We call F an *eventual equivalence of categories* if there exist a functor $G: \mathcal{B}_2 \longrightarrow \mathcal{B}_1$, called an *eventual pseudo-inverse* for F, and natural equivalences $\mu: GF \longrightarrow C_1^N$ and $\upsilon: FG \longrightarrow C_2^N$ for some integer $N \geq 0$. The functor G is called a *strong eventual pseudo-inverse* for F if there exists a natural equivalence $\delta: GC_2 \longrightarrow C_1G$ for which the diagram

$$\begin{array}{ccc} G & \xrightarrow{G(\tau_2)} & GC_2 \\ & \searrow^{\tau_{1G}} & \downarrow \delta \\ & & C_1G \end{array}$$

commutes and F is called a *strong eventual equivalence of categories* if it has a strong eventual pseudo-inverse.

Lemma 6.1 *Let* $(\mathcal{B}_i, C_i, \tau_i)$ $(i=1,2)$ *be categories with endomorphism,* $F: \mathcal{B}_1 \longrightarrow \mathcal{B}_2$ *be a functor that commutes with* C *and* τ, *and* $\mathcal{C}$ *be a co-complete category. If* F *is a strong eventual equivalence of categories, then we have the following equivalences of categories:*

(i) $F_!: \mathcal{B}_1(\Sigma^{-1}) \longrightarrow \mathcal{B}_2(\Sigma^{-1})$ *and*

(ii) $F_!: \mathcal{C}^{\mathcal{B}_1}(\Sigma^{-1}) \longrightarrow \mathcal{C}^{\mathcal{B}_2}(\Sigma^{-1})$

The proof is given later in this section.

Let $F: \mathcal{B}_1 \longrightarrow \mathcal{B}_2$ be a functor between categories with endomorphism as above. We say that F is *eventually onto* if there exists an integer $r \geq 0$ such that for every $B_2 \in |\mathcal{B}_2|$, there exists a $B_1 \in |\mathcal{B}_1|$ and an isomorphism $g: F(B_1) \longrightarrow C_2^r B_2$. We say, F is *eventually full* if there exists an integer $s \geq 0$ such that given any morphism $h: F(B_1) \longrightarrow F(B_1')$ $(B_1, B_1' \in |\mathcal{B}_1|)$, there exists a morphism $f: C_1^s B_1 \longrightarrow C_1^s B_1'$ such that $F(f) = C_2^s(h)$; i.e. the diagram

$$\begin{array}{ccc} F(C_1^s B_1) & \xrightarrow{F(f)} & F(C_1^s B_1') \\ \| & & \| \\ C_2^s F(B_1) & \xrightarrow{C_2^s(h)} & C_2^s F(B_1') \end{array}$$

commutes. We shall call the morphism f a *lift* of h. Finally, we say F is *eventually faithful* if there exists an integer $t \geq 0$ such that if $f_1', f_1'': B_1 \longrightarrow B_1''$ are two morphisms in $\mathcal{B}_1$ for which $F(f_1') = F(f_1'')$, then $C_1^t(f_1') = C_1^t(f_1'')$.

Theorem 6.2 *Let* $(\mathcal{B}_i, C_i, \tau_i)$ $(i=1,2)$ *be categories with endomorphism and* $F: \mathcal{B}_1 \longrightarrow \mathcal{B}_2$ *be a functor that commutes with* C *and* τ. *If* F *is eventually onto, eventually full, and eventually faithful, then* F *is a strong eventual equivalence of categories. In particular, for any co-complete category* $\mathcal{C}$, $F_!: \mathcal{C}^{\mathcal{B}_1}(\Sigma^{-1}) \longrightarrow \mathcal{C}^{\mathcal{B}_2}(\Sigma^{-1})$ *is an equivalence of categories.*

The proof is given later in this section.

Proof of 6.1: Let $G: \mathcal{B}_2 \longrightarrow \mathcal{B}_1$ be a strong eventual pseudo-inverse for F. It follows from 3.2 that G induces a functor $G_!: \mathcal{B}_2(\Sigma^{-1}) \longrightarrow \mathcal{B}_1(\Sigma^{-1})$ and from 3.3 that $\mu: GF \longrightarrow C_1^N$ and $\nu: FG \longrightarrow C_2^N$ induce natural equivalences $\mu_!: G_! F_! \longrightarrow (C_1^N)_!$ and $\nu_!: F_! G_! \longrightarrow (C_2^N)_!$. Since $(\tau_i^N)_!$ is a natural equi-valence from the identity on $\mathcal{B}_i(\Sigma^{-1})$ to $(C_i^N)_!$ by 3.4, part (i) follows.

To prove (ii), we note first that if $F_1, F_2: \mathcal{B}_1 \longrightarrow \mathcal{B}_2$ are functors and $\alpha: F_1 \longrightarrow F_2$ is a natural transformation (respectively, equivalence), it is easy to see that if $F_i^*: \mathcal{C}^{\mathcal{B}_2} \longrightarrow \mathcal{C}^{\mathcal{B}_1}$ $(i=1,2)$ is the functor of 5.2, then $\alpha^*: F_1^* \longrightarrow F_2^*$ given by $\alpha^*(H) = H(\alpha)$ for $H \in |\mathscr{C}^{\mathscr{B}_2}|$ is also a natural transformation (respectively, equivalence). Now if $F: \mathscr{B}_1 \longrightarrow \mathscr{B}_2$ is a strong eventual equivalence with strong eventual pseudo-inverse G, the proof of 5.2 shows that F^* commutes with C and τ and the preceeding remark then shows that F^* is a strong eventual pseudo-equivalence with strong eventual pseudo-inverse G^*. It then follows from (i) that $F^!: \mathscr{C}^{\mathscr{B}_2}(\Sigma^{-1}) \longrightarrow \mathscr{C}^{\mathscr{B}_1}(\Sigma^{-1})$ is an equivalence of categories with pseudo-inverse $G_!$. By symmetry $G^!$ is an equivalence of categories with pseudo-inverse $F^!$.

Since it is a general fact from category theory that the pseudo-inverse of an equivalence is a left adjoint for the equivalence (cf. [Ma, p. 91]), $G^!$ is a left adjoint for $F^!$. Since $F_!$ is also a left adjoint for $F^!$, $F_!$ is naturally equivalent to $G^!$. It now follows easily that $F_!$ is an equivalence of categories since $G^!$ is and (ii) is established.

Proof of 6.2: We begin by constructing a functor $G: \mathscr{B}_2 \longrightarrow \mathscr{B}_1$ that will be a strong eventual pseudo-inverse for F. To do this, use the fact that F is eventually onto to choose for each $B_2 \in |\mathscr{B}_2|$, an object $B_1 \in |\mathscr{B}_1|$ and an isomorphism $g: F(B_1) \longrightarrow C_2^r B_2$. These choices will remain fixed for the remainder of this proof. We set $G(B_2) = C_1^{s+t}(B_1)$ and, for $h: B_2 \longrightarrow B_2'$, $G(h) = C_1^t f$ where $f: C_1^s B_1 \longrightarrow C_1^s B_1'$ is any lift of $(g')^{-1}(C_2^r h)g$. That f exists follows from the fact that F is eventually full. That $C_1^t(f)$ is well defined follows from the fact that F is eventually faithful. The reader may check that G is a functor.

We let $N = r+s+t$ and define $\nu: FG \longrightarrow C_2^N$ by noting first that for every $B_2 \in |\mathscr{B}_2|$, $FG(B_2) = F(C_1^{s+t}B_1) = C_2^{s+t}F(B_1)$, observing that $C_2^{s+t}(g): C_2^{s+t}F(B_1) \longrightarrow C_2^{r+s+t}B_2$, and setting $\nu_{B_2} = C_2^{s+t}(g)$. Since g is an isomorphism, so is ν_{B_2}. Hence ν is a natural equivalence if it is a natural transformation. To see that ν is a natural transformation recall that if $h\colon B_2 \longrightarrow B_2'$, then $G(h) = C_1^t f$ where $f\colon C_1^s B_1 \longrightarrow C_1^s B_1'$ is a lift of $(g')^{-1}(C_2^r f)g$. That is, $F(f) = C_2^s(g')^{-1}C_2^{s+t}(h)C_2^s(g)$ or $C_2^s(g')F(f) = C_2^{s+t}(h)C_2^s(g)$. But then the following diagram commutes

$$\begin{array}{ccccccc}
FG(B_2) & = & FC_1^{s+t}(B_1) & = & C_2^t F(C_1^s B_1) & \xrightarrow{C_2^{s+t}(g)} & C_2^{r+s+t}(B_1) \\
\downarrow{\scriptstyle FG(h)} & & \downarrow{\scriptstyle F(C_1^t f)} & & \downarrow{\scriptstyle C_2^t F(f)} & & \downarrow{\scriptstyle C_2^{r+s+t}(h)} \\
FG(B_2') & = & FC_1^{s+t}(B_1') & = & C_2^t F(C_1^s B_1') & \xrightarrow{C_2^{s+t}(g')} & C_2^{r+s+t}(B_2')
\end{array}$$

Since the rows are ν_{B_2} and $\nu_{B_2'}$, respectively, this shows that ν is a natural transformation.

We recall that $N = r+s+t$ and define $\mu\colon GF \longrightarrow C_1^N$ as follows: Let $\bar{B}_1 \in |\mathcal{B}_1|$ and $B_1 = GF(\bar{B}_1)$. By the way G is defined, there is a preferred isomorphism $g\colon F(B_1) \longrightarrow C_2^r F(\bar{B}_1) = F(C_1^r \bar{B}_1)$. Lift g to $u\colon C_1^s B_1 \longrightarrow C_1^{r+s}\bar{B}_1$ using the fact that F is eventually full, note that then $F(u) = C_2^s(g)$, and set $\mu_{\bar{B}_1} = C_1^t u$.

Let $\bar{f}\colon \bar{B}_1 \longrightarrow \bar{B}_1'$ and consider the following diagram in which f is a lift of $(g')^{-1}C_2^r(\bar{f})g$

$$\begin{array}{ccc}
C_1^s B_1 & \xrightarrow{u} & C_1^{r+s}\bar{B}_1 \\
{\scriptstyle f}\downarrow & & \downarrow{\scriptstyle C_1^{r+s}\bar{f}} \\
C_1^s B_1' & \xrightarrow{u'} & C_1^{r+s}\bar{B}_1'
\end{array}$$

Then $F(f) = C_2^s(g')^{-1}C_2^{r+s}F(\bar{f})C_2^s(g) = F(u')^{-1}F(C_1^{r+s}\bar{f})F(u)$ from which it follows that $F(u'f) = F(C_1^{r+s}\bar{f}u)$. Since F is eventually faithful, we now have that $\mu_{\bar{B}_1'}GF(f) = C_1^t(u')C_1^t(f) = C_1^{r+s+t}(\bar{f})C_1^t(u) = C_1^N(\bar{f})\mu_{\bar{B}_1}$ which shows that μ is a natural transformation. That μ is a natural equivalence follows by letting g and u, respectively, play the roles of h and f, respectively, in the following lemma:

Lemma 6.3 *Let* $B_1, B_1' \in |\mathcal{B}_1|$, $h\colon F(B_1) \longrightarrow F(B_1')$ *be an isomorphism in* $\mathcal{B}_2$, *and* $f\colon C_1^s B_1 \longrightarrow C_1^s B_1'$ *be a lift of* h. *Then* $C_1^t(f)$ *is an isomorphism.*

Proof: Let $h_1: F(B_1') \longrightarrow F(B_1)$ be an inverse for h_1 and $f_1: C_1^s B_1' \longrightarrow C_1^s B_1$ be a lift of h_1. Then $F(f_1 f) = F(f_1)F(f) = C_2^s(h_1)C_2^s(h) = C_2^s(h_1 h) = C_2^s(1_{FB_1}) = F(1_{C_1^s B_1})$. Since F is eventually faithful, it follows that $C_1^t(f_1)C_1^t(f) = 1_{C_1^{s+t}B_1}$. Similarly, $C_1^t(f)C_1^t(f_1) = 1_{C_1^{s+t}B_1'}$ and the lemma follows.

It remains only to construct a natural equivalence $\delta: GC_2 \longrightarrow C_1G$ for which the diagram

$$\begin{array}{ccc} G & \xrightarrow{G(\tau_2)} & GC_2 \\ & \searrow^{\tau_{1G}} & \downarrow \delta \\ & & C_1G \end{array}$$

commutes. To do this, let $B_2 \in |\mathcal{B}_2|$, $g: F(B_1) \longrightarrow C_2^r B_2$ and $\bar{g}: F(\bar{B}_1) \longrightarrow C_2^r(C_2B_2)$ be the isomorphisms chosen at the outset of this proof, $d_1: C_1^s\bar{B}_1 \longrightarrow C_1^{s+1}B_1$ be a lift of the composite

$$F(\bar{B}_1) \xrightarrow{\bar{g}} C_2^r C_2 B_2 \xrightarrow{C_2(g)^{-1}} C_2F(B_1) = F(C_1B_1),$$

and set $\delta_{B_2} = C_1^t(d_1)$. Since $C_2(g)^{-1}\bar{g}$ is an isomorphism, so is δ_{B_2} by 6.3.

Let $\tau_2: B_2 \longrightarrow C_2B_2$ and $t_1: C_1^sB_1 \longrightarrow C_1^s\bar{B}_1$ be a lift of $(\bar{g})^{-1}C_2^r(\tau_2)g$. Then $G(\tau_2) = C_1^t(t_1)$ and $F(t_1) = (C_2^s\bar{g})^{-1}C_2^{r+s}(\tau_2)C_2^s g$. But then $F(d_1t_1) = C_2^{s+1}(g)^{-1}C_2^s(\bar{g})(C_2^s\bar{g})^{-1}C_2^{r+s}(\tau_2)C_2^s(g) = C_2^{s+1}(g^{-1})C_2^{r+s}(\tau_2)C_2^s(g)$. On the other hand, since F commutes with C and τ and since τ_2 is a natural transformation, the diagram

$$\begin{array}{ccccc} F(C_1^sB_1) & = & C_2^sF(B_1) & \xrightarrow{C_2^s(g)} & C_2^{r+s}B_2 \\ {\scriptstyle F(C_1^s\tau_1)}\downarrow & & \downarrow{\scriptstyle C_2^s\tau_2} & & \downarrow{\scriptstyle C_2^s(\tau_{2C^r})=C_2^{r+s}(\tau_2)} \\ F(C_1^{s+1}B_1) & = & C_2^{s+1}F(B_1) & \xrightarrow{C_2^{s+1}(g)} & C_2^{r+s+1}B_2 \end{array}$$

commutes. Thus $F(C_1^s\tau_1) = C_2^{s+1}(g)^{-1}C_2^{r+s}(\tau_2)C_2^s(g) = F(d_1t_1)$ and since F is eventually faithful $\tau_{1GB_2} = C_1^{s+t}\tau_{B_1} = C_1^t(C_1^s\tau_1) = C_1^t(d_1)C_1^t(t_1) = \delta_{B_2}G(\tau_2)$. Thus

$$\begin{array}{ccc} G(B_2) & \xrightarrow{G(\tau_2)} & GC_2(B_2) \\ & \searrow^{\tau_{1GB_2}} & \downarrow \delta_{B_2} \\ & & C_1G(B_2) \end{array}$$

commutes as required.

We now show that δ is a natural transformation. To do this, let $h\colon B_2 \longrightarrow B_2'$ and choose lifts $f_1\colon C_1^sB_1 \longrightarrow C_1^sB_1'$ and $f_2\colon C_1^s\bar{B}_1 \longrightarrow C_1^s\bar{B}_1'$ of $(g')^{-1}(C_2^rh)g$ and $(\bar{g}')^{-1}C_2^r(C_2h)\bar{g}$ respectively. Consider the following diagram

$$\begin{array}{ccc} C_1^s\bar{B}_1 & \xrightarrow{d_1} & C_1^{s+1}B_1 \\ f_2\downarrow & & \downarrow C_1f_1 \\ C_1^s\bar{B}_1' & \xrightarrow{d_1'} & C_1^{s+1}B_1' \end{array}$$

in which d_1 and d_1' are the lifts of $C_2(g)^{-1}\bar{g}$ and $C_2(g')^{-1}\overline{g'}$, respectively, used to define δ_{B_2} and $\delta_{B_2'}$, respectively. Since

$$\begin{aligned} F(C_1f_1)F(d_1) &= C_2F(f_1)F(d_1) \\ &= [(C_2^{s+1}g')^{-1}(C_2^{r+s+1}h)(C_2^{s+1}g)][(C_2^{s+1}g)^{-1}(C_2^s\bar{g})] \\ &= (C_2^{s+1}g')^{-1}(C_2^{r+s+1}h)(C_2^s\bar{g}) \\ &= [(C_2^{s+1}g')^{-1}C_2^s(\overline{g'})][C_2^s(\overline{g'})^{-1}(C^{r+s+1}h)(C_2^s\bar{g})] \\ &= F(d_1')F(f_2), \end{aligned}$$

we have $F(C_1f_1{\circ}d_1) = F(d_1'f_2)$. Since F is eventually faithful it follows that $C_1^t(C_1f_1{\circ}d_1) = C_1^t(d_1'f_2)$ and

$$C_1G(h)\delta_{B_2} = (C_1^{t+1}f_1)(C_1^td_1) = (C_1^td_1')(C_1^tf_2) = \delta_{B_2'}GC_2(h).$$

Thus δ is a natural transformation and the proof of 6.2 is complete.

CHAPTER II

THE ALGEBRAIC TOPOLOGY OF BOUNDEDLY CONTROLLED SPACES

This chapter develops the basic algebraic topological tools needed to study spaces in the categories $\mathcal{T}op^c/Z$ and $\mathcal{CW}^c/Z$. We again urge that this chapter be read in conjunction with Chapter I. The main ideas in Chapter I developed out of necessity as the authors undertook the study of spaces in $\mathcal{T}op^c/Z$ and the development of their algebraic topology. Thus the motivation for most of the results in Chapter I lies in the developments in this chapter and they are probably most easily understood in that context.

In section 1, we define the notion of a boundedness control structure, introduce the concepts of bc spaces and bc CW-complexes, and define the categories $\mathcal{T}op^c/Z$ and $\mathcal{CW}^c/Z$. Several examples of boundedness control structures are given which illustrate the generality of this idea.

The basic idea underlying the definition of homology or homotopy of a bc space is to view the space as an assembled jigsaw puzzle made up of overlapping pieces called "fragments". The algebraic invariants for the whole space are built out of standard algebraic invariants for the fragments that are assembled into a whole following the pattern of the puzzle. This is encoded by the boundedness control structure.

In section 2 we introduce the ideas of a fragmented space and of a fragmentation of a bc space or bc CW-complex. A basic invariant of a fragmented space, its fundamental groupoid, is defined and used in introducing the universal cover of a fragmented space in Example 2.3. Some basic results on fragmentations are also established.

The homology of fragmented spaces or pairs is defined in section 3 and its basic properties are developed. Similarly, the main homotopy theoretic invariants of fragmented spaces and pairs are defined and developed in section 4. These include the appropriate analogues of the usual homotopy groups. Section 5 explores the relationship between the fundamental groupoid and homotopy of a bc space.

Analogues, for fragmented spaces and pairs, of the absolute and relative Hurewicz theorems are developed in sections 6 through 8. The main results are Theorems 6.1, 7.1, and 8.2. The reader may well want to skip the proofs of 6.1 and 7.1 since they follow the standard proofs fairly closely.

The homology and homotopy of bc spaces and pairs are defined in section 9 essentially by composing the fragmentation functors of section 2 with the homology (or homotopy) of a fragmented space or pair. The main properties of these invariants are described briefly and the Hurewicz theorems (Theorems 9.1 and 9.2) are recorded.

Basic results needed for studying bc maps between bc CW-complexes are proven in section 10. These include the Cellular Approximation Theorem (Corollary 10.3) and two versions of the Whitehead Theorem, one involving homotopy (Corollary 10.4) and one involving homology (Corollary 10.5).

1. The categories $\mathcal{Top}^c/Z$ and $\mathcal{CW}^c/Z$.

This section describes, in their full generality, the categories of spaces we wish to study. We also give several of the main examples of such categories.

Definition 1.1 *Let* Z *be a space and regard its power set* 2^Z *as a partially ordered set (poset). A* boundedness control structure *on* Z *is a pair* (P,C) *such that*

(i) P *is a subposet of* 2^Z *and* $C: P \longrightarrow P$ *is an order preserving function such that for all* $K \in P$, $K \subseteq C(K)$.

(ii) *For all* $K \in P$, $Z = \cup_n C^n(K)$.

(iii) *For all* $K \in P$, *there exists a minimal element* $K_0 \in P$ *with* $K_0 \subseteq K$.

(iv) *There exists a function* $\theta\colon \mathbb{Z}_+ \longrightarrow \mathbb{Z}_+$ *such that if* $K_0 \in P$ *is minimal,* $L \in P$, *and* $C^nK_0 \cap L \neq \emptyset$, *then* $K_0 \subseteq C^{\theta(n)}(L)$.

It follows from (i) and I.1.2 that (P,C) determines a category with endomorphism (P,C,τ). In the sequel, when we regard P as a category, we shall denote it by $\mathcal{P}$ and we call C the *boundedness control functor.*

A triple (Z,P,C) consisting of a space Z and a boundedness control structure (P,C) on Z is called a *boundedness control space.* When no confusion arises, we shall denote a boundedness control space simply by Z and suppress mention of the boundedness control structure.

Let Z be a boundedness control space and $A \subseteq Z$. The *radius* of A, radA, is defined to be

$$\text{radA} = \inf\{n \in \mathbb{Z}_+ \cup \{\infty\} \mid A \subseteq C^nK_0 \text{ for some minimal element } K_0 \in P\}.$$

A collection $\mathcal{A}$ of subsets of Z is called *bounded* if $\{\text{radA} \mid A \in \mathcal{A}\}$ is a bounded subset of $\mathbb{Z}_+$.

Let (Z,P,C) be a boundedness control space. A *boundedly controlled space* (or, simply, a *bc space*) over Z is a pair (X,p) where X is a space and $p\colon X \longrightarrow Z$ is a continuous map. Let (X_i,p_i) (i=1,2) be bc spaces over Z. A map $f\colon X_1 \longrightarrow X_2$ is *boundedly controlled* (or simply *bc*) if there exists an integer $m \geq 0$ such that for all $K \in P$, $f(p_1^{-1}(K)) \subseteq p_2^{-1}(C^m(K))$. Clearly the composite of bc maps is bc and we let $\mathcal{T}op^c/Z$ denote the category of bc spaces over Z and bc maps.

Let (X,p) be a bc space over Z and $A \subseteq X$. Then the *radius* of *A*, radA, is defined by setting radA = rad p(A).

A boundedly controlled CW-complex (or simply a *bc CW-complex*) over the boundedness control space (Z,P,C) is a pair (X,p) where (X,p) is a bc space over Z with X a *finite dimensional* CW-complex and $\{p(e) \mid e \text{ is a cell in } X\}$ a bounded collection of subsets of Z. Note that this means that there exists an integer $d \geq 0$ such that for every cell $e \in X$, there exists a minimal element $K_e \in P$ with $e \subseteq p^{-1}(C^d K_e)$. Let $\mathcal{CW}^c/Z$ be the category whose objects are bc CW-complexes over Z and whose morphisms are bc maps $f\colon (X_1,p_1) \longrightarrow (X_2,p_2)$.

A bc CW-complex (X,p) is *finite* if, for each $K \in P$, $p^{-1}(K)$ is contained in a finite subcomplex of X. Let $\mathcal{CW}^c_f/Z$ be the full subcategory of $\mathcal{CW}^c/Z$ whose objects are the finite bc CW-complexes over Z.

The categories of pairs of bc spaces (or bc CW-complexes, respectively) over Z, denoted $\mathcal{PTop}^c/Z$ (or $\mathcal{PCW}^c/Z$, respectively), are defined similarly. An object in $\mathcal{PTop}^c/Z$ is a triple (X,Y,p) where (X,p) and $(Y,p|Y)$ are bc spaces (or bc CW-complexes, respectively) over Z; while a morphism $f\colon (X,Y,p) \longrightarrow (V,W,q)$ is required to restrict to bc maps $f\colon (X,p) \longrightarrow (V,q)$ and $f|Y\colon (Y,p|Y) \longrightarrow (W,q|W)$.

To help the reader develop some intuition, we give several examples of how boundedness control structures arise.

Example 1.2 Let Z be any space. The *indiscrete boundedness control structure* (P,C) on Z is given by $P = \{Z\}$ and $C(Z) = Z$. In this case, $\mathcal{CW}^c/Z$ and $\mathcal{CW}^c_f/Z$, respectively, are the (usual) categories of CW complexes and finite CW complexes, respectively.

Let Z be a metric space. A metric $\rho\colon Z \times Z \longrightarrow \mathbb{R}_+$ is called *proper* if for all $z \in Z$, $\rho(z,\ )\colon Z \longrightarrow \mathbb{R}_+$ is a proper map. Note that since B(z,r) is then compact for all r, Z is locally compact, hence complete.

Example 1.3 Let (Z,ρ) be a metric space with proper metric ρ and suppose ρ satisfies

Condition A If $B(z,r) \subseteq B(u,s)$, then $B(z,r+1) \subseteq B(u,s+1)$.

The *metric boundedness control structure* (P,C) on Z is defined by setting $P = \{B(z,r) \mid z\in Z,\ r\in\mathbb{R}_+\}$ and $C(B(z,r)) = B(z,r+1)$. We note that the technical Condition A is imposed to insure that $C\colon P \longrightarrow P$ is well defined and satisfies (i) of 1.1. We also note that in this case a map $f\colon (X_1,p_1) \longrightarrow (X_2,p_2)$ of bc spaces over Z is bc if and only if there exists an integer $m\geq 0$ such that for all $x \in X_1$, $\rho(p_2f(x),p_1(x)) \leq m$. In the sequel, such maps will often simply be called *bounded*.

The reader may wish to examine the special case of 1.3 in which (Z,ρ) is $\mathbb{R}^k$ with the box metric $\rho(x,y) = \max_i\{|x_i-y_i| \mid i=1,\ldots,k\}$ where $x = (x_1,\ldots,x_k)$ and $y = (y_1,\ldots,y_k)$. It was this special case of the theory presented here that motivated much of our work as well as that of Pedersen [Pe1], [Pe2], and [Pe3] and Pedersen and Weibel [PW1].

Example 1.4 Let (Z,ρ) be a metric space with proper metric ρ. For any compact subset $K \subseteq Z$, we define the *halo* of K to be

$$H(K) = \{z\in Z \mid \rho(z,K) \leq 1\}.$$

Since H(K) is again compact, we can define $H^n(K)$ inductively. Suppose ρ satisfies

Condition B For any two points $x,y \in Z$ there exist points $x = x_o,x_1,\ldots,x_n = y$ such that $\rho(x_{i-1},x_i) \leq 1$ $(i=1,\ldots,n)$.

For example, since any path $w\colon I \longrightarrow Z$ is uniformly continuous, one sees easily that any path connected metric space satisfies Condition B.

The *halo boundedness control structure* (P,C) on Z is defined by setting $P = \{H^n(B(z,m)) \mid z\in Z,\ n,m\in\mathbb{Z}_+\}$ and $C(H^n(B(z,m))) = H^{n+1}(B(z,m))$. It is easy to verify that (P,C) is a boundedness control structure.

Now let (X_1,p_1) and (X_2,p_2) be bc spaces over Z with the halo boundedness control structure. It is clear that any halo-bc map

$f\colon (X_1,p_1) \longrightarrow (X_2,p_2)$ is also bounded. In general, however, bounded maps need not be halo-bc. We leave it to the reader to check that the following Condition C guarantees that any bounded map $f\colon (X_1,p_1) \longrightarrow (X_2,p_2)$ is indeed halo-bc.

Condition C There exists a function $\sigma\colon \mathbb{Z}_+ \longrightarrow \mathbb{Z}_+$ such that for every $z \in Z$, $B(z,n) \subseteq H^{\sigma(n)}(B(z,0))$.

We note that for a path connected space (Z,ρ), Condition C holds if one assumes that any two points x,y with $\rho(x,y) \leq n$ can be joined by a path $w\colon I \longrightarrow Z$ having $\rho\left(w\left(\frac{i-1}{\sigma(n)}\right), w\left(\frac{i}{\sigma(n)}\right)\right) \leq 1$ for all $i=1,2,\dots,\sigma(n)$.

Example 1.5 Let $Z = \mathbb{R}^k$ and endow $\mathbb{R}^k$ with the structure of a cubical complex by cutting it along the hyperplanes $x_i = n$ ($i=1,\dots,k$; $n \in \mathbb{Z}$). Let $P(\mathbb{R}^k) = \{K \subseteq \mathbb{R}^k \mid K$ is a finite, connected, cubical subcomplex$\}$ and set $C(K) = \{x \in \mathbb{R}^k \mid \rho(x,K) \leq 1\}$ where ρ is the box metric. It is easy to see $(P(\mathbb{R}^k),C)$ is a boundedness control structure on $\mathbb{R}^k$. It is called the *cubical boundedness control structure.*

Example 1.6 Let Z be a connected simplicial complex, $P(Z) = \{K \subseteq Z \mid K$ is a finite, connected subcomplex$\}$, and $C(K)$ be the closed simplicial neighborhood of K (i.e. $C(K) = \{\tau \in Z \mid \tau$ is a face of a simplex $\sigma \in Z$ with $\sigma \cap K \neq \emptyset\}$). Then $(P(Z),C)$ is a boundedness control structure on Z. It will be called the *simplicial boundedness control structure* on Z.

Example 1.7 Let $\mathcal{U} = \{U_\alpha \mid \alpha \in A\}$ be an open cover of the arcwise connected space Z. For any open set $V \subset Z$, let $St(V;\mathcal{U}) = \cup\{U_\beta \in \mathcal{U} \mid U_\beta \cap V \neq \emptyset\}$. $St(V;\mathcal{U})$ is called the *star of* V *relative to* $\mathcal{U}$. Since $St(V;\mathcal{U})$ is again open, we may define the *n-fold star of* V *relative to* $\mathcal{U}$ for $n \geq 2$ inductively by setting $St^n(V;\mathcal{U}) = St(St^{n-1}(V;\mathcal{U});\mathcal{U})$. Let $St^0(V;\mathcal{U}) = V$. It is now easily seen that $P(\mathcal{U}) = \{St^n(U_\alpha;\mathcal{U}) \mid \alpha \in A, n \in \mathbb{Z}_+\}$ with $C(V) = St(V;\mathcal{U})$ defines a boundedness control structure $(P(\mathcal{U}),C)$ on Z. We will call this the *star enlargement boundedness control structure.*

Let Z be a boundedness control space. A morphism $f\colon (X,p) \longrightarrow (Y,q)$ in $\mathcal{T}op^c/Z$ is an isomorphism exactly when f is bc and a homeomorphism whose inverse is also bc. We shall call the isomorphisms in $\mathcal{T}op^c/Z$ *bc homeomorphisms*. The reader should be warned that there are examples of bc maps that are homeomorphisms whose inverses are not bc. Such maps therefore fail to be bc homeomorphisms. The following lemma shows that this anomaly does not occur in $\mathcal{CW}^c/Z$.

Lemma 1.8 *Let* $f\colon (X,p) \longrightarrow (Y,q)$ *be a map in* $\mathcal{CW}^c/Z$ *that is a homeomorphism of spaces. Then* f *is a bc homeomorphism.*

Corollary 1.9 *Let* $(X,p), (Y,q) \in |\mathcal{CW}^c/Z|$. *A map* $f\colon X \longrightarrow Y$ *is bc if and only if* $(X,qf) \in |\mathcal{CW}^c/Z|$ and $1_X\colon (X,p) \longrightarrow (X,qf)$ *is a bc homeomorphism.*

Proof of 1.8: To show f^{-1} is bc, let $K \in |\mathcal{P}|$ where (P,C) is the boundedness control structure on Z, suppose $y \in q^{-1}K$, and write $y = f(x)$. Then x is in the interior of some cell $e \in X$. Since $(X,p) \in |\mathcal{CW}^c/Z|$, there exists an integer m, independent of e, and a minimal element $K_e \in P$ such that $p(e) \subseteq C^m(K_e)$. In particular, $p(x) \in C^m(K_e)$. Since f is bc, there exists an integer n such that for all $L \in P$, $fp^{-1}(L) \subseteq q^{-1}(C^nL)$. In particular, $y = f(x) \in q^{-1}(C^{n+m}(K_e))$. Hence $q(y) \in K \cap C^{n+m}(K_e)$. It follows that $K_e \subseteq C^{\theta(n+m)}(K)$ where $\theta\colon \mathbb{Z}_+ \longrightarrow \mathbb{Z}_+$ is the function given by of 1.1.(iv). But then $x = f^{-1}(y) \in p^{-1}(C^m(K_e)) \subseteq p^{-1}(C^{m+\theta(m+n)}(K_e))$. Hence $f^{-1}q^{-1}(K) \subseteq p^{-1}(C^d(K))$ for $d = m+\theta(m+n)$ and f^{-1} is bc. The lemma follows.

Proof of 1.9: Suppose f is bc and that $d_1 \geq 0$ is a bound for f (so that then $fp^{-1}(K) \subseteq q^{-1}(C^{d_1}K)$ for every $K \in |\mathcal{P}|$). Let $d_2 \geq 0$ be a bound for {rad p(e) | e ∈ X}. It is easily verified that then d_1+d_2 is a bound for {rad qf(e) | e ∈ X}. Hence $(X,qf) \in |\mathcal{CW}^c/Z|$. It is also easily verified that $1_X: (X,p) \longrightarrow (X,qf)$ is bc. Hence 1_X is a bc homomorphism by 1.8.

If $1_X: (X,p) \longrightarrow (X,qf)$ is a bc homeomorphism, then $f: (X,p) \longrightarrow (Y,q)$ is the composite $(X,p) \xrightarrow{1_X} (X,qf) \xrightarrow{f} (Y,q)$ of the bc map 1_X and the obviously bc map $f: (X,qf) \longrightarrow (Y,q)$. Hence f is bc and 1.9 follows.

2. Fragmented spaces and fragmentations.

Let $(\mathcal{B},C,\tau)$ be a category with endomorphism. We begin this section by introducing the category of fragmented spaces over $\mathcal{B}$, $\mathcal{FTop}/\mathcal{B}$. In the case when $\mathcal{P}$ is the category with endomorphism associated with a boundedness control structure (P,C) on Z, we also introduce a fragmentation functor $Fr: \mathcal{Top}^c/Z \longrightarrow \mathcal{FTop}/\mathcal{P}$. Several important examples of fragmented spaces, including universal covers, will also be given.

Let $(\mathcal{B},C,\tau)$ be a category with encomorphism and $\mathcal{Top}$ be the category of topological spaces and continuous maps. Then the category $\mathcal{Top}^{\mathcal{B}}(\Sigma^{-1})$ (cf. I.4) is called the *category of fragmented spaces over* $\mathcal{B}$ and will be denoted by $\mathcal{FTop}/\mathcal{B}$. An object in $\mathcal{FTop}/\mathcal{B}$ is just a functor $\underline{X}: \mathcal{B} \longrightarrow \mathcal{Top}$ and is called a *fragmented space*. It is useful to think of a fragmented space as a family of spaces $\underline{X} = \{\underline{X}(K) \mid K \in |\mathcal{B}|\}$ together with a family of maps $\underline{X}(g): \underline{X}(K) \longrightarrow \underline{X}(L)$ whenever $g \in \mathcal{B}(K,L)$. To simplify notation, in the sequel the space $\underline{X}(K)$ will be denoted by X_K. It is called the *fragment* of $\underline{X}$ over K.

Let $\underline{X},\underline{Y} \in |\mathcal{FTop}/\mathcal{B}|$. By I.1.5 a morphism $F: \underline{X} \longrightarrow \underline{Y}$ in $\mathcal{FTop}/\mathcal{B}$ is represented by a natural transformation $f_n: \underline{X} \longrightarrow C^n\underline{Y}$ where (C,τ) is the endomorphism structure on $\mathcal{Top}^{\mathcal{B}}$ induced from the endomorphism structure on $\mathcal{B}$ (cf. I.4). Since $(C^n\underline{Y})(K) = Y_{C^nK}$, f_n is just a family of maps

$\{f_{nK}: X_K \longrightarrow Y_{C^n K} \mid K \in |\mathcal{B}|\}$ for which $f_{nL}\underline{X}(g) = \underline{Y}(C^n g)f_{nK}$ whenever $g \in \mathcal{B}(K,L)$.

Similarly, if we replace $\mathcal{T}op$ in the above discussion with any of the categories $\mathcal{C}$ listed below, the category $\mathcal{C}^{\mathcal{B}}(\Sigma^{-1})$ is called the *category of fragmented* $\mathcal{C}$ *objects over* B. The main examples of such categories $\mathcal{C}$ that arise in this book are the following:

(i) $\mathcal{T}op_*$ - the category of pointed spaces and pointed maps.

(ii) $\mathcal{CW}$ - the category of CW complexes and continuous maps.

(iii) $\mathcal{CW}_f$ - the category of finite CW complexes and continuous maps.

(iv) $\mathcal{PC}$ - the category of pairs of objects in $\mathcal{C}$ for $\mathcal{C}$ any of the above categories.

Thus $\mathcal{FCW}_f/\mathcal{B}$ is called the category of fragmented, finite CW complexes over $\mathcal{B}$, etc.

Let (Z,P,C) be a boundedness control space, $(X,p) \in |\mathcal{T}op^c/Z|$ and $\mathcal{X}$ be the subcategory of $\mathcal{T}op$ consisting of all subspaces of X and their inclusion maps. A *fragmentation* of (X,p) is a functor $F: \mathcal{P} \longrightarrow \mathcal{X}$ such that $\cup F(K) = X$ where the union runs over all $K \in |\mathcal{P}|$. We often denote F(K) by X_K. It is convenient to think of a fragmentation of X as being a family $\{X_K \mid K \in |\mathcal{P}|\}$ of subspaces of X, covering X, together with an inclusion $X_K \longrightarrow X_L$ whenever $K \subseteq L$. The subspace X_K is again called the *fragment of* X *over* K.

Example 2.1 Let $(X,p) \in |\mathcal{T}op^c/Z|$. Then $K \longrightarrow p^{-1}(K)$ defines a fragmentation of X called the *inverse image fragmentation*.

Example 2.2 Let $(X,p) \in |\mathcal{CW}^c/Z|$. The correspondence K $\longrightarrow$ smallest subcomplex of K containing $p^{-1}(K)$ defines a fragmentation of X called the *smallest subcomplex fragmentation*. This has values in the category of subcomplexes of X and inclusion maps.

Since one of the objectives of this book is to study Whitehead torsion in the categories $\mathcal{CW}^c_\ell/Z$, it is necessary to introduce universal covers of such spaces. It turns out $(\tilde{X},pq)$ is not the right object to use as the universal cover of $(X,p) \in |\mathcal{CW}^c_\ell/Z|$ where $q\colon \tilde{X} \longrightarrow X$ is the universal cover of X. This is due mainly to the fact that $(\tilde{X},pq)$ fails to detect adequately and to control boundedly local changes in the fundamental group.

To circumvent this problem, we might try to view (X,p) as a fragmented space (say via 2.1 or 2.2) and to think of the universal cover of X as the fragmented space $K \longrightarrow \tilde{X}_K$. Unfortunately, since the operation "take universal covers" is not a functor on $\mathcal{Top}$ (there is no natural choice of a map of universal covers covering a map of spaces), $K \longrightarrow \tilde{X}_K$ is not a fragmented space and this does not quite work. However, since the operation "take universal covers" is a functor on $\mathcal{Top}_*$, we are lead to find a way for introducing basepoints.

The problem now is that since different fragments of X may well have empty intersection, there is no single choice of basepoint for X that will work. We resolve this problem by making, in some sense, all possible choices of basepoint and assembling these into the fundamental groupoid. The details are as follows:

Let $\underline{X}\colon \mathfrak{B} \longrightarrow \mathcal{Top}$ be a fragmented space over the category with endomorphism $\mathfrak{B}$ and $G_1(\underline{X})$ be the composite functor $\mathfrak{B} \xrightarrow{\underline{X}} \mathcal{Top} \xrightarrow{G_1} \mathcal{Gpoid}$ where G_1 is the fundamental groupoid functor (cf. [Sp, Chapter 1]). The category $\mathfrak{B}G_1(\underline{X})$ of I.1.3 is called the *fundamental groupoid* of $\underline{X}$ (although this is an abuse of language since $\mathfrak{B}G_1(\underline{X})$ is not a groupoid in the usual sense). We recall that an object in $\mathfrak{B}G_1(\underline{X})$ is a pair (x,K), where $K \in |\mathfrak{B}|$ and $x \in X_K$, and a morphism $(x,K) \longrightarrow (y,L)$ is a pair (ω,i) where $i \in \mathfrak{B}(K,L)$ and ω is a homotopy class of paths (relative to endpoints) in X_L from y to $\underline{X}(i)(x)$.

Example 2.3 Let $\underline{X}\colon \mathcal{B} \longrightarrow \mathcal{T}op$ be a fragmented space over $\mathcal{B}$. The *universal cover* of $\underline{X}$ is the fragmented space $\tilde{\underline{X}}\colon \mathcal{B}G_1(\underline{X}) \longrightarrow \mathcal{T}op$ defined by setting

$$\tilde{X}_{(x,K)} = \tilde{\underline{X}}(x,K) = P(X_K,x)/\sim$$

for $(x,K) \in |\mathcal{B}(G_1(\underline{X}))|$ and

$$\tilde{\underline{X}}(\omega,i)(\alpha) = \omega\underline{X}(i)(\alpha)$$

for $(\omega,i)\colon (x,K) \longrightarrow (y,L)$ and $\alpha \in \tilde{\underline{X}}(x,K)$. Here $P(X_K,x)/\sim$ is the space of paths in X_K with initial point x, modulo the relation of homotopy relative to endpoints, and the juxtaposition $\omega\underline{X}(i)(\alpha)$ denotes concatenation of paths. We let ε_x be the homotopy class of the constant path to $x \in X_K$ and $p_{(x,K)}\colon (\tilde{X}_{(x,K)},\varepsilon_x) \longrightarrow (X_K,x)$ be the terminal point projection.

We note that if $\underline{X}\colon \mathcal{B} \longrightarrow \mathcal{CW}$ is a fragmented CW complex over $\mathcal{B}$, then for all $(x,K) \in |\mathcal{B}G_1(\underline{X})|$, $p_{(x,K)}\colon (\tilde{X}_{(x,K)},\varepsilon_x) \longrightarrow (X_K,x)$ is a universal cover of the component of X_K containing x and inherits a cellular structure from X_K in the usual fashion. Thus, $\tilde{\underline{X}}$ is a fragmented CW complex over $\mathcal{B}G_1(\underline{X})$.

Let $F_1,F_2\colon \mathcal{F} \longrightarrow \mathcal{K}$ be two fragmentations of $(X,p) \in |\mathcal{T}op^c/Z|$. We say that F_1 *is equivalent to* F_2 if there exist integers $m,n \geq 0$ such that for all $K \in |\mathcal{F}|$, $F_1(K) \subseteq F_2(C^mK)$ and $F_2(K) \subseteq F_1(C^nK)$. Clearly this is an equivalence relation.

Lemma 2.4 *Let* $F_1,F_2\colon \mathcal{F} \longrightarrow \mathcal{K}$ *be equivalent fragmentations of* (X,p). *Then there is a preferred isomorphism from* $Q(F_1)$ *to* $Q(F_2)$ *in* $\mathcal{FT}op/\mathcal{F}$.

In this lemma $Q\colon \mathcal{T}op^{\mathcal{F}} \longrightarrow \mathcal{T}op^{\mathcal{F}}(\Sigma^{-1}) = \mathcal{FT}op/\mathcal{F}$ is the obvious functor.

The proof, which is an exercise in using the definitions, is left to the reader.

Lemma 2.5 *Let* $(X,p) \in |\mathcal{CW}_{\ell}^{c}/Z|$ *and* $K \longrightarrow X_K$ *be the smallest subcomplex fragmentation of* (X,p). *Then there exists an integer* n *such that for all* $K \in |\mathcal{F}|$, $X_K \subseteq p^{-1}(C^n K)$. *In particular,*

(i) $\operatorname{rad} X_K \leq \operatorname{rad} K + n$; *and*

(ii) *The inverse image fragmentation of* (X,p) *is equivalent to the smallest subcomplex fragmentation.*

Proof: Let $d = \max\{\operatorname{rad}(e) \mid e \text{ is a cell of } X\}$ and $n = (\dim X + 1)(d + \theta(d))$ where θ is the function of 1.1.(iv). We claim that this is the desired integer n.

To establish the claim, note that an i-cell $e^i \in X_K$ if and only if there exists a chain of cells $e^i < e^{i(1)} < \cdots < e^{i(j)}$ where $i = i(0) < i(1) < \cdots < i(j)$, where $e^{i(j)} \cap p^{-1}(K) \neq \emptyset$, and where $e^{i(k)} < e^{i(k+1)}$ $(k=0,\ldots,j-1)$ means that $\bar{e}^{i(k)} \cap \bar{e}^{i(k+1)} \neq \emptyset$. Let K_j be a minimal element with $p(e^{i(j)}) \subseteq C^d(K_j)$. Then $C^d(K_j) \cap K \neq \emptyset$ and by 1.1.(iv), $K_j \subseteq C^{\theta(d)}(K)$. But then $p(e^{i(j)}) \subseteq C^d(K_j) \subseteq C^{d+\theta(d)}(K)$; that is, $e^{i(j)} \subseteq p^{-1}(C^{d+\theta(d)}(K))$. By decreasing induction on k, $e^{i(k)} \subseteq p^{-1}(C^{n_k}K)$ where $n_k = (j-k+1)(d+\theta(d))$. In particular, $e^i \subseteq p^{-1}(C^{n_0}K) \subseteq p^{-1}(C^n K)$ since $n_0 = (j+1)(d+\theta(d)) \leq (\dim X + 1)(d+\theta(d))$. Since every cell of X_K is contained in $p^{-1}(C^n K)$ so is X_K and the claim follows.

Statement (i) follows easily from the first part of 2.5 and the definition of the radius; while (ii) follows from the first part of 2.5 and the fact that $p^{-1}(K) \subseteq X_K$. This completes the proof of 2.5.

Lemma 2.6 *Let* (Z,P,C) *be a boundedness control space. Then*

(i) *The inverse image fragmentation defines a functor* $Fr_1\colon \mathcal{T}op^c/Z \longrightarrow \mathcal{FT}op/\mathcal{F}$.

(ii) *The smallest subcomplex fragmentation defines a functor* $Fr_2\colon \mathcal{CW}_{\ell}^{c}/Z \longrightarrow \mathcal{FCW}_{\ell}/\mathcal{F}$.

Proof: Let (X,p), $(Y,q) \in |\mathcal{T}op^c/Z|$. We let $Fr_1(X,p)$ be the functor $\mathcal{P} \longrightarrow \mathcal{T}op$ that sends K to $p^{-1}(K)$. If $f\colon (X,p) \longrightarrow (Y,q)$, then there exists an integer $m \geq 0$ such that for all $K \in |\mathcal{P}|$, $f(p^{-1}(K)) \subseteq q^{-1}(C^mK)$. Let $f_{mK} = f|p^{-1}(K)\colon p^{-1}(K) \longrightarrow q^{-1}(C^mK)$. Then $f_m = \{f_{mK} \mid K \in P\}$ defines a natural transformation $f_m\colon Fr_1(X,p) \longrightarrow C^m Fr_1(Y,q)$ (i.e. a morphism in $\mathcal{T}op^{\mathcal{P}}$) and we let $Fr_1(f)\colon Fr_1(X,p) \longrightarrow Fr_1(Y,q)$ be the morphism in $\mathcal{F}\mathcal{T}op/\mathcal{P}$ represented by f_m. It follows easily from I.1.5 and the discussion following it that $Fr_1(f)$ is independent of the choice of m and that Fr_1 is a functor. Part (i) of 2.6 follows.

If (X,p), $(Y,q) \in |\mathcal{CW}^c_\ell/Z|$, we let $Fr_2(X,p)$ be represented by the functor $\mathcal{P} \longrightarrow \mathcal{CW}_\ell$ that sends K to the smallest subcomplex X_K of X containing $p^{-1}(K)$. If $f\colon (X,p) \longrightarrow (Y,q)$ is a bc map, choose m as in the proof of part (i), let n be the integer of 2.5, and $\bar{f}_{n+mK}\colon X_K \longrightarrow Y_{C^{n+m}K}$ be the composite

$$X_K \subseteq p^{-1}(C^nK) \xrightarrow[f_{mK}]{} q^{-1}(C^{m+n}K) \subseteq Y_{C^{n+m}K} .$$

Then $\bar{f}_{n+m} = \{\bar{f}_{n+mK} \mid K \in |\mathcal{P}|\}\colon Fr_2(X,p) \longrightarrow C^{n+m}Fr_2(Y,q)$ is a morphism in $\mathcal{CW}^{\mathcal{P}}_\ell$ and we let $Fr_2(f)\colon Fr_2(X,p) \longrightarrow Fr_2(Y,q)$ be the morphism in $\mathcal{F}\mathcal{CW}_\ell/\mathcal{P}$ that it represents. The remaining details of part (ii) are similar to those of part (i).

It follows from 2.6 that there are two fragmentation functors $\mathcal{CW}^c_\ell/Z \longrightarrow \mathcal{F}\mathcal{T}op/\mathcal{P}$ given by the two composites in the diagram

$$\begin{array}{ccc}
\mathcal{CW}^c_\ell/Z & \xrightarrow{Fr_2} & \mathcal{F}\mathcal{CW}_\ell/\mathcal{P} \\
\Big\downarrow{\scriptstyle J_1} & & \Big\downarrow{\scriptstyle J_2} \\
\mathcal{T}op^c/Z & \xrightarrow{Fr_1} & \mathcal{F}\mathcal{T}op/\mathcal{P}
\end{array}$$

where J_1 and J_2 are the functors obtained by regarding a CW complex as a topological space.

Lemma 2.7 *The functors* Fr_1J_1 *and* J_2Fr_2 *are naturally equivalent.*

Proof: Let $(X,p) \in |\mathcal{CW}_{\ell}^{c}/Z|$ and $K\in|\mathcal{P}|$. Then $Fr_1J_1(X,p)(K) = p^{-1}(K)$, $J_2Fr_2(X,p)(K) = X_K$ is the smallest subcomplex containing $p^{-1}(K)$, and we let $\nu_{(X,p)K}\colon p^{-1}(K) \longrightarrow X_K$ be the inclusion. Clearly

$$\nu_{(X,p)} = \{\nu_{(X,p)K} \mid K\in|\mathcal{P}|\}$$

is a natural transformation and represents a morphism

$$\nu_{(X,p)}\colon Fr_1J_1(X,p) \longrightarrow J_2Fr_2(X,p)$$

in $\mathcal{FTop}/\mathcal{P}$. We claim that $\nu = \{\nu_{(X,p)} \mid (X,p) \in |\mathcal{CW}_{\ell}^{c}/Z|\}$ is the desired natural equivalence $\nu\colon Fr_1J_1 \longrightarrow J_2Fr_2$.

We check first that ν is a natural transformation. To do this, let $f\colon (X,p) \longrightarrow (Y,q)$ be a morphism in $\mathcal{CW}_{\ell}^{c}/Z$ and choose m as in 2.6 and n as in 2.5 such that for all $K\in|\mathcal{P}|$, $f(p^{-1}(K)) \subseteq q^{-1}(C^mK)$ and $X_K \subseteq p^{-1}(C^nK)$. Then the following diagram of objects in $\mathcal{Top}^{\mathcal{P}}$ commutes

$$\begin{array}{ccccccccc}
p^{-1}(\ ) & \xrightarrow{f_m} & q^{-1}C^m(\ ) & & \xleftarrow{\qquad\tau^m\qquad} & & & & q^{-1}(\ ) \\
\downarrow{\scriptstyle \nu_{(X,p)}} & \searrow{\scriptstyle \tau^n} & & \searrow{\scriptstyle \tau^n} & & & & & \downarrow{\scriptstyle \nu_{(Y,q)}} \\
X_{(\ )} & \xrightarrow{j} & p^{-1}C^n(\ ) & \xrightarrow{f_m} & q^{-1}C^{m+n}(\ ) & \xrightarrow{k} & Y_{C^{n+m}(\ )} & \xleftarrow{\tau^n} & Y_{(\ )}
\end{array}$$

where $p^{-1}(\ )$ and $X_{(\ )}$ denote the functors $K \longrightarrow p^{-1}(K)$ and $K \longrightarrow X_K$, respectively, etc.; and j and k are the inclusions. Since $Fr_1J_1(f) = Q(\tau^m)^{-1}Q(f_m)$ and $J_2Fr_2(f) = Q(\tau^n)^{-1}Q(kf_mj)$ by 2.6, an easy calculation using this diagram shows that

$$J_2Fr_2(f)\nu_{(X,p)} = \nu_{(Y,q)}Fr_1J_1(f).$$

Hence ν is a natural transformation.

It remains to be shown that ν is a natural equivalence; i.e. that $\nu_{(X,p)}\colon Fr_1J_1(X,p) \longrightarrow J_2Fr_2(X,p)$ is an isomorphism for all $(X,p) \in |\mathcal{CW}_{\ell}^{c}/Z|$. This, however, follows immediately from the facts that the diagram in $\mathcal{T}op^{\mathcal{P}}$

$$\begin{array}{ccc} p^{-1}(\ ) & \xrightarrow{\ \tau^n\ } & p^{-1}C^n(\ ) \\ {\scriptstyle \nu_{(X,p)}}\big\downarrow & \searrow^{j} & \big\downarrow{\scriptstyle \nu_{(X,p)}} \\ X_{(\ )} & \xrightarrow[\ \tau^n\ ]{} & X_{C^n(\ )} \end{array}$$

commutes where n is the integer of 2.5, and that $Q(\tau^n)$ is an isomorphism. This completes the proof of 2.7.

Remark 2.8 It is clear that the results of this section carry over to pairs of bc spaces or bc CW complexes virtually without change. Thus, the inverse image construction defines a functor $FR_1\colon \mathcal{PT}op^{c}/Z \longrightarrow \mathcal{FPT}op/\mathcal{P}$, the least subcomplex construction defines a functor $Fr_2\colon \mathcal{PCW}_{\ell}^{c}/Z \longrightarrow \mathcal{FPCW}_{\ell}/\mathcal{P}$, and Fr_1J_1 is naturally equivalent to J_2Fr_2. A similar remark applies to triples.

3. Homology of fragmented spaces and pairs.

In this section we define the homology of a fragmented space (or pair) and establish its basic properties.

Let $\mathcal{A}b$ be the category of abelian groups and for $n \in \mathbb{Z}_+$, let $H_n\colon \mathcal{T}op \longrightarrow \mathcal{A}b$ (respectively, $H_n\colon \mathcal{PT}op \longrightarrow \mathcal{A}b$) be the usual n-th singular homology of a space (respectively, of a pair). Let $(\mathcal{B},C,\tau)$ be a category with endomorphism. Then by I.5.1 composition on the left with H_n induces functors H_{n_*}, $H_{n_!}$ such that the diagrams

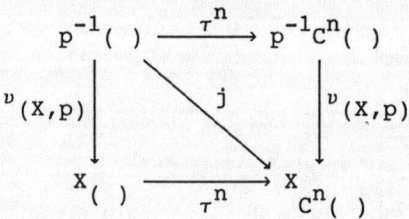

$$\begin{array}{ccc} \mathcal{T}op^{\mathcal{B}} & \xrightarrow{H_{n_*}} & \mathcal{A}b^{\mathcal{B}} \\ \Big\downarrow Q & & \Big\downarrow Q \\ \mathcal{T}op^{\mathcal{B}}(\Sigma^{-1}) & \xrightarrow{H_{n_!}} & \mathcal{A}b^{\mathcal{B}}(\Sigma^{-1}) \end{array} \qquad \begin{array}{ccc} \mathcal{P}\mathcal{T}op^{\mathcal{B}} & \xrightarrow{H_{n_*}} & \mathcal{A}b^{\mathcal{B}} \\ \Big\downarrow Q & & \Big\downarrow Q \\ \mathcal{P}\mathcal{T}op^{\mathcal{B}}(\Sigma^{-1}) & \xrightarrow{H_{n_!}} & \mathcal{A}b^{\mathcal{B}}(\Sigma^{-1}) \end{array}$$

commute. Since $\mathcal{F}\mathcal{T}op/\mathcal{B}$ (respectively $\mathcal{F}\mathcal{P}\mathcal{T}op/\mathcal{B}$) is another notation for $\mathcal{T}op^{\mathcal{B}}(\Sigma^{-1})$ (respectively, $\mathcal{P}\mathcal{T}op^{\mathcal{B}}(\Sigma^{-1})$), we make a similar change of notation and let H_n^F denote either of the functors $H_{n_!}$. If $\underline{X} \in |\mathcal{F}\mathcal{T}op/\mathcal{B}|$ (respectively, $(\underline{X},\underline{Y}) \in |\mathcal{F}\mathcal{P}\mathcal{T}op/\mathcal{B}|$), we call $H_n^F(\underline{X})$ (respectively, $H_n^F(\underline{X},\underline{Y})$ the n-th *homology of the fragmented space* (respectively, *pair*) $\underline{X}$ (respectively, $(\underline{X},\underline{Y})$).

The following more concrete description of H_n^F is used repeatedly: Let $\underline{X}$: $\mathcal{B} \longrightarrow \mathcal{T}op$ be a fragmented space over $\mathcal{B}$ and let X_K denote $\underline{X}(K)$. For $n \in \mathbb{Z}_+$, define a functor $H_n(\underline{X})$: $\mathcal{B} \longrightarrow \mathcal{A}b$ by setting $H_n(\underline{X})(K) = H_n(X_K)$, the n-th singular homology group of X_K for $K \in |\mathcal{B}|$; and setting $H_n(\underline{X})(g) = H_n(\underline{X}(g)) = \underline{X}(g)_*$: $H_n(X_K) \longrightarrow H_n(X_L)$ for $g \in \mathcal{B}(K,L)$. Then $Q(H_n(\underline{X})) = H_n^F(\underline{X})$. (The reader will note a slight abuse of notation on the right. The more precise, but less suggestive, notation would be $H_n^F(Q(\underline{X}))$. We shall frequently follow this convention in the sequel and not distinguish between an object K' in a category with endomorphism $\mathcal{B}'$ and its image under Q in $\mathcal{B}'(\Sigma^{-1})$.)

Let F: $\underline{X} \longrightarrow \underline{Y}$ be a map of fragmented spaces, and let f_m: $\underline{X} \longrightarrow C^m\underline{Y}$ be a natural transformation representing f (cf. I.1.5). Then

$$f_m = \{f_{mK}: X_K \longrightarrow C^mY_K = Y_{C^mK} \mid K\in|\mathcal{B}|\}$$

and the family

$$\{f_{mK_*}: H_n(X_K) \longrightarrow H_n(Y_{C^mK}) \mid K\in|\mathcal{B}|\}$$

defines a natural transformation f_{m_*}: $H_n(\underline{X}) \longrightarrow H_n(\underline{Y}C^m) = C^mH_n(\underline{Y})$ which represents $H_n^F(f)$. In the sequel we denote this morphism by f_*.

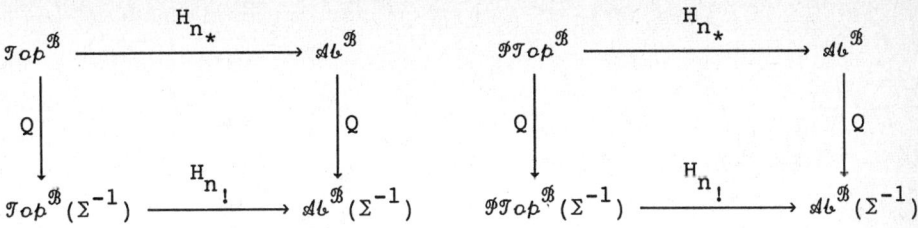

Suppose now that $(\underline{X},\underline{Y})$: $\mathcal{B} \longrightarrow \mathcal{P}\mathcal{T}op$ is a pair of fragmented spaces over $\mathfrak{B}$. It is convenient to view $(\underline{X},\underline{Y})$ as a pair of functors $\mathfrak{B} \longrightarrow \mathcal{T}op$ such that the inclusion maps $\{i_K\colon Y_K \longrightarrow X_K \mid K\in|\mathfrak{B}|\}$ define a natural transformation $i\colon \underline{Y} \longrightarrow \underline{X}$ and hence, a morphism $i_*\colon H_n^F(\underline{Y}) \longrightarrow H_n^F(\underline{X})$. It is also easy to see that the families

$$\{j_{K_*}\colon H_n(X_K) \longrightarrow H_n(X_K,Y_K) \mid K\in|\mathfrak{B}|\}$$

and

$$\{\partial_{K_*}\colon H_n(X_K,Y_K) \longrightarrow H_{n-1}(Y_K) \mid K\in|\mathfrak{B}|\}$$

$(n\geq 1)$, define morphisms $j_*\colon H_n^F(\underline{X}) \longrightarrow H_n^F(\underline{X},\underline{Y})$ and $\partial_*\colon H_n^F(\underline{X},\underline{Y}) \longrightarrow H_{n-1}^F(\underline{Y})$ $(n\geq 1)$ in $\mathcal{A}b^{\mathfrak{B}}(\Sigma^{-1})$. Then the following lemma is an immediate corollary of I.4.2 and I.2.4.

Lemma 3.1 *For any pair* $(\underline{X},\underline{Y})$ *of fragmented spaces over* $\mathfrak{B}$*, the following sequence is exact*

$$\cdots \longrightarrow H_{n+1}^F(\underline{X},\underline{Y}) \xrightarrow{\partial_*} H_n^F(\underline{Y}) \xrightarrow{i_*} H_n^F(\underline{X}) \xrightarrow{j_*} H_n^F(\underline{X},\underline{Y}) \longrightarrow \cdots$$

It is easy to see that this exact sequence is natural: the morphisms induced by a map $f\colon (\underline{X}_1,\underline{Y}_1) \longrightarrow (\underline{X}_2,\underline{Y}_2)$ of pairs of fragmented spaces become the rungs in a commutative ladder whose sides are the exact sequences for $(\underline{X}_i,\underline{Y}_i)$ $(i=1,2)$.

Remark 3.2 A result similar to 3.1 holds for a triple $(\underline{X},\underline{Y},\underline{Z})$ of fragmented spaces over $\mathfrak{B}$. Namely the inclusions $i\colon (\underline{Y},\underline{Z}) \longrightarrow (\underline{X},\underline{Z})$ and $j\colon (\underline{X},\underline{Z}) \longrightarrow (\underline{X},\underline{Y})$ induce morphisms $i_*\colon H_n^F(\underline{Y},\underline{Z}) \longrightarrow H_n^F(\underline{X},\underline{Z})$ and $j_*\colon H_n^F(\underline{X},\underline{Z}) \longrightarrow H_n^F(\underline{X},\underline{Y})$ such that the following sequence is exact

$$\cdots \longrightarrow H_{n+1}^F(\underline{X},\underline{Y}) \xrightarrow{\partial_*} H_n^F(\underline{Y},\underline{Z}) \xrightarrow{i_*} H_n^F(\underline{X},\underline{Z}) \xrightarrow{j_*} H_n^F(\underline{X},\underline{Y}) \longrightarrow \cdots$$

where $\partial_* = \{\partial_{*K}\colon H_{n+1}(X_K,Y_K) \longrightarrow H_n(Y_K,Z_K) \mid K\in|\mathfrak{B}|\}$ and ∂_{*K} is the connecting homomorphism in the exact sequence of the triple (X_K,Y_K,Z_K). Furthermore, this sequence is natural with respect to maps of triples of fragmented spaces.

Let $I = [0,1]$ and for $(\underline{X},\underline{Y}) \in |\mathcal{FPT}op/\mathcal{B}|$, let $(\underline{X},\underline{Y})\times I \in |\mathcal{FPT}op/\mathcal{B}|$ be represented by the functor $[(\underline{X},\underline{Y})\times I](K) = (X_K\times I, Y_K\times I)$ for $K\in|\mathcal{B}|$, and $[(\underline{X},\underline{Y})\times I](g) = \underline{X}(g)\times 1$ for $g\in\mathcal{B}(K,L)$, where $\underline{X}(g)\colon (X_K,Y_K) \longrightarrow (X_L,Y_L)$ and 1 is the identity on I. Then the inclusions $(X_K,Y_K) \longrightarrow (X_K\times\{j\},Y_K\times\{j\})$ $(j=0,1)$ define morphisms $i_j\colon (\underline{X},\underline{Y}) \longrightarrow (\underline{X},\underline{Y})\times I$ in $\mathcal{FPT}op/\mathcal{B}$.

Let $f_0,f_1\colon (\underline{X},\underline{Y}) \longrightarrow (\underline{V},\underline{W})$ be maps in $\mathcal{FPT}op/\mathcal{B}$. Then f_0 *is homotopic to* f_1 if there exists $F\colon (\underline{X},\underline{Y})\times I \longrightarrow (\underline{V},\underline{W})$ in $\mathcal{FPT}op/\mathcal{B}$ such that $fi_j = f_j$ $(j=0,1)$. If F is represented by the family of maps

$$F_n = \{F_{nK}\colon (X_K\times I,Y_K\times I) \longrightarrow (V_{C^n K},W_{C^n K}) \mid K\in|\mathcal{B}|\},$$

then clearly $F_{nK}|\colon (X_K\times\{j\},Y_K\times\{j\}) \longrightarrow (V_{C^n K},W_{C^n K})$ represents f_j $(j=0,1)$. Thus F is, in effect, a (compatible) family of ordinary homotopies.

The proof of the following lemma is a simple exercise:

Lemma 3.3 *Let* $f_0,f_1\colon (\underline{X},\underline{Y}) \longrightarrow (\underline{V},\underline{W})$ *be homotopic maps of fragmented pairs over* $\mathcal{B}$. *Then* $f_{0_*} = f_{1_*}\colon H^F_n(\underline{X},\underline{Y}) \longrightarrow H^F_n(\underline{V},\underline{W})$.

The reader will note that both the definition of homotopy and 3.3 specialize to fragmented spaces by taking $\underline{Y}$ and $\underline{W}$ to be the empty space functors $Y_K = \emptyset = W_K$ for all $K\in|\mathcal{B}|$.

In the sequel, we will occasionally want to view the homology of a fragmented space or pair as arising from a chain complex (e.g. either the singular or cellular chains). Thus we digress somewhat and examine this situation in the abstract.

Lemma 3.4 *Let* $(\mathcal{A},C,\tau)$ *be an abelian category* $\mathcal{A}$ *with endomorphism and suppose* C *preserves kernels. Then for any chain complex*

$C_* = \{C_n, \partial_n\}$ *in* $\mathcal{A}$, $QC_* = \{Q(C_n), Q(\partial_n)\}$ *is a chain complex in* $\mathcal{A}(\Sigma^{-1})$ *and for every* p,

$$QH_p(C_*) = H_p(QC_*)$$

where $Q\colon \mathcal{A} \longrightarrow \mathcal{A}(\Sigma^{-1})$ *is the obvious functor.*

Proof: Since Q preserves initial and terminal objects by I.2.1 and $\partial_n\partial_{n+1}=0$, $Q(\partial_n)Q(\partial_{n+1}) = 0$ and QC_* is a chain complex. By definition $H_p(C_*) = \operatorname{coker} d_{p+1}$ where d_{p+1} is the unique morphism such that the following diagram commutes

$$\begin{array}{ccccc} C_{p+1} & \xrightarrow{\quad \partial_{p+1} \quad} & C_p & \xrightarrow{\ \partial_p\ } & C_{p-1} \\ & {\scriptstyle d_{p+1}} \searrow \quad \nearrow {\scriptstyle i_p} & & & \\ & \ker\partial_p & & & \end{array}$$

If we apply Q to this diagram, we get the diagram

$$\begin{array}{ccccc} Q(C_{p+1}) & \xrightarrow{\quad Q(\partial_{p+1}) \quad} & Q(C_p) & \xrightarrow{\ Q(\partial_p)\ } & Q(C_{p-1}) \\ & {\scriptstyle Q(d_{p+1})} \searrow \quad \nearrow {\scriptstyle Q(i_p)} & & & \\ & Q(\ker\partial_p) & & & \end{array}$$

Since C preserves kernels, by I.2.4 so does Q. Thus, $(Q(\ker\partial_p), Q(i_p))$ is the kernel of $Q(\partial_p)$. Since Q always preserves cokernels by I.2.1 and I.2.2, we have

$$H_p(QC_*) = \operatorname{coker} Q(d_{p+1}) = Q(\operatorname{coker} d_{p+1}) = QH_p(C_*)$$

and 3.4 follows.

Corollary 3.5 *Let* $\mathfrak{D}$ *be an abelian category and* $(\mathfrak{B}, C, \tau)$ *be a category with endomorphism. Then for any chain complex* $C_* = \{C_n, \partial_n\}$ *in* $\mathfrak{D}^{\mathfrak{B}}$ *and for every* p,

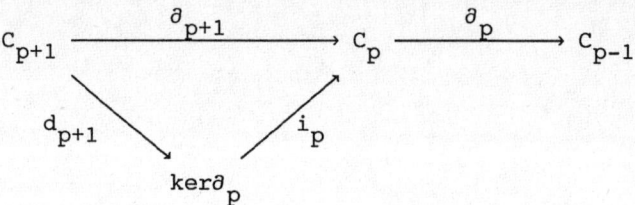

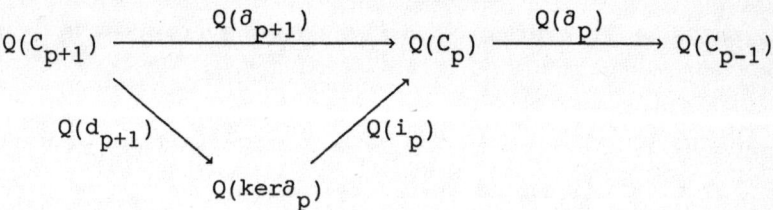

(i) $H_p(C_*) \in |\mathcal{D}^{\mathcal{B}}|$ *is the functor* $K \longrightarrow H_p(C_*(K))$ *where* $C_*(K) = \{C_n(K), \partial_{nK}\}$; *and*

(ii) $QH_p(C_*) = H_p(QC_*)$.

Proof: Part (i) follows easily from the fact that kernels and cokernels in $\mathcal{D}^{\mathcal{B}}$ are given pointwise (cf. the proof of I.4.2); while part (ii) is an immediate consequence of I.4.1, the proof of I.2.4, and 3.4.

Example 3.6 Let $(\underline{X},\underline{Y}) \in \mathcal{FPTop}/\mathcal{B}$ and let $\Delta_*^F(\underline{X},\underline{Y}) = \{\Delta_n(\underline{X},\underline{Y}),\partial_n\}$ where $\Delta_n\colon \mathcal{PTop} \longrightarrow \mathcal{Ab}$ is the n-th singular chain group functor. Then by 3.5, for every p, $H_p^F(\underline{X},\underline{Y}) = QH_p(\Delta_*^F(\underline{X},\underline{Y})) = H_p(Q\Delta_*^F(\underline{X},\underline{Y}))$.

Example 3.7 Let $(\underline{X},\underline{Y}) \in |\mathcal{FPCW}_f/\mathcal{B}|$ and view $(\underline{X},\underline{Y})$ as a functor that sends $K \longrightarrow (X_K,Y_K)$. Let $(\underline{X},\underline{Y})^{(n)}$ be the functor with

$$(\underline{X},\underline{Y})^{(n)}(K) = X_K^{(n)} \cup Y_K,$$

where $X_K^{(n)}$ is the n-skeleton of X_K. Let $C_*^{F,\mathrm{cell}}(\underline{X},\underline{Y})$ be the chain complex in $\mathcal{Ab}^{\mathcal{B}}$ with

$$C_n^F(\underline{X},\underline{Y})(K) = H_n^F((\underline{X},\underline{Y})^{(n)}, (\underline{X},\underline{Y})^{(n-1)})(K) = H_n(X_K^{(n)} \cup Y_K, X_K^{(n-1)} \cup Y_K)$$

and $\partial_n = \{\partial_{nK}\}$ where ∂_{nK} is the boundary operator of the triple $(X_K^{(n)} \cup Y_K, X_K^{(n-1)} \cup Y_K, X_K^{(n-2)} \cup Y_K)$. Then for every $K \in |\mathcal{B}|$, $C_*^{F,\mathrm{cell}}(\underline{X},\underline{Y})(K)$ is the usual cellular chain complex for the pair (X_K,Y_K). Hence by 3.5, for every p

$$H_p^F(\underline{X},\underline{Y}) = QH_p(C_*^{F,\mathrm{cell}}(\underline{X},\underline{Y})) = H_p(QC_*^{F,\mathrm{cell}}(\underline{X},\underline{Y})) \text{ in } \mathcal{Ab}^{\mathcal{B}}(\Sigma^{-1}).$$

4. Homotopy of fragmented spaces and pairs.

In this section we define the homotopy of fragmented spaces and pairs and establish its basic properties. The main new difficulty that arises here is the failure of homotopy to be functorial, at least when the action of the fundamental groups is taken into account. We describe this difficulty more completely.

If $f\colon (X,x_0) \longrightarrow (Y,y_0)$, then f certainly induces a group homomorphism $f_*\colon \pi_n(X,x_0) \longrightarrow \pi_n(Y,y_0)$ for all $n \geq 1$. Since $\pi_n(X,x_0)$ is a module over $Z\pi_1(X,x_0)$ and $\pi_n(Y,y_0)$ is a module over $Z\pi_1(Y,y_0)$ and $f_*\colon \pi_1(X,x_0) \longrightarrow \pi_1(Y,y_0)$ is usually not an isomorphism, $f_*\colon \pi_n(X,x_0) \longrightarrow \pi_n(Y,y_0)$ cannot be a homomorphism of modules. To circumvent this one uses the homomorphism $f_*\colon \pi_1(X,x_0) \longrightarrow \pi_1(Y,y_0)$ to endow $\pi_n(Y,y_0)$ with a $Z\pi_1(X,x_0)$-module structure, denoted $f^!\pi_n(Y,y_0)$, and then observes that $f_*\colon \pi_n(X,x_0) \longrightarrow f^!\pi_n(Y,y_0)$ is a $Z\pi_1(X,x_0)$-module homomorphism. The construction of our induced morphisms involves a similar kind of pullback construction.

Let $\mathscr{C}$ be the category of pointed sets, groups, or abelian groups respectively, if n=0, n=1, or $n \geq 2$ respectively. For $n \in \mathbb{Z}_+$, define a functor

$$\pi_n(\underline{X})\colon \mathscr{B}G_1(\underline{X}) \longrightarrow \mathscr{C}$$

by setting $\pi_n(\underline{X})(x,K) = \pi_n(X_K,x)$ for $(x,K) \in |\mathscr{B}G_1(\underline{X})|$ and $\pi_n(\omega,i)$ equal to the composite

$$\pi_n(X_K,x) \xrightarrow{(\underline{X}(i))_*} \pi_n(X_L,\underline{X}(i)(x)) \xrightarrow{\omega_*} \pi_n(X_L,y)$$

for $(\omega,i)\colon (x,K) \longrightarrow (y,L)$ where ω_* is the change of basepoint isomorphism induced by ω.

Definition 4.1 *For* $n \geq 0$, $\pi_n^F(\underline{X}) = Q(\pi_n(\underline{X})) \in |\mathscr{C}^{\mathscr{B}G_1(\underline{X})}(\Sigma^{-1})|$ *is called the* n-th homotopy *of the fragmented space* $\underline{X}$.

Note that since the group of automorphisms of (x,K) in $\mathcal{B}G_1(\underline{X})$ is isomorphic to $\pi_1(X_K,x)$, the usual action of the fundamental group is built into $\pi_n^F(\underline{X})$. In a sense, $\pi_n^F(\underline{X})$ is a "module over $\mathcal{B}G_1(\underline{X})$".

Proposition 4.2 *Let* $f\colon \underline{X} \longrightarrow \underline{Y}$ *be a map of fragmented spaces. For* $\mathcal{C}$ *the category of pointed sets, groups, or abelian groups,* f *induces a functor*

$$f^!\colon \mathcal{C}^{\mathcal{B}G_1(\underline{Y})}(\Sigma^{-1}) \longrightarrow \mathcal{C}^{\mathcal{B}G_1(\underline{X})}(\Sigma^{-1})$$

and a morphism $f_*\colon \pi_n^F(\underline{X}) \longrightarrow f^!\pi_n^F(\underline{Y})$. *The functor* $f^!$ *and the morphism* f_* *are unique up to a canonical natural equivalence.*

Proof: Suppose that $f\colon \underline{X} \longrightarrow \underline{Y}$ is represented by the natural transformation $f_m\colon \underline{X} \longrightarrow C^m\underline{Y}$. Then $f_m = \{f_{mK}\colon X_K \longrightarrow Y_{C^mK} \mid K\in|\mathcal{B}|\}$ and induces a functor $F_m\colon \mathcal{B}G_1(\underline{X}) \longrightarrow \mathcal{B}G_1(\underline{Y})$ that sends (x,K) to $(f_{mK}(x),C^mK)$ and (ω,i) to $(f_{mL}(\omega),C^m(i))$ for $i\in \mathcal{B}(K,L)$. The maps $\{f_{mK_*}\colon \pi_n(X_K,x) \longrightarrow \pi_n(Y_{C^mK},f_{mK}(x))\}$ are now seen to constitute a natural transformation $f_{m_*}\colon \pi_n^F(X) \longrightarrow F_m^*\pi_n^F(Y)$ where $F_m^*\colon \mathcal{C}^{\mathcal{B}G_1(\underline{Y})} \longrightarrow \mathcal{C}^{\mathcal{B}G_1(\underline{X})}$ is induced by F_m. We set $f^! = F_m^!$ and $f_* = Q(f_{m_*})$. The uniqueness of $f^!$ and f_* is an immediate consequence of the following lemma:

Lemma 4.3 *Let* $f_i\colon \underline{X} \longrightarrow C^i\underline{Y}$ $(i=m_1,m_2)$ *both represent* $f\colon \underline{X} \longrightarrow \underline{Y}$ *in* $\mathcal{FT}op/\mathcal{B}$. *Then the functors* $F_i^*\colon \mathcal{C}^{\mathcal{B}G_1(\underline{Y})} \longrightarrow \mathcal{C}^{\mathcal{B}G_1(\underline{X})}$ *are eventually equal. Hence there is a canonical natural equivalence* $\sigma\colon F_{m_1}^! \longrightarrow F_{m_2}^!$. *Furthermore, the following diagram commutes*

$$
\begin{array}{ccc}
 & \nearrow^{f_{m_{1*}}} & F^!_{m_1}\pi^F_n(\underline{Y}) \\
\pi^F_n(\underline{X}) & & \downarrow \sigma \\
 & \searrow_{f_{m_{2*}}} & F^!_{m_2}\pi^F_n(\underline{Y})
\end{array}
$$

Proof: Let i be either m_1 or m_2. We note first that $F_i\colon \mathfrak{B}G_1(\underline{X}) \longrightarrow \mathfrak{B}G_1(\underline{Y})$ commutes with C and τ. Furthermore, it suffices to consider the case when $m_2 \geq m_1$ and $f_{m_2} = \tau^{m_2-m_1} f_{m_1}$. Then a calculation shows that $F_{m_1} C^{m_2-m_1} = F_{m_2}$ from which the equation $C^{m_2-m_1} F^*_{m_1} = F^*_{m_2}$ follows easily. Hence $F^*_{m_1}$ is eventually equal to $F^*_{m_2}$ (cf. Section I.3 for definitions) and by I.3.5 there is a canonical natural equivalence $\sigma\colon F^!_{m_1} \longrightarrow F^!_{m_2}$. That the diagram at the end of 4.3 commutes follows easily from the description of σ given in the proof of I.3.5.

Let $F\colon \underline{X} \longrightarrow \underline{Y}$ and $g\colon \underline{Y} \longrightarrow \underline{Z}$ be represented by $f_m\colon \underline{X} \longrightarrow C^m\underline{Y}$ and $g_p\colon \underline{Y} \longrightarrow C^p\underline{Z}$. Let $F_m\colon \mathfrak{B}G_1(\underline{X}) \longrightarrow \mathfrak{B}G_1(\underline{Y})$ and $G_p\colon \mathfrak{B}G_1(\underline{Y}) \longrightarrow \mathfrak{B}G_1(\underline{Z})$ be the functors induced by f_m and g_p as in the proof above. We note that then $(gf)_{m+p} = (C^m g_p) f_m\colon \underline{X} \longrightarrow C^{m+p}\underline{Z}$ represents gf and induces the functor $G_pF_m\colon \mathfrak{B}G_1(\underline{X}) \longrightarrow \mathfrak{B}G_1(\underline{Z})$. It is easy to check that the following diagram commutes

$$
\begin{array}{ccc}
\pi^F_n(\underline{X}) & \xrightarrow{\quad f_{m_*}\quad} & F^!_m\pi^F_n(\underline{Y}) \\
{\scriptstyle (gf)_{m+p_*}}\downarrow & & \downarrow{\scriptstyle F^!_m(g_{p_*})} \\
(G_pF_m)^!\pi^F_n(\underline{Z}) & = & F^!_m G^!_p \pi^F_n(\underline{Z})
\end{array}
$$

The next lemma is an immediate consequence of these remarks and 4.3.

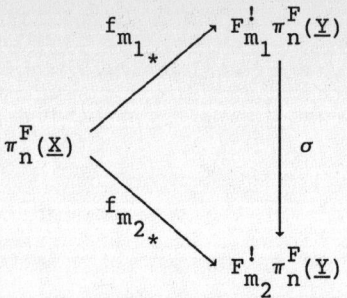

Lemma 4.4 *Let* $f\colon \underline{X} \longrightarrow \underline{Y}$ *and* $g\colon \underline{Y} \longrightarrow \underline{Z}$ *be maps of fragmented spaces over* $\mathfrak{B}$. *Then there is a canonical natural equivalence* η *such that the following diagram commutes*

$$\begin{array}{ccc} \pi_n^F(\underline{X}) & \xrightarrow{\quad f_* \quad} & f^!\pi_n^F(\underline{Y}) \\ {\scriptstyle (gf)_*}\big\downarrow & & \big\downarrow{\scriptstyle f^!(g_*)} \\ (gf)^!\pi_n^F(\underline{Z}) & \xrightarrow{\quad \eta \quad} & f^!g^!\pi_n^F(\underline{Z}) \end{array}$$

Now let $(\underline{X},\underline{Y})$ be a pair of fragmented spaces over $\mathfrak{B}$. Let $\mathcal{C}$ be the category of pointed sets, groups, or abelian groups, respectively, for $n=0,1$, $n=2$, or $n\geq 3$, respectively, and define a functor

$$\pi_n(\underline{X},\underline{Y})\colon \mathfrak{B}G_1(\underline{Y}) \longrightarrow \mathcal{C}$$

by setting $\pi_n(\underline{X},\underline{Y})\ (y,K) = \pi_n(X_K,Y_K,y)$ for $(y,K) \in |\mathfrak{B}G_1(\underline{Y})|$ and if $(\omega,i)\colon (y,K) \longrightarrow (z,L)$ in $\mathfrak{B}G_1(\underline{Y})$, by setting $\pi_n(\omega,i)$ equal to the composite

$$\pi_n(X_K,Y_K,y) \xrightarrow{\underline{X}(i)_*} \pi_n(X_L,Y_L,X(i)(y)) \xrightarrow{\omega_*} \pi_n(X_L,Y_L,z)\ .$$

(For any pair of spaces (X,A) we let $\pi_0(X,A,a)$ be the pointed set $\mathrm{coker}\{i_*\colon \pi_0(A,a) \longrightarrow \pi_0(X,a)\}$. Thus, $\pi_0(X,A,a)$ is the set whose objects are the path components of X not meeting A together with the distinguished point consisting of the union of all the path components of X that meet A. Notice that the sequence of pointed sets

$$\cdots \longrightarrow \pi_0(A,a) \longrightarrow \pi_0(X,a) \longrightarrow \pi_0(X,A,a) \longrightarrow 0$$

is exact where 0 is the pointed set with one object.)

Definition 4.5 *For* $n\geq 0$, $\pi_n^F(\underline{X},\underline{Y}) = Q(\pi_n(\underline{X},\underline{Y})) \in |\mathcal{C}^{\mathfrak{B}G_1(\underline{Y})}(\Sigma^{-1})|$ *is called the* n-th relative homotopy *of the fragmented pair* $(\underline{X},\underline{Y})$.

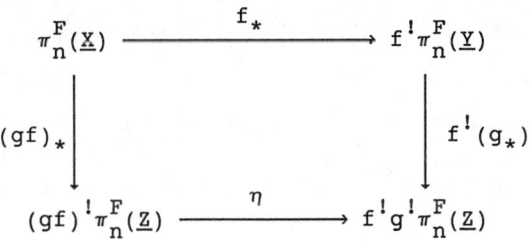

We note the analogy with the classical homotopy groups: $\pi_n(X,Y,y_0)$ is a module over $\mathbb{Z}\pi_1(Y,y_0)$; while for fragmented pairs, $\pi_n^F(\underline{X},\underline{Y})$ is a "module over $\mathcal{B}G_1(\underline{Y})$".

If $f\colon (\underline{X},\underline{Y}) \longrightarrow (\underline{V},\underline{W})$ is a map of fragmented pairs, then f restricts to a map $\underline{Y} \longrightarrow \underline{W}$ which we also denote by f (with only a slight ambiguity). Then, as before f induces a functor

$$f^!\colon \mathcal{C}^{\mathcal{B}G_1(\underline{W})}(\Sigma^{-1}) \longrightarrow \mathcal{C}^{\mathcal{B}G_1(\underline{Y})}(\Sigma^{-1})$$

and a morphism $f_*\colon \pi_n^F(\underline{X},\underline{Y}) \longrightarrow f^!\pi_n^F(\underline{V},\underline{W})$ unique up to a canonical natural equivalence. The details, which are entirely similar to the absolute case, are left to the reader. The proof of the following lemma is also left to the reader:

Lemma 4.6 *Let* $f\colon (\underline{X},\underline{Y}) \longrightarrow (\underline{V},\underline{W})$ *and* $g\colon (\underline{V},\underline{W}) \longrightarrow (\underline{S},\underline{T})$ *be maps of fragmented pairs over* $\mathcal{B}$. *Then there is a canonical natural equivalence* η *such that the following diagram commutes*

$$\begin{array}{ccc}
\pi_n^F(\underline{X},\underline{Y}) & \xrightarrow{\quad f_* \quad} & f^!\pi_n^F(\underline{V},\underline{W}) \\
{\scriptstyle (gf)_*}\big\downarrow & & \big\downarrow{\scriptstyle f^!(g_*)} \\
(gf)^!\pi_n^F(\underline{S},\underline{T}) & \xrightarrow{\quad \eta \quad} & f^!g^!\pi_n^F(\underline{S},\underline{T})
\end{array}$$

Let $(\underline{X},\underline{Y})$ be a fragmented pair over $\mathcal{B}$. Following the discussion preceeding 3.1, we let $i\colon \underline{Y} \longrightarrow \underline{X}$ be the inclusion map and note that $i^!\pi_n^F(\underline{X})$ is represented essentially by the restriction of $\pi_n^F(\underline{X})$ to $\mathcal{B}G_1(\underline{Y})$. For $n \geq 0$, we let $j_*\colon i^!\pi_n^F(\underline{X}) \longrightarrow \pi_n^F(\underline{X},\underline{Y})$ be represented by the natural transformation $j_*\colon i^*\pi_n^F(\underline{X}) \longrightarrow \pi_n^F(\underline{X},\underline{Y})$ given by the family

$$\{j(y,K)_* = j_{K*}\colon \pi_n(X_K,y) \longrightarrow \pi_n(X_K,Y_K,y) \mid (y,K) \in |\mathcal{B}G_1(\underline{Y})|\}$$

and, for $n \geq 1$, we let $\partial_*\colon \pi_n^F(\underline{X},\underline{Y}) \longrightarrow \pi_{n-1}^F(\underline{Y})$ be represented by the natural transformation given by the family

$$\{\partial(y,K)_* = \partial_{K*}\colon \pi_n(X_K,Y_K,y) \longrightarrow \pi_{n-1}(Y_K,y) \mid (y,K) \in |\mathcal{B}G_1(\underline{Y})|\}.$$

The following lemma is an immediate consequence of I.4.2 and I.2.4

Lemma 4.7 *For any pair* $(\underline{X},\underline{Y})$ *of fragmented spaces over* $\mathfrak{B}$, *the following sequence is exact*

$$\cdots \longrightarrow \pi^F_{n+1}(\underline{X},\underline{Y}) \xrightarrow{\partial_*} \pi^F_n(\underline{Y}) \xrightarrow{i_*} i^!\pi^F_n(\underline{X}) \xrightarrow{j_*} \pi^F_n(\underline{X},\underline{Y}) \longrightarrow \cdots$$

Remarks 4.8 (i) This sequence terminates in

$$\cdots\ i^!\pi^F_1(\underline{X}) \longrightarrow \pi^F_1(\underline{X},\underline{Y}) \longrightarrow \pi^F_0(\underline{Y}) \longrightarrow i^!\pi^F_0(\underline{X}) \longrightarrow \pi^F_0(\underline{X},\underline{Y}) \rightarrow 0.$$

(ii) The term "exact" in this lemma is somewhat misleading as the objects lie in different categories. In particular,

$$\cdots \xrightarrow{j_*} \pi^F_3(\underline{X},\underline{Y}) \xrightarrow{\partial_*} \pi^F_2(\underline{Y}) \xrightarrow{i_*} i^!\pi^F_2(\underline{X})$$

is exact in $\mathcal{Ab}^{\mathfrak{B}G_1(\underline{Y})}$; while

$$\pi^F_2(\underline{Y}) \longrightarrow i^!\pi^F_2(\underline{X}) \longrightarrow \pi^F_2(\underline{X},\underline{Y}) \longrightarrow \pi^F_1(\underline{Y}) \longrightarrow i^!\pi^F_1(\underline{X})$$

and

$$\pi^F_1(\underline{Y}) \longrightarrow i^!\pi^F_1(\underline{X}) \longrightarrow \pi^F_1(\underline{X},\underline{Y}) \longrightarrow \pi^F_0(\underline{Y}) \longrightarrow i^!\pi^F_0(\underline{X}) \longrightarrow \pi^F_0(\underline{X},\underline{Y})$$

are exact in $\mathcal{Gp}^{\mathfrak{B}G_1(\underline{Y})}(\Sigma^{-1})$ and $\mathcal{Sets}_*^{\mathfrak{B}G_1(\underline{Y})}(\Sigma^{-1})$, respectively. (Strictly speaking, the forgetful functors $\mathcal{Ab} \longrightarrow \mathcal{Gp}$ and $\mathcal{Gp} \longrightarrow \mathcal{Sets}_*$ should be applied to the first two terms on the left in the last two sequences.)

(iii) The sequence of 4.7 is natural with respect to maps of pairs.

Lemma 4.9 *Let* $f_j\colon \underline{X} \longrightarrow \underline{Y}$ $(j=0,1)$ *be maps of fragmented spaces over* $\mathfrak{B}$ *and* $F\colon \underline{X}\times I \longrightarrow \underline{Y}$ *be a homotopy from* f_0 *to* f_1. *Then for all* $n\geq 0$, F *induces a natural equivalence* ω_F *such that the following diagram commutes*

$$\begin{array}{ccc} & \overset{f_{1_*}}{\nearrow} & f_1^! \, \pi_n^F(\underline{Y}) \\ \pi_n^F(\underline{X}) & & \big\downarrow \omega_F \\ & \underset{f_{0_*}}{\searrow} & f_0^! \, \pi_n^F(\underline{Y}) \end{array}$$

Proof: Let F be represented by $F_m\colon \underline{X}\times I \longrightarrow C^m\underline{Y}$. Then $F_m|X\times\{j\} = f_{jm}$ $(j=0,1)$ represents f_j and $F_{mK}\colon X_K\times I \longrightarrow Y_{C^m K}$. If $(x,K) \in |\mathcal{B}G_1(\underline{X})|$, we let $\omega_F(x,K) = [F_{mK}|x\times I]$. Then $\omega_F(x,K)$ is a path class from $f_{0m}(x)$ to $f_{1m}(x)$ in $Y_{C^m K}$. Since it is easily seen that for all $(x,K) \in |\mathcal{B}G_1(\underline{X})|$, the following diagram commutes

$$\begin{array}{ccc} & \overset{f_{1m_*}}{\nearrow} & \pi_n(Y_{C^m K}, f_{1m}(x)) \\ \pi_n(X_K, x) & & \big\downarrow \omega_F(x,K)_* \\ & \underset{f_{0m_*}}{\searrow} & \pi_n(Y_{C^m K}, f_{0m}(x)) \end{array}$$

and that $\omega_F = \{\omega_F(x,K)_* \mid (x,K) \in |\mathcal{B}G_1(\underline{X})|\}$ is a natural equivalence, the lemma follows.

We remark that although ω_F above depends on the choice of F_m representing F, any two such choices for ω_F are canonically naturally equivalent.

Corollary 4.10 *Let* $f\colon \underline{X} \longrightarrow \underline{Y}$ *be a homotopy equivalence of fragmented spaces over* $\mathcal{B}$. *Then for all* $n\geq 0$, $f_*\colon \pi_n^F(\underline{X}) \longrightarrow f^!\pi_n^F(\underline{Y})$ *is an isomorphism.*

Proof: Let $g\colon \underline{Y} \longrightarrow \underline{X}$ be a homotopy inverse for f and $F\colon \underline{X}\times I \longrightarrow \underline{Y}$ be a homotopy from $1_{\underline{X}}$ to gf. By 4.6 and 4.8 the following diagram

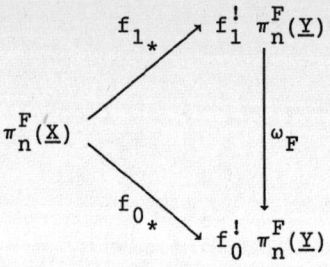

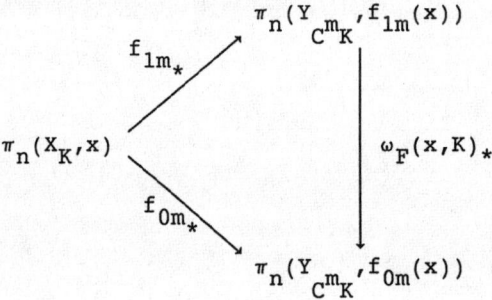

commutes

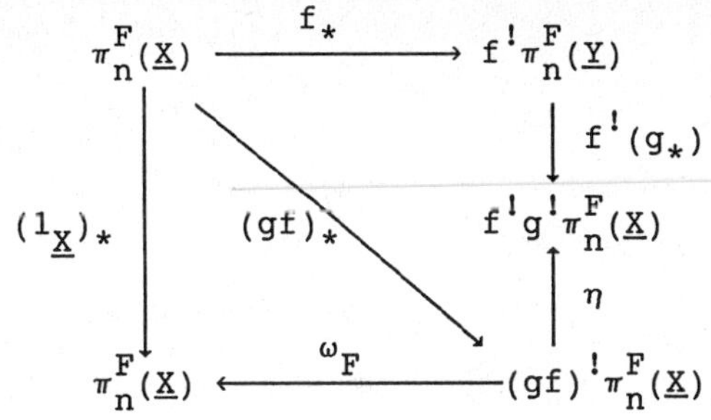

where η is the natural equivalence of 4.6. Hence, $(gf)_*$ and $f^!(g_*)f_*$ are isomorphisms. Similarly, $(fg)_*$ and $g^!(f_*)g_*$ are isomorphisms. But then the composites of the first two and the last two morphisms in the diagram

$$\pi_n^F(\underline{X}) \xrightarrow{f_*} f^!\pi_n^F(\underline{Y}) \xrightarrow{f^!(g_*)} f^!g^!\pi_n^F(\underline{X}) \xrightarrow{f^!g^!(f_*)} f^!g^!f^!\pi_n^F(\underline{Y})$$

are isomorphisms. It follows easily that all the morphisms are isomorphisms.

Remark 4.11 Since 4.8 easily extends to cover the case of a homotopy $F\colon (\underline{X},\underline{Y})\times I \longrightarrow (\underline{V},\underline{W})$ between the maps of fragmented pairs $f_j\colon (\underline{X},\underline{Y}) \longrightarrow (\underline{V},\underline{W})$ $(j=0,1)$, the proof of 4.9 can be adapted to prove that a homotopy equivalence $f\colon (\underline{X},\underline{Y}) \longrightarrow (\underline{V},\underline{W})$ of fragmented pairs over $\mathfrak{B}$ induces an isomorphism $f_*\colon \pi_n^F(\underline{X},\underline{Y}) \longrightarrow f^!\pi_n^F(\underline{V},\underline{W})$ for all $n\geq 0$.

Remark 4.12 Arguments similar to those of 4.7, show that for any triple $(\underline{X},\underline{Y},\underline{Z})$ of fragmented spaces over $\mathfrak{B}$ there is an exact sequence

$$\cdots \longrightarrow j^!\pi_{n+1}^F(\underline{X},\underline{Y}) \xrightarrow{\partial_*} \pi_n^F(\underline{Y},\underline{Z}) \xrightarrow{i_*} \pi_n^F(\underline{X},\underline{Z}) \xrightarrow{j_*} j^!\pi_n^F(\underline{X},\underline{Y}) \longrightarrow \cdots$$

where $i\colon (\underline{Y},\underline{Z}) \longrightarrow (\underline{X},\underline{Z})$ and $j\colon (\underline{X},\underline{Z}) \longrightarrow (\underline{X},\underline{Y})$ are the inclusions and ∂_* is represented by the maps $\partial_{(z,K)*}\colon \pi_{n+1}(X_K,Y_K,z) \longrightarrow \pi_n(Y_K,Z_K,z)$ for $z\in Z_K$. This sequence terminates when $n=0$. Furthermore, results analogous to 4.8 and 4.9 hold for such triples. We leave the details to the reader.

5. The fundamental groupoid.

Let $\underline{X}$ be a fragmented space over the category with endomorphism $\mathfrak{B}$. Then the fundamental groupoid $\mathfrak{B}G_1(\underline{X})$ can itself be regarded as an algebraic invariant of $\underline{X}$. It is the purpose of this section to compare this invariant with other invariants of $\underline{X}$. In making this comparison, it is convenient to work at a more general level. In particular, we shall study certain functors between the categories $\mathfrak{B}G$ introduced in I.1.3.

Let $\mathfrak{B}$ be a category with endomorphism and $G: \mathfrak{B} \longrightarrow \mathcal{G}poid$ be a functor into the category of groupoids. We note that the correspondence $G \longrightarrow \mathfrak{B}G$ of I.1.3 extends to a functor

$$\mathfrak{B}(\ ): \mathcal{G}poid^{\mathfrak{B}} \longrightarrow \mathcal{C}at$$

into the category of U-small categories. Specifically, if $\nu: G \longrightarrow H$ is a morphism in $\mathcal{G}poid^{\mathfrak{B}}$ (i.e. a natural transformation), let $\mathfrak{B}\nu: \mathfrak{B}G \longrightarrow \mathfrak{B}H$ be the functor given by setting $\mathfrak{B}\nu(x,K) = (\nu_K(x),K)$ and $\mathfrak{B}\nu(\omega,i) = (\nu_L(\omega),i)$ when $(x,K) \in |\mathfrak{B}G|$ and $(\omega,i): (x,K) \longrightarrow (y,L)$ is a morphism in $\mathfrak{B}G$ and where $\nu_J: G(J) \longrightarrow H(J)$ $(J=K,L)$ is the value of ν on J. We note further that $\mathfrak{B}\nu$ commutes with C and τ.

Some conditions sufficient to guarantee that $\mathfrak{B}\nu$ is a strong eventual equivalence of categories (cf. I.6) are given in Theorem 5.1 below. The first of these is the following: We say that G *is coextensive with* H if there exists an integer $m \geq 0$ such that for every $K \in |\mathfrak{B}|$, if $|H(K)| \neq \emptyset$, then $|G(C^mK)| \neq \emptyset$ and if $|G(K)| \neq \emptyset$, then $|H(C^mK)| \neq \emptyset$. The other conditions involve the homotopy of $\mathfrak{B}G$, which we now describe. Let $\pi_0(\mathfrak{B}G): \mathfrak{B}G \longrightarrow \mathcal{S}ets_*$ be the functor that assigns to $(x,K) \in |\mathfrak{B}G|$ the set of path components of $G(K)$, with that of $x \in |G(K)|$ distinguished, and assigns to $(\omega,i)\colon (x,K) \longrightarrow (y,L)$ the function that sends the path component of $z \in |G(K)|$ to that of $G(i)(z) \in |G(L)|$. Since $\omega\colon G(i)(x) \longrightarrow y$ in $G(L)$, this function preserves the distinguished element. We think of $\pi_0(\mathcal{B}G)$ as being an object in $\mathcal{S}ets_*^{\mathcal{B}G}(\Sigma^{-1})$.

In a similar vein, let $\pi_1(\mathcal{B}G)\colon \mathcal{B}G \longrightarrow \mathcal{G}p$ be the functor given by $\pi_1(\mathcal{B}G)(x,K) = G(K)(x,x)$, the $G(K)$-automorphisms of x, and $\pi_1(\mathcal{B}G)(\omega,i)(\alpha) = \omega G(i)(\alpha)\omega^{-1}$ for $(\omega,i)\colon (x,K) \longrightarrow (y,L)$. We regard $\pi_1(\mathcal{B}G)$ as an object in $\mathcal{G}p^{\mathcal{B}G}(\Sigma^{-1})$.

A natural transformation $\nu\colon G \longrightarrow H$ induces morphisms

$$\pi_i(\mathcal{B}\nu)\colon \pi_i(\mathcal{B}G) \longrightarrow \mathcal{B}\nu^!\pi_1(\mathcal{B}H),$$

$(i=0,1)$ by letting $\pi_0(\mathcal{B}\nu)(x,K)$ be the function that sends the component of z in $G(K)$ to that of $\nu_K(z)$ in $H(K)$ and letting $[\pi_1(\mathcal{B}\nu)(x,K)](\alpha) = \nu_K(\alpha)$. (Note that if $\alpha \in G(K)(x,x) = \pi_1(\mathcal{B}G)(x,K)$, then $\nu_K(\alpha) \in H(K)(\nu_K(x),\nu_K(x)) = \pi_1(\mathcal{B}H)(\nu_K(x),K) = \mathcal{B}\nu^!\pi_1(\mathcal{B}H)(x,K)$.)

Theorem 5.1 *Let $\mathcal{B}$ be a category with endomorphism, $G,H\colon \mathcal{B} \longrightarrow \mathcal{G}poid$ be functors, and $\nu\colon G \longrightarrow H$ be a natural transformation. If G is coextensive with H and $\pi_i(\mathcal{B}\nu)\colon \pi_i(\mathcal{B}G) \longrightarrow \mathcal{B}\nu^!\pi_i(\mathcal{B}H)$ is an isomorphism for $i=0,1$, then $\mathcal{B}\nu\colon \mathcal{B}G \longrightarrow \mathcal{B}H$ is a strong eventual equivalence of categories. In particular, for any cocomplete category $\mathcal{C}$, $\mathcal{B}\nu_!\colon \mathcal{C}^{\mathcal{B}G}(\Sigma^{-1}) \longrightarrow \mathcal{C}^{\mathcal{B}H}(\Sigma^{-1})$ is an equivalence of categories.*

The proof is given later in this section.

Let $\underline{X}$ be a fragmented space over $\mathcal{B}$. We say that *$\underline{X}$ is coextensive with $\mathcal{B}$* if there exists an $m \geq 0$ such that $X_{C^mK} \neq \emptyset$ for every $K \in |\mathcal{B}|$. We say that $\underline{X}$ is *connected* if $\pi_0^F(\underline{X}) = 0$ and that $\underline{X}$ is *simply connected* if $\pi_i^F(\underline{X}) = 0$ $(i=0,1)$.

Corollary 5.2 *Let $\underline{X}$ be a simply connected fragmented space over $\mathfrak{B}$. If $\underline{X}$ is coextensive with $\mathfrak{B}$, then ρ: $\mathfrak{B}G_1(\underline{X}) \longrightarrow \mathfrak{B}$ is a strong eventual equivalence of categories.*

Let $\underline{X}$ and $\underline{Y}$ be fragmented spaces over $\mathfrak{B}$. We say $\underline{X}$ *is coextensive with* $\underline{Y}$ if there exists an integer $m \geq 0$ such that for every $K \in |\mathfrak{B}|$, if $Y_K \neq \emptyset$, then $X_{C^m K} \neq 0$ and if $X_K \neq \emptyset$, then $Y_{C^m K} \neq \emptyset$.

Corollary 5.3 *Let f: $\underline{X} \longrightarrow \underline{Y}$ be a map of fragmented spaces over $\mathfrak{B}$. If $\underline{X}$ is coextensive with $\underline{Y}$ and f_*: $\pi_i^F(\underline{X}) \longrightarrow f^!\pi_i^F(\underline{Y})$ is an isomorphism for i=0,1, then f induces a strong eventual equivalence f_*: $\mathfrak{B}G_1(\underline{X}) \longrightarrow \mathfrak{B}G_1(\underline{Y})$. Hence for any cocomplete category $\mathscr{C}$, $f_!$: $\mathscr{C}^{\mathfrak{B}G_1(\underline{X})}(\Sigma^{-1}) \longrightarrow \mathscr{C}^{\mathfrak{B}G_1(\underline{Y})}(\Sigma^{-1})$ is an equivalence of categories.*

The following result, weaker than 5.1, has an entirely elementary proof:

Proposition 5.4 *Let f: $\underline{X} \longrightarrow \underline{Y}$ be an isomorphism of fragmented spaces over $\mathfrak{B}$. Then f induces a strong eventual equivalence f_*: $\mathfrak{B}G_1(\underline{X}) \longrightarrow \mathfrak{B}G_1(\underline{Y})$. Hence for every cocomplete category $\mathscr{C}$, $f_!$: $\mathscr{C}^{\mathfrak{B}G_1(\underline{X})} \longrightarrow \mathscr{C}^{\mathfrak{B}G_1(\underline{Y})}$ is an equivalence of categories.*

Proof of 5.2: Let H: $\mathfrak{B} \longrightarrow \mathscr{G}poid$ be the constant functor with value the trivial groupoid $\mathscr{T}$ (i.e. the groupoid with only one object x and one morphism, the identity of x). For any groupoid $\mathfrak{D}$, there is a unique functor $\nu_{\mathfrak{D}}$: $\mathfrak{D} \longrightarrow \mathscr{T}$ and $\nu = \{\nu_{G_1(\underline{X})(K)} \mid K \in |\mathfrak{B}|\}$ defines a natural transformation ν: $G_1(\underline{X}) \longrightarrow H$. Clearly $G_1(\underline{X})$ is coextensive with H. Since it is easily seen that if i=0,1, $\pi_i(\mathfrak{B}G_1(\underline{X})) = \pi_i^F(\underline{X}) = 0$ and that $\pi_i(\mathfrak{B}H) = 0$, $\pi_i(\mathfrak{B}\nu)$ is an isomorphism for i=0,1. Hence $\mathfrak{B}\nu$ is a strong eventual equivalence of categories by 5.1.

On the other hand the functor F: $\mathfrak{B} \longrightarrow \mathfrak{B}H$ given by F(K) = (x,K) and

$F(i) = (1,i)$ for $i \in \mathcal{B}(K,L)$ is an isomorphism of categories making the diagram

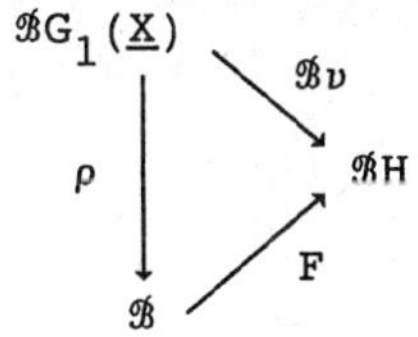

commute. Corollary 5.2 is now clear.

Proof of 5.3: Let f be represented by $f_m\colon \underline{X} \longrightarrow C^m\underline{Y}$ and $F_m\colon \mathcal{B}G_1(\underline{X}) \longrightarrow \mathcal{B}G_1(\underline{Y})$ be the functor given by $F_m(x,K) = (f_{mK}(x), C^mK)$ and $F_m(\omega,i) = (f_{mL}(\omega), C^m i)$ for $(\omega,i)\colon (x,K) \longrightarrow (y,L)$ (cf. the proof of 4.2). We shall show that F_m is a strong eventual equivalence of categories. Setting $f_* = F_m$ will then complete the proof of 5.3.

To do this, we note first that f_m induces a natural transformation $\upsilon\colon G_1(\underline{X}) \longrightarrow G_1(C^m\underline{Y})$ by setting $\upsilon_K(x) = f_{mK}(x)$ for $x \in G_1(X_K)$ and $\upsilon_K(\omega) = f_{mK}(\omega)$ for $\omega\colon x \longrightarrow y$ in $G_1(X_K)$. In addition, we define a functor $J\colon \mathcal{B}(G_1(C^m\underline{Y})) \longrightarrow \mathcal{B}(G_1(\underline{Y}))$ by $J(x,K) = (x, C^mK)$ and $J(\omega,i) = (\omega, C^m i)$. It is easy to see that the following diagram commutes

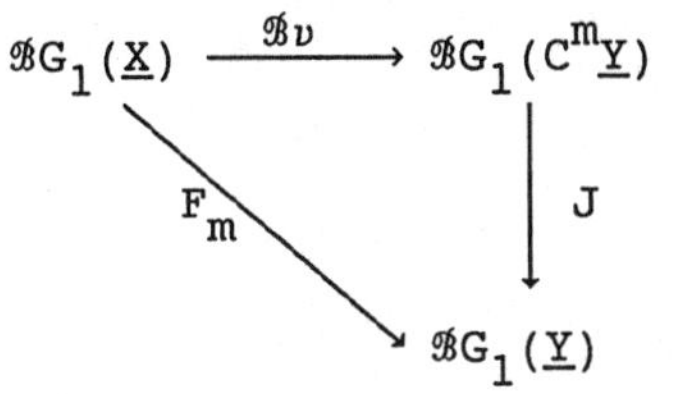

Thus it suffices to prove that each of J and $\mathcal{B}\upsilon$ is a strong eventual equivalence of categories.

That J is a strong eventual equivalence of categories is straightforward. For simple calculations show that J commutes with C and τ and that $\mathcal{B}G_1(\underline{Y})(\tau^m)$ is a strong eventual pseudo inverse for J.

Since $\underline{X}$ is coextensive with $\underline{Y}$, it is easy to see that $G_1(\underline{X})$ is coextensive with $G_1(\underline{Y})$. Let $1_*\colon \pi_i^F(C^m\underline{Y}) \longrightarrow J^*\pi_i^F(\underline{Y})$ be given by

$1_* = \{\text{id}: \pi_i(Y_{C^m K}, y) \longrightarrow \pi_i(Y_{C^m K}, y)\}$. Since $\pi_i(Y_{C^m K}, y)$ can be regarded as either $\pi_i^F(C^m\underline{Y})(y,K)$ or as $\pi_i^F(\underline{Y})(y,C^m K) = J^*\pi_i^F(\underline{Y})(y,K)$, this makes sense. Clearly 1_* is an isomorphism for all i. Now consider the diagram

$$\begin{array}{ccc} \pi_i^F(\underline{X}) & \xrightarrow{\pi_i(\mathcal{B}\nu)} & \mathcal{B}\nu^!\pi_i^F(C^m\underline{Y}) \\ \Big\downarrow f_{m_*} & & \Big\downarrow \mathcal{B}\nu^!(1_*) \\ F_m^!\pi_i^F(\underline{Y}) & =\!=\!=\!= & \mathcal{B}\nu^!J^!\pi_i^F(F) \end{array}$$

which is easily seen to commute. By hypothesis, f_{m_*} is an isomorphism for i=0,1. Hence $\pi_i(\mathcal{B}\nu)$ is an isomorphism for i=0,1 and by 5.1, $\mathcal{B}\nu$ is a strong eventual equivalence of categories. This completes the proof of 5.3.

Proof of 5.4: Let $g: \underline{Y} \longrightarrow \underline{X}$ be an inverse for f and represent f and g by $f_m: \underline{X} \longrightarrow C^m\underline{Y}$ and $g_n: \underline{Y} \longrightarrow C^n\underline{X}$ as in the proof of 5.3. Since $gf = 1_{\underline{X}}$ and $fg = 1_{\underline{Y}}$, we may assume m and n chosen so that for every $K \in |\mathcal{B}|$ the composites

$$X_K \xrightarrow{f_m} Y_{C^m K} \xrightarrow{g_n} X_{C^{n+m}K} \quad \text{and} \quad Y_K \xrightarrow{g_n} X_{C^n K} \xrightarrow{f_m} Y_{C^{n+m}K}$$

are $\underline{X}(\tau^{n+m})$ and $\underline{Y}(\tau^{n+m})$ respectively. Let $F_m: \mathcal{B}G_1(\underline{X}) \longrightarrow \mathcal{B}G_1(\underline{Y})$ be the functor defined in the proof of 5.3 and define $G_n: \mathcal{B}G_1(\underline{Y}) \longrightarrow \mathcal{B}G_1(\underline{X})$ similarly. An easy calculation using the fact that $g_n f_m = \underline{X}(\tau^{n+m})$ shows that $G_n F_m = C_1^{n+m}$ where C_1 is the expansion functor in $\mathcal{B}G_1(\underline{X})$. Similarly $F_m G_n = C_2^{n+m}$. If we now let $f_* = F_m$ and $g_* = G_n$, the first part of 5.4 follows (cf. I.6). The second part is immediate from I.6.1.

Proof of 5.1: We shall show that $\mathcal{B}\nu$ is eventually onto, eventually full, and eventually faithful. Thus 5.1 will follow from I.6.2.

To show $\mathcal{B}\nu$ is eventually onto, let $(y,K) \in \mathcal{B}H$. Then $|H(K)| \neq \emptyset$.

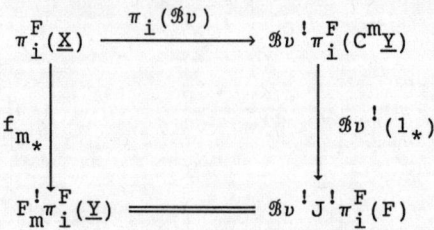

Since G is coextensive with H, there exists an integer $m \geq 0$ such that for every $L \in |\mathcal{B}|$, if $|H(L)| \neq 0$, then $|G(C^m L)| \neq \emptyset$. In particular, $|G(C^m K)| \neq \emptyset$ and we pick $x \in |G(C^m K)|$. We now regard $H(\tau^m)(y)$ as an element $[H(\tau^m)(y)] \in \pi_0(\mathcal{B}H)(\upsilon_{C^m K}(x), C^m K) = \mathcal{B}\upsilon^! \pi_0(\mathcal{B}H)(x, C^m K)$. Since $\pi_0(\mathcal{B}\upsilon)\colon \pi_0(\mathcal{B}G) \longrightarrow \pi_0(\mathcal{B}H)$ is an isomorphism, by I.4.7 $\mathcal{B}\upsilon_*$ is eventually epimorphic. Hence there exists an integer $r \geq 0$ and an object $z \in |G(C^{m+r}K)|$ such that $\pi_0(\mathcal{B}\upsilon)[z] = H(\tau^r)_*[H(\tau^m)(y)] = [H(\tau^{m+r})(y)]$ in $\pi_0(\mathcal{B}H)(H(\tau^r)\upsilon_{C^m K}(x),\ C^{m+r}K)$. In particular there is an isomorphism $\omega\colon \upsilon_{C^{m+r}K}(z) \longrightarrow H(\tau^{m+r})(y)$ in $H(C^{m+r}K)$. But then $(\omega,1)\colon (\upsilon_{C^{m+r}}(z), C^{m+r}K) \longrightarrow (H(\tau^{m+r})(y), C^{m+r}K)$ is an isomorphism from $\mathcal{B}\upsilon(z, C^{m+r}K)$ to $C^{m+r}(y,K)$. It follows that $\mathcal{B}\upsilon$ is eventually onto.

To show that $\mathcal{B}\upsilon$ is eventually full, let $(\omega,i)\colon \mathcal{B}\upsilon(x_1,K_1) \longrightarrow \mathcal{B}\upsilon(x_2,K_2)$ be a morphism in $\mathcal{B}H$. Since $\mathcal{B}\upsilon(x_j,K_j) = (\upsilon_{K_j}(x_j),K_j)$ $(j=1,2)$, $\omega\colon H(i)\upsilon_{K_1}(x_1) \longrightarrow \upsilon_{K_2}(x_2)$ in $H(K_2)$. Since $H(i)\upsilon_{K_1}(x_1) = \upsilon_{K_2}G(i)(x_1)$, this says that $\upsilon_{K_2}G(i)(x_1)$ lies in the same component of $H(K_2)$ as $\upsilon_{K_2}(x_2)$; i.e. both objects represent the same element of $\pi_0(\mathcal{B}H)(x_2,K_2)$. Since $\pi_0(\mathcal{B}\upsilon)$ is an isomorphism, it is eventually monomorphic by I.4.7. Hence there exists $m \geq 0$ such that $G(\tau^m)[G(i)(x_1)] = G(\tau^m)[x_2]$ in $\pi_0(\mathcal{B}G)(G(\tau^m)x_2), C^m K_2)$. Since $G(C^m K_2)$ is a groupoid, this means that there exists an isomorphism $\omega'\colon G(\tau^m)G(i)(x_1) \longrightarrow G(\tau^m)(x_2)$ in $G(C^m K_2)$. Since $C^m(x_j,K_j) = (G(\tau^m)(x_j), C^m K_j)$ $(j=1,2)$ and $G(\tau^m)G(i) = G(C^m i)G(\tau^m)$ for $i\colon K_1 \longrightarrow K_2$, $(\omega', C^m i)\colon C^m(x_1,K_1) \longrightarrow C^m(x_2,K_2)$ in $\mathcal{B}G$.

Since

$$\upsilon_{C^m K_2} G(\tau^m i)(x_1) = H(\tau^m i)\upsilon_{K_1}(x_1) \text{ and } \upsilon_{C^m K_2} G(\tau^m)(x_2) = H(\tau^m)\upsilon_{K_2}(x_2),$$

$\upsilon_{C^m K_2}(\omega')$ and $H(\tau^m)(\omega)$ are both morphisms from $H(\tau^m i)\upsilon_{K_1}(x_1)$ to $H(\tau^m)\upsilon_{K_2}(x_2)$ in $H(C^m K_2)$. Let $\alpha = [\upsilon_{C^m K_2}(\omega')]^{-1} \circ H(\tau^m)(\omega)$. Then $\alpha \in \pi_1(\mathcal{B}H)(H(\tau^m i)\upsilon_{K_1}(x_1), C^m K_2)$. Since $\pi_1(\mathcal{B}\upsilon)\colon \pi_1\mathcal{B}(G) \longrightarrow \pi_1\mathcal{B}(H)$ is an isomorphism, it is eventually epimorphic by I.4.7. Hence there exists an integer $n\geq 0$ and a class $\alpha' \in \pi_1(\mathfrak{B}G)(G(\tau^m i)(x_1), C^n(C^m K_2))$ such that $\nu_{C^{n+m}K_2}(\alpha') = \pi_1(\mathfrak{B}\nu)(\alpha') = H(\tau^n)(\alpha)$. Let

$$\omega'' = G(\tau^n)(\omega')\alpha' \colon G(\tau^{n+m} i)(x_1) \longrightarrow G(\tau^{n+m} x_2)$$

in $G(C^{n+m}K_2)$. Then

$$\begin{aligned}
\nu(\omega'') &= \nu[G(\tau^n)(\omega')\alpha'] \\
&= \nu[G(\tau^n)(\omega')]\ \nu(\alpha') \\
&= H(\tau^n)\ \nu(\omega')\ H(\tau^n)(\alpha) \\
&= H(\tau^n)\nu(\omega')\ [H(\tau^n)\nu(\omega')]^{-1}\ H(\tau^n)H(\tau^m)(\omega) \\
&= H(\tau^{n+m})(\omega).
\end{aligned}$$

It now follows that $(\omega'', C^{n+m} i)\colon C^{n+m}(x_1,K_1) \longrightarrow C^{n+m}(x_2,K_2)$ has $\mathfrak{B}\nu(\omega'', C^{n+m} i) = C^{n+m}(\omega,i)$. Since m and n are independent of (ω,i), this shows that $\mathfrak{B}\nu$ is eventually full.

To show that $\mathfrak{B}\nu$ is eventually faithful, let $(\omega_j,i_j)\colon (x_1,K_1) \longrightarrow (x_2,K_2)$ $(j=1,2)$ be morphisms in $\mathfrak{B}G$ such that $\mathfrak{B}\nu(\omega_1,i_1) = \mathfrak{B}\nu(\omega_2,i_2)$. Then $(\nu_{K_2}(\omega_1),i_1) = (\nu_{K_2}(\omega_2),i_2)$ from which it follows that $i_1 = i_2\colon K_1 \longrightarrow K_2$ and $\nu_{K_2}(\omega_1) = \nu_{K_2}(\omega_2)$. Let $\alpha = \omega_2\omega_1^{-1} \in \pi_1(\mathfrak{B}G)(x_2,K_2)$. Then $\pi_1(\mathfrak{B}\nu)(\alpha) = \nu_{K_2}(\alpha) = 1$ in $\pi_1(\mathfrak{B}H)(\nu_{K_2}(x_2),K_2)$. Since $\pi_1(\mathfrak{B}\nu)$ is an isomorphism, it is eventually monomorphic by I.4.7. Hence there exists a $t\geq 0$, independent of (x_2,K_2), such that $G(\tau^t)(\alpha) = 1$ in $\pi_1(\mathfrak{B}G)(G(\tau^t)(x_2),C^t K_2)$. Since $1 = G(\tau^t)(\alpha) = [G(\tau^t)(\omega_2)][G(\tau^t)(\omega_1)]^{-1}$, it follows that $G(\tau^t)(\omega_1) = G(\tau^t)(\omega_2)$. But then $C^t(\omega_1,i_1) = (G(\tau^t)(\omega_1),C^t i_1) = (G(\tau^t)(\omega_2),C^t i_2) = C^t(\omega_2,i_2)$ and $\mathfrak{B}\nu$ is eventually faithful. This completes the proof of 5.1.

6. The absolute Hurewicz theorem.

In this section and the next we shall state and prove absolute and relative Hurewicz theorems for fragmented spaces. We begin by setting the context for the absolute Hurewicz theorem.

Let $\mathfrak{B}$ be a category with endomorphism, $\underline{X}$ be a fragmented space over $\mathfrak{B}$, $\mathfrak{B}G_1(\underline{X})$ be the fundamental groupoid of $\underline{X}$, and $\rho\colon \mathfrak{B}G_1(\underline{X}) \longrightarrow \mathfrak{B}$ be the forgetful functor. For any $n\geq 1$, the Hurewicz homomorphisms

$$\varphi_{(x,K)}\colon \pi_n(X_K,x) \longrightarrow H_n(X_K)$$

define a natural transformation $\varphi\colon \pi_n^F(\underline{X}) \longrightarrow \rho^* H_n^F(\underline{X})$ and hence, by adjointness, a natural transformation $\psi\colon \rho_!\pi_n^F(\underline{X}) \longrightarrow H_n^F(\underline{X})$. When $n\geq 2$, ψ is called the *Hurewicz transformation*. When $n=1$, the *Hurewicz transformation* is the natural transformation $\psi^{ab}\colon (\rho_!\pi_1^F(\underline{X}))^{ab} \longrightarrow H_1^F(\underline{X})$ induced by ψ above where $(\ )^{ab}\colon \mathcal{G}p^{\mathfrak{B}G_1(\underline{X})}(\Sigma^{-1}) \longrightarrow \mathcal{A}b^{\mathfrak{B}G_1(\underline{X})}(\Sigma^{-1})$ is induced by the abelianization functor ab: $\mathcal{G}p \longrightarrow \mathcal{A}b$. It is the object of this section to prove the following theorem:

Theorem 6.1 *(Absolute Hurewicz Theorem) Let $\underline{X}$ be a fragmented space over $\mathfrak{B}$. Then*

(i) $\psi^{ab}\colon (\rho_!\pi_1^F(\underline{X}))^{ab} \longrightarrow H_1^F(\underline{X})$ *is an isomorphism in* $\mathcal{A}b^{\mathfrak{B}}(\Sigma^{-1})$; *and*

(ii) *If for some* $n\geq 2$, $\pi_i^F(\underline{X}) = 0$ *for* $1 \leq i \leq n-1$, *then* $\psi\colon \rho_!\pi_n^F(\underline{X}) \longrightarrow H_n^F(\underline{X})$ *is an isomorphism in* $\mathcal{A}b^{\mathfrak{B}}(\Sigma^{-1})$.

Proof: To prove (i), we note that by I.5.6 for any $K \in |\mathfrak{B}|$, $\rho_!\pi_1^F(\underline{X})(K) = \coprod_\alpha ab(\pi_1(X_K,x_\alpha))$ where the free product runs over the set $\{\alpha\}$ of path components of X_K. Hence $(\rho_!\pi_1^F(\underline{X}))^{ab} = \sum_\alpha ab(\pi_1(X_K,x_\alpha))$. But also $H_1^F(\underline{X})(K) = \sum_\alpha H_1(X_K,x_\alpha)$ and there is a commutative diagram

$$\begin{array}{ccc}
(\rho_! \pi_1^F(\underline{X}))^{ab}(K) & \xrightarrow{\quad \psi_K \quad} & H_1^F(\underline{X})(K) \\
\| & & \| \\
\sum_\alpha ab(\pi_1(X_K, x_\alpha)) & \xrightarrow{\quad \Sigma_\alpha \psi_\alpha \quad} & \sum_\alpha H_1(X_K, x_\alpha)
\end{array}$$

where $\psi_\alpha\colon\ ab(\pi_1(X_K,x_\alpha)) \longrightarrow H_1(X_K,x_\alpha)$ is the classical Hurewicz isomorphism. Thus ψ_K is an isomorphism for all K and part (i) follows.

The proof of part (ii) of 6.1 is based on the following easy lemmas whose proofs are sketched at the end of this section:

Lemma 6.2 *Let* $n \geq 2$ *and suppose* $\pi_1^F(\underline{X}) = 0$ *for* $1 \leq i \leq n-1$. *Then there exists an* $m \geq 0$ *such that for any* CW *complex* W, *any* $K \in |\mathfrak{B}|$ *and any map* $f_0\colon W \longrightarrow X_K$, *there exists a map* $f_1\colon W \longrightarrow X_{C^m K}$ *homotopic to* $X(\tau_K^m)f_0$ *such that*

(i) *For any path component* W_γ *of* W, $f_1(W_\gamma^{(n-1)}) = x_\beta$ *is any preferred basepoint of the path component of* $X_{C^m K}$ *containing* $f_1(V_\gamma)$.

(ii) *If* $V \subseteq W$ *is a path connected subcomplex such that* $f_0(V^{(n-1)}) = x_\alpha$ *and* $\tau_K^m(x_\alpha) = x_\beta$, *then* $f_1|V = X(\tau_k^m)f_0|V$ *and the homotopy restricted to* $V \times I$ *is the composite* $V \times I \xrightarrow{p_1} V \xrightarrow{f_1|V} X_{C^m K}$.

Lemma 6.3 *Let* U *be a space. Then*

(i) *For any* $u \in H_n(U)$, *there exists a finite n-dimensional* CW *com plex* V, *a class* $v \in H_n(V)$, *and a map* $f\colon V \longrightarrow U$ *with* $f_* v = u$.

(ii) *If* V, v, *and* f *are as in (i) and* $f_* v = 0$, *then there exists a finite* (n+1)-*dimensional* CW *complex* W *containing* V, *a class* w $\in H_{n+1}(W,v)$ *with* $\partial w = v$ *where* $\partial\colon H_{n+1}(W,V) \longrightarrow H_n(V)$, *and a* *map* g: W ⟶ Z *extending* f. *If* V *is connected, then* W *may be taken to be connected.*

We complete the proof of part (ii) of 6.1 assuming 6.2 and 6.3. Let $K \in |\mathfrak{B}|$ and $\xi \in H_n(X_K)$. By 6.3, there exists a finite n-complex V, a class $v \in H_n(V)$, and a map $f\colon V \longrightarrow X_K$ such that $f_*(v) = \xi$. By 6.2, there exists m, independent of K, such that $X(\tau_K^m)f$ is homotopic to a map $\bar{g}$ such that for any path component V_γ of V, $\bar{g}(V_\gamma^{(n-1)})$ is a preferred basepoint x_β of the path component of X_{C^mK} that contains $\bar{g}(V_\gamma)$. It follows that there is a homotopy commutative diagram

$$\begin{array}{ccccc}
V & \xrightarrow{\;p\;} & \coprod_\gamma V_\gamma/V_\gamma^{(n-1)} & & \\
{\scriptstyle f}\downarrow & \searrow^{g} & & \searrow^{\coprod g_\gamma} & \\
X_K & \xrightarrow[X(\tau_K^m)]{} & X_{C^mK} & = & \coprod_\beta X_{C^mK,\beta}
\end{array}$$

where g is induced by $\bar{g}$, $g_\gamma = g|V_\gamma/V_\gamma^{(n-1)}$, and p is the obvious collapse map. Since $V_\gamma/V_\gamma^{(n-1)}$ is a bouquet of n-spheres,

$$\varphi_V = \Sigma_\gamma \varphi_\gamma\colon \Sigma_\gamma \pi_n(V_\gamma/V_\gamma^{(n-1)}, v_\gamma) \longrightarrow H_n(\coprod V_\gamma/V_\gamma^{(n-1)})$$

is an isomorphism. A simple chase of the diagram

$$\begin{array}{ccccc}
 & & \Sigma_\gamma \pi_n(V_\gamma/V_\gamma^{(n-1)}, v_\gamma) & \xrightarrow{\;g_*\;} & \\
 & \Sigma_\alpha \pi_n(X_K, x_\alpha) & \xrightarrow{\;X(\tau_K^m)_*\;} & & \Sigma_\beta \pi_n(X_{C^mK}, x_\beta) \\
 & & \downarrow{\scriptstyle \varphi_V} & & \\
H_n(V) & \xrightarrow{\;p_*\;} & H_n(\coprod V_\gamma/V_\gamma^{(n-1)}) & \xrightarrow{\;g_*\;} & \downarrow{\scriptstyle \varphi_{C^mK}} \\
{\scriptstyle f_*}\searrow & \downarrow{\scriptstyle \varphi_K} & & & \\
 & H_n(X_K) & \xrightarrow[X(\tau_K^m)_*]{} & & H_n(X_{C^mK})
\end{array}$$

now shows that $X(\tau_K^m)_*(\xi) \in \operatorname{Im} \varphi_{C^mK}$. Since φ_{C^mK} factors as

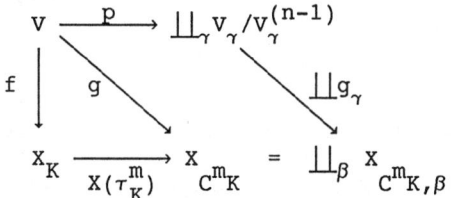

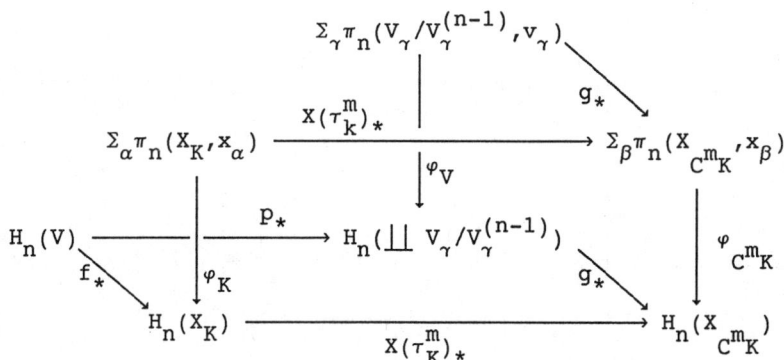

$$\sum_{\beta} \pi_n(X_{C^m K}, x_\beta) \longrightarrow \rho_! \pi_n^F(X)(C^m K) \longrightarrow H_n(X_{C^m K})$$

and the latter map is $\psi_{C^m K}$, $X(\tau_K^m)_*(\xi) \in \operatorname{Im} \psi_{C^m K}$. It follows from I.4.6 that $\psi\colon \rho_!\pi_n^F(X) \longrightarrow H_n^F(X)$ is epic.

Now let $\xi \in \ker\{\psi_K\colon \rho_!\pi_n^F(X)(K) \longrightarrow H_n^F(X_K)\}$. Since the following diagram commutes

$$\begin{array}{ccc}
\psi_K\colon \rho_!\pi_n^F(X)(K) & \longrightarrow & H_n^F(X)(K) \\
\| & & \| \\
\sum_\alpha \pi_n'(X_K, x_\alpha) & \longrightarrow & \Sigma_\alpha H(X_K, x_\alpha) \\
\nwarrow & & \nearrow \Sigma_\alpha \varphi(x_\alpha, K) \\
& \Sigma_\alpha \pi_n(X_K, x_\alpha) &
\end{array}$$

where $\pi_n'(X_K, x_\alpha)$ denotes $\pi_n(X_K, x_\alpha)$ modulo the action of $\pi_1(X_K, x_\alpha)$, it follows that for any $\xi = (\xi_\alpha) \in \Sigma_\alpha \pi_n(X_K, x_\alpha)$ with $\eta(\xi) = \xi'$, $\varphi_{(x_\alpha, K)}(\xi_\alpha) = 0$.

Let $f_\alpha\colon (S_\alpha^n, *) \longrightarrow (X_K, x_\alpha)$ represent ξ_α. Since $0 = \varphi_{(x_\alpha, K)}(\xi_\alpha) = f_{\alpha_*}(i_\alpha)$ where $i_\alpha \in H_n(S_\alpha^n)$ is a generator, by 6.3 there exists a connected $W \supseteq S_\alpha^n$, a class $w \in H_{n+1}(W, S_\alpha^n)$ with $\partial w = i_\alpha$, and $\bar{g}\colon W \longrightarrow X_K$ extending f_α. By 6.2 there is an $m \geq 0$ independent of K such that $X(\tau_K^m)\bar{g}$ is homotopic relative to S_α^n to a map $g'\colon W \longrightarrow X_{C^m K}$ such that $g'(W^{(n-1)}) = x_\beta = X(\tau_K^m)(x_\alpha)$. Hence, there is a homotopy commutative diagram

$$\begin{array}{ccccc}
W & \xrightarrow{p} & W/W^{(n-1)} & \longleftarrow & S_\alpha^n \\
\bar{g} \downarrow & \searrow^{g'} & \downarrow g & & \downarrow f_\alpha \\
X_K & \xrightarrow[X(\tau_K^m)]{} & X_{C^m K} & \xleftarrow[X(\tau_K^m)]{} & X_K
\end{array}$$

where g is induced by g'.

Let $V = W/W^{(n-1)}$ and notice that (V, S_α^n) is (n-1) connected. Since

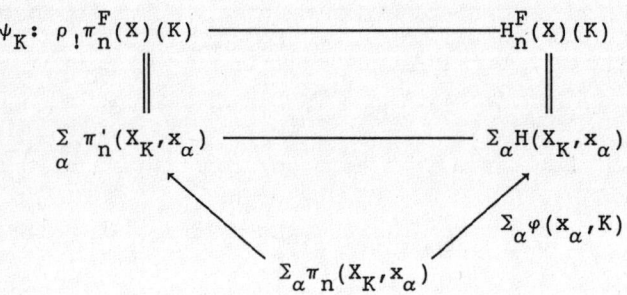

$n \geq 2$, and S^n_α is (n-1) connected, the map k_* in the commutative diagram

$$\begin{array}{ccccc}
\pi_n(S^n_\alpha) & \xleftarrow{\partial_*} & \pi_{n+1}(V,S^n_\alpha) & \xrightarrow{k_*} & \pi_{n+1}(V/S^n_\alpha,*) \\
\downarrow{\scriptstyle\varphi''} & & \downarrow{\scriptstyle\varphi'} & & \downarrow{\scriptstyle\varphi} \\
H_n(S^n_\alpha) & \xleftarrow{\partial'} & H_{n+1}(V,S^n_\alpha) & \xrightarrow{\ell_*} & H_{n+1}(V/S^n_\alpha) \\
& {\scriptstyle\partial}\nwarrow & & \nearrow{\scriptstyle p_*} & \\
& & H_{n+1}(W,S^n_\alpha) & &
\end{array}$$

is onto. Since V/S^n_α is (n-1) connected, φ is also onto and since ℓ_* is an isomorphism, φ' is onto. Furthermore, if $w' \in \pi_{n+1}(V,S^n_\alpha)$ satisfies $\varphi'(w') = p_*w$, then $\partial_* w'$ is the preferred generator of $\pi_n(S^n_\alpha)$. Since $g\colon (V,S^n_\alpha) \longrightarrow (X_{C^m K}, X_K)$ and $g|S^n_\alpha = f_\alpha$, a chase of the commutative diagram

$$\begin{array}{ccccc}
\pi_{n+1}(V,S^n_\alpha) & \xrightarrow{\partial_*} & \pi_n(S^n_\alpha,*) & & \\
\downarrow{\scriptstyle g_*} & & \downarrow{\scriptstyle f_{\alpha *}} & & \\
\pi_{n+1}(X_{C^{m}K},X_K) & \xrightarrow{\partial_*} & \pi_n(X_K,x_\alpha) & \xrightarrow{X(\tau^m_K)_*} & \pi_n(X_{C^{m}K},X(\tau^m_K)(x_\alpha))
\end{array}$$

shows that $\partial_* g_*(\omega') = \xi_\alpha$ and that $X(\tau^m_K)_*(\xi_\alpha) = 0$. It now follows from I.4.6 that $\psi\colon \rho_! \pi^F_n(\underline{X}) \longrightarrow H^F_n(\underline{X})$ is monic. Since $\mathcal{Ab}^{\mathcal{B}}(\Sigma^{-1})$ is abelian and ψ is both monic and epic, φ is an isomorphism and part (ii) of 6.1 follows.

We now give the proofs of 6.2 and 6.3.

Proof of 6.2: Let $1 \leq i \leq n-1$. Since $\pi^F_i(\underline{X}) = 0$, there exists an integer $m(i) \geq 0$ such that for all $(x,K) \in |\mathcal{B}G_1(\underline{X})|$,

$$\underline{X}(\tau^{m(i)}_K)_* = 0\colon \pi_i(X_K,x) \longrightarrow \pi_i(X_{C^{m(i)}K}, X(\tau^{m(i)}_K)(x)).$$

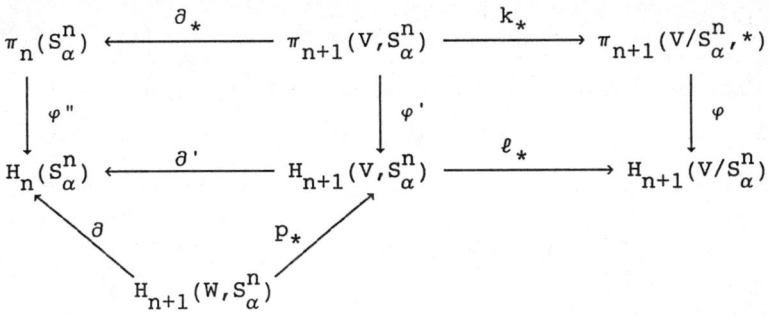

Let $p_0 = 0$ and $p(i) = m(1) + \cdots + m(i)$ $(1 \leq i \leq n-1)$ and let $f_0\colon W \longrightarrow X_K$. Using standard obstruction theory arguments, it is easy to construct maps $f^{(i)}: W \longrightarrow X_{C^{p(i)}K}$ $(0 \le i \le n-1)$, inductively, such that

(i) $f^{(i)}$ is homotopic to $\underline{X}(\tau_K^{p(i)})f_0$;

(ii) For each component W_γ of W, $f^{(i)}(W^{(i)})$ is a point.

(iii) If $V \subset W$ is path connected and $f_0(V^{(n-1)}) = x_\alpha$, then $f^{(i)}|V = \underline{X}(\tau_K^{p(i)})f_0|V$ and the homotopy restricted to $V\times I$ is the composite $V\times I \xrightarrow{p_1} V \xrightarrow{f^{(i)}} \underline{X}_{C^{p(i)}K}$.

The map $f^{(n-1)}$ is the desired map, except that $f^{(n-1)}(W^{(n-1)}) = x'_\beta$ may not be the preferred basepoint. A simple homotopy along a path from x'_β to x_β corrects this, however. The details are left to the reader.

Proof of 6.3: This lemma follows from the facts that the obvious map $f: |\Delta(Z)| \longrightarrow Z$ from the geometric realization of the singular complex of Z is a weak homotopy equivalence and that $|\Delta(Z)|$ is a CW complex and from standard facts about the homology of CW complexes.

7. The relative Hurewicz theorem.

Let $\mathcal{B}$ be a category with endomorphism, $(\underline{X},\underline{Y})$ be a fragmented pair over $\mathcal{B}$, $\mathcal{B}G_1(\underline{Y})$ be the fundamental groupoid of $\underline{Y}$, and $\rho: \mathcal{B}G_1(\underline{Y}) \longrightarrow \mathcal{B}$ be the forgetful functor. The Hurewicz homomorphism

$$\varphi_{(y,K)}: \pi_n(X_K,Y_K,y) \longrightarrow H_n(X_K,Y_K)$$

define a natural transformation $\varphi: \pi_n^F(\underline{X},\underline{Y}) \longrightarrow \rho^* H_n^F(\underline{X},\underline{Y})$ if $n\ge 3$ and a natural transformation $\varphi: \pi_2^F(\underline{X},\underline{Y}) \longrightarrow \rho^* U_* H_2^F(\underline{X},\underline{Y})$, where $U_*: \mathcal{A}b^{\mathcal{B}} \longrightarrow \mathcal{G}p^{\mathcal{B}}$ is induced by composition on the left with the forgetful functor $U: \mathcal{A}b \longrightarrow \mathcal{G}p$ (cf. I.5.1). By adjointness, φ induces a natural transformation $\psi: \rho_! \pi_n^F(\underline{X},\underline{Y}) \longrightarrow H_n^F(\underline{X},\underline{Y})$ if $n\ge 3$ and a natural transformation $\psi: \rho_! \pi_2^F(\underline{X},\underline{Y}) \longrightarrow U_! H_2^F(\underline{X},\underline{Y})$. If $n\ge 3$, we call ψ the *Hurewicz transformation*. If $n=2$, the transformation

$\psi^{ab}: (\rho_! \pi_2^F(\underline{X},\underline{Y}))^{ab} \longrightarrow (U_! H_2^F(\underline{X},\underline{Y}))^{ab} = H_2^F(\underline{X},\underline{Y})$ is called the *Hurewicz transformation* where $(\ )^{ab}$ is induced by the abelianization functor. It is the object of this section to prove the following theorem:

Theorem 7.1 *(Relative Hurewicz Theorem)* *Suppose that the inclusion* $i: \underline{Y} \longrightarrow \underline{X}$ *induces an isomorphism* $i_*: \pi_0^F(\underline{Y}) \longrightarrow i^! \pi_0^F(\underline{X})$ *and that for some* $n \geq 2$, $\pi_i^F(\underline{X},\underline{Y}) = 0$ *for* $0 \leq i \leq n-1$.

(i) *If* $n=2$, *then* $\psi^{ab}: (\rho_! \pi_2^F(\underline{X},\underline{Y}))^{ab} \longrightarrow H_2^F(\underline{X},\underline{Y})$ *is an isomorphism.*

(ii) *If* $n \geq 3$, *then* $\psi: \rho_! \pi_n^F(\underline{X},\underline{Y}) \longrightarrow H_n^F(\underline{X},\underline{Y})$ *is an isomorphism.*

In the following proof we assume $n \geq 3$. The changes needed when $n=2$ are left to the reader.

For any pair (V,W,v_0) with basepoint $v_0 \in W$, we set

$$\pi_n'(V,W,v_0) = \pi_n(V,W,v_0)/\{\xi - \alpha\xi \mid \xi \in \pi_n(V,W,v_0),\ \alpha \in \pi_1(W,v_0)\}.$$

The following lemma is a corollary of I.5.6.

Lemma 7.2 *For any* $K \in |\mathcal{B}|$, $\rho_! \pi_n^F(\underline{X},\underline{Y})(K) = \Sigma \pi_n'(X_K, Y_K, y_\gamma)$ *where the summation runs over the set of path components* $\{Y_\gamma\}$ *of* Y *and* $y_\gamma \in Y_\gamma$.

We now construct a commutative diagram that forms the basis of the proof of 7.1. Let (V,W) be any pair of spaces and $n \geq 1$. Then $\Delta^{(n-1)}(V,W)$ will denote the subcomplex of the singular chain complex $\Delta(V)$ generated by singular simplices $\sigma: (\Delta^q, (\Delta^q)^{n-1}) \longrightarrow (V,W)$. Similarly $H_p^{(n-1)}(V,W)$ will denote the p-th homology of the chain complex $\Delta^{(n-1)}(V,W)/\Delta(W)$. Then $H_p^{(n-1)}: \mathcal{PT}op \longrightarrow \mathcal{A}b$ is a functor and there is a natural transformation $\eta: H_p^{(n-1)} \longrightarrow H_p$ induced by the chain inclusion $\Delta^{(n-1)}(V,W)/\Delta(W) \longrightarrow \Delta(V)/\Delta(W)$. It follows from I.5.1 and I.3.2 that there is an induced functor $H_p^{(n-1)}: \mathcal{FPT}op/\mathcal{B} \longrightarrow \mathcal{A}b^{\mathcal{B}}(\Sigma^{-1})$ (which we will denote simply by $H_p^{(n-1)}$) and an induced natural transformation $\eta_!: H_p^{(n-1)} \longrightarrow H_p^F$.

We observe now that for all $(y,K) \in |\mathcal{B}G_1(Y)|$, there is a commutative diagram of homomorphisms

$$\begin{array}{ccc} & \nearrow^{\varphi'_{(Y,K)}} & H_n^{(n-1)}(X_K,Y_K) \\ \pi_n(X_K,Y_K,Y) & & \downarrow \eta(X_K,Y_K) \\ & \searrow_{\varphi_{(Y,K)}} & H_n(X_K,Y_K) \end{array}$$

which yields the commutative diagram in $\mathcal{Ab}^{\mathcal{B}}(\Sigma^{-1})$

$$\begin{array}{ccc} & \nearrow^{\psi'} & H_n^{(n-1)}(\underline{X},\underline{Y}) \\ \rho_!\pi_n^F(\underline{X},\underline{Y}) & & \downarrow \eta(\underline{X},\underline{Y}) \\ & \searrow_{\psi} & H_n^F(\underline{X},\underline{Y}) \end{array}$$

Theorem 7.1 follows immediately from chasing this diagram using the following propositions:

Proposition 7.3 *Let* $(\underline{X},\underline{Y}) \in |\mathcal{FPT}op/\mathcal{B}|$ *and let* $n \geq 3$. *Then* $\psi': \rho_!\pi_n^F(\underline{X},\underline{Y}) \longrightarrow H_n^{(n-1)}(\underline{X},\underline{Y})$ *is an isomorphism in* $\mathcal{Ab}^{\mathcal{B}}(\Sigma^{-1})$.

Proposition 7.4 *Let* $(\underline{X},\underline{Y}) \in |\mathcal{FPT}op/\mathcal{B}|$ *and suppose that* $i_*: \pi_0^F(\underline{Y}) \longrightarrow i^!\pi_0^F(\underline{X})$ *is an isomorphism. Let* $n \geq 1$. *If* $\pi_i^F(\underline{X},\underline{Y}) = 0$ *for* $0 \leq i \leq n-1$, *then for all* $p \geq 0$, $\eta_{(X,Y)}: H_p^{(n-1)}(\underline{X},\underline{Y}) \longrightarrow H_p^F(\underline{X},\underline{Y})$ *is an isomorphism in* $\mathcal{Ab}^{\mathcal{B}}(\Sigma^{-1})$.

Proof of 7.3: Since $n \geq 2$,

$$H_n^{(n-1)}(\underline{X},\underline{Y})(K) = H_n^{(n-1)}(X_K,Y_K) = \Sigma\, H_n^{(n-1)}(X_K,(Y_K)_\gamma),$$

where the direct sum runs over the set of path components $\{(Y_K)_\gamma\}$ of Y_K. Furthermore, ψ'_K induces homomorphisms

$$\psi'_{K,\gamma}: \pi'_n(X_K,(Y_K)_\gamma,x_\gamma) \longrightarrow H_n^{(n-1)}(X_K,(Y_K)_\gamma)$$

such that the following diagram commutes

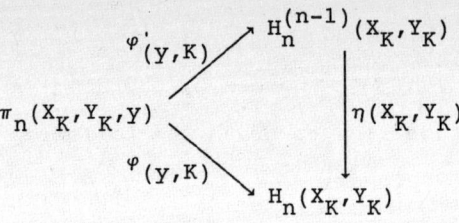

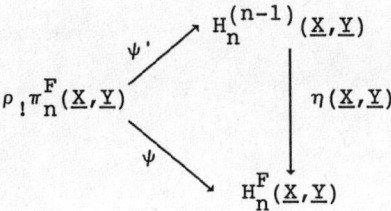

$$\begin{array}{ccc} \rho_! \pi_n^F(\underline{X},\underline{Y})(K) & \xrightarrow{\psi'_K} & H_n^{(n-1)}(\underline{X},\underline{Y})(K) \\ \| & & \| \\ \Sigma\, \pi'_n(X_K,(Y_K)_\gamma,x_\gamma) & \xrightarrow[\Sigma\psi'_{K,\gamma}]{} & \Sigma\, H_n^{(n-1)}(X_K,(Y_K)_\gamma) \end{array}$$

It suffices therefore to show that $\psi'_{K,\gamma}$ is an isomorphism for all K,γ.

Now for any pair (V,W,v_0) with basepoint $v_0 \in W$, let $\Delta_n^{(n-1)}(V,W,v_0)$ be the subcomplex of $\Delta(V)$ generated by singular simplices σ with $\sigma\colon (\Delta_n,(\Delta_n)^{(n-1)},(\Delta_n)^{(0)}) \longrightarrow (V,W,v_0)$. We define $H_p^{(n-1)}(V,W,v_0)$ to be the p-th homology of the chain complex

$$\Delta^{(n-1)}(V,W,v_0)/\Delta(W) \cap \Delta^{(n-1)}(V,W,v_0).$$

The inclusion $i\colon \Delta^{(n-1)}(V,W,v_0) \longrightarrow \Delta^{(n-1)}(V,W)$ induces a homomorphism $i_*\colon H_p^{(n-1)}(V,W,v_0) \longrightarrow H_p^{(n-1)}(V,W)$.

It is now easy to see that for any K,γ, the homomorphism $\psi'_{K,\gamma}$ can be factored as a composite

$$\pi'_n(X_K,(Y_K)_\gamma,x_\gamma) \xrightarrow{\psi''_{K,\gamma}} H_n^{(n-1)}(X_K,(Y_K)_\gamma,x_\gamma) \xrightarrow{i_*} H_n^{(1-n)}(X_K,(Y_K)_\gamma) .$$

In [Sp, p.397] it is shown that $\psi''_{K,\gamma}$ is an isomorphism. Thus the proof of 7.3 is completed by proving the following lemma:

Lemma 7.5 *Let $n\geq 2$. Let (V,W,v_0) be a pair of spaces with W path connected and basepoint $v_0 \in W$. Then for all $p\geq 0$, $i_*\colon H_p^{(n-1)}(V,W,v_0) \longrightarrow H_p^{(n-1)}(V,W)$ is an isomorphism.*

Proof: The homomorphism i_* fits into an exact sequence

$$\cdots \longrightarrow H_{p+1}(C_*) \longrightarrow H_p^{(n-1)}(V,W,v_0) \xrightarrow{i_*} H_p^{(n-1)}(V,W) \longrightarrow H_p(C_*) \longrightarrow \cdots$$

Where C_* is the chain complex $\Delta^{(n-1)}(V,W)/\Delta^{(n-1)}(V,W,v_0)+\Delta(W)$. Thus, it suffices to show that the inclusion $\Delta^{(n-1)}(V,W,v_0)+\Delta(W) \longrightarrow \Delta^{(n-1)}(V,W)$ is a chain equivalence. However, this follows easily by the techniques given in [Sp, pp. 392-3].

Proof of 7.4: We note first that for any pair (V,W) there is an exact sequence of chain complexes

$$0 \longrightarrow \Delta^{(n-1)}(V,W)/\Delta(W) \longrightarrow \Delta(V)/\Delta(W) \longrightarrow \Delta(V)/\Delta^{(n-1)}(V,W) \longrightarrow 0$$

and, hence, an exact homology sequence. Thus by I.4.2 and I.2.4, there is an exact sequence in $\mathcal{A}b^{\mathfrak{B}}(\Sigma^{-1})$

$$\cdots \longrightarrow \mathcal{H}_{p+1}(\underline{X},\underline{Y}) \longrightarrow H_p^{(n-1)}(\underline{X},\underline{Y}) \longrightarrow H_p^F(\underline{X},\underline{Y}) \longrightarrow \mathcal{H}_p(\underline{X},\underline{Y}) \longrightarrow \cdots$$

where for any $K \in |\mathfrak{B}|$, $\mathcal{H}_p(\underline{X},\underline{Y})(K)$ is the p-th homology of $\Delta(X_K)/\Delta^{(n-1)}(X_K,Y_K)$. To prove 7.4, it now suffices to prove $\mathcal{H}_p(\underline{X},\underline{Y}) = 0$ for all p. By I.4.5 this is an immediate consequence of the following lemma:

Lemma 7.6 *Under the hypothesis of* 7.4 *there exists an integer* $m \geq 0$ *such that for all* $K \in |\mathfrak{B}|$, $\underline{X}(\tau_K^m)_*\colon \mathcal{H}_p(\underline{X},\underline{Y})(K) \longrightarrow \mathcal{H}_p(\underline{X},\underline{Y})(C^mK)$ *is the zero map.*

Proof: To simplify notation in this proof, we shall write τ_K^m for $\underline{X}(\tau_K^m)\colon (X_K,Y_K) \longrightarrow (X_{C^mK},Y_{C^mK})$ and if $y \in Y_K$ we shall denote its image under τ_K^m again by y rather than by $\underline{X}(\tau_K^m)(y)$. Since $i_*\colon \pi_0^F(Y) \longrightarrow i^!\pi_0^F(X)$ is an isomorphism, it is epic. Hence, by an obvious modification of I.4.6, there exists an $r \geq 0$ such that for all $(y,K) \in |\mathfrak{B}G_1(\underline{Y})|$, $\mathrm{Im}(\tau_K^r)_* \subseteq \mathrm{Im}(i_{C^rK})_*$ where $\tau_{K*}^r\colon \pi_0(X_K,y) \longrightarrow \pi_0(X_{C^rK},y)$ and $(i_{C^rK})_*\colon \pi_0(Y_{C^rK},y) \longrightarrow \pi_0(X_{C^rK},y)$. Similarly, since $\pi_i^F(\underline{X},\underline{Y}) = 0$ for $0 \leq i \leq n-1$, there exists an $r_i \geq 0$ such that for all $(y,K) \in |\mathfrak{B}G_1(Y)|$, $(\tau_K^{r_i})_*\colon \pi_i(X_K,Y_K,y) \longrightarrow \pi_i(X_{C^{r_i}K},Y_{C^{r_i}K},y)$ is the zero map.

Let $m = r+r_0+\cdots+r_{n-1}$. By using the techniques of [Sp, pp. 392-393], it is easy to construct a chain map $f_K\colon \Delta(X_K) \longrightarrow \Delta^{(n-1)}(X_K,Y_K)$ such that the top triangle in the diagram

$$\begin{array}{ccc}
\Delta^{(n-1)}(X_K,Y_K) & \longrightarrow & \Delta^{(n-1)}(X_{C^m K},Y_{C^m K}) \\
\downarrow & \nearrow f_K & \downarrow \\
\Delta(X_K) & \xrightarrow[\tau_K^m]{} & \Delta(X_{C^m K})
\end{array}$$

commutes and a chain homotopy $H\colon \tau_K^m \simeq if_K$ such that $H(\Delta^{(n-1)}(X_K,Y_K)) \subseteq \Delta^{(n-1)}(X_{C^m K},Y_{C^m K})$. From this it follows easily that

$$\tau_K^m\colon \Delta(X_K)/\Delta^{(n-1)}(X_K,Y_K) \longrightarrow \Delta(X_{C^m K})/\Delta^{(n-1)}(X_{C^m K},Y_{C^m K})$$

is chain homotopic to the zero chain map; and, hence, that $(\tau_K^m)_* = 0$. The details are left to the reader.

8. Variations on the Hurewicz theorem.

In this section we establish some variations on the Hurewicz theorems that are useful in applications.

Let $\underline{X}$ be a fragmented space over the category with endomorphism $\mathcal{B}$. Let $\rho\colon \mathcal{B}G_1(\underline{X}) \longrightarrow \mathcal{B}$ be the forgetful functor and $\varphi\colon \pi_n^F(\underline{X}) \longrightarrow \rho^! H_n^F(\underline{X})$ be the morphism defined by the Hurewicz homomorphisms in section 6.

Theorem 8.1 *Let $\underline{X}$ be a fragmented space over the category with endomorphism $\mathcal{B}$. Let $n \geq 2$. If $\underline{X}$ is simply connected and $H_i^F(\underline{X}) = 0$ for $i \leq n-1$, then $\pi_i^F(\underline{X}) = 0$ for $i \leq n-1$ and $\varphi\colon \pi_n^F(\underline{X}) \longrightarrow \rho^! H_n(\underline{X})$ is an isomorphism.*

Similarly if $(\underline{X},\underline{Y})$ is a pair of fragmented spaces over $\mathcal{B}$, let $\varphi\colon \pi_2^F(\underline{X},\underline{Y}) \longrightarrow \rho^! U_! H_2^F(\underline{X},\underline{Y})$ and $\varphi\colon \pi_n^F(\underline{X},\underline{Y}) \longrightarrow \rho^! H_n^F(\underline{X},\underline{Y})$ $(n \geq 3)$ be the morphisms defined in section 7.

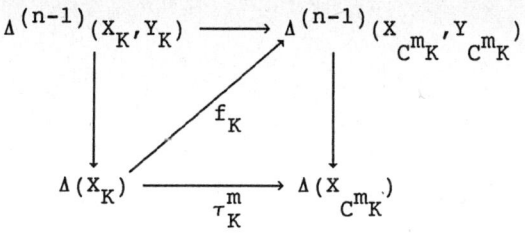

Theorem 8.2 *Let* $(\underline{X},\underline{Y})$ *be a pair of fragmented spaces over* $\mathfrak{B}$ *such that* $\underline{Y}$ *is coextensive with* $\underline{X}$ *and suppose both* $\underline{X}$ *and* $\underline{Y}$ *are simply connected. Then*

(i) $\varphi\colon \pi_2^F(\underline{X},\underline{Y}) \longrightarrow \rho^! U_! H_2^F(\underline{X},\underline{Y})$ *is an isomorphism, and*

(ii) *If for some* $n\geq 3$, $H_i^F(\underline{X},\underline{Y}) = 0$ *for* $i\leq n-1$, *then* $\pi_i^F(\underline{X},\underline{Y}) = 0$ *for* $i\leq n-1$ *and* $\varphi\colon \pi_n^F(\underline{X},\underline{Y}) \longrightarrow \rho^! H_n^F(\underline{X},\underline{Y})$ *is an isomorphism.*

Proof of 8.1: **Case 1.** We assume that X is coextensive with $\mathfrak{B}$. In this case the proof is by induction on k ($1\leq k\leq n$) with the inductive hypothesis being that $\pi_i^F(\underline{X}) = 0$ for $i\leq k-1$ and that $\varphi\colon \pi_k^F(\underline{X}) \longrightarrow \rho^! H_k^F(\underline{X})$ is an isomorphism. By definition (cf. section 5), since $\underline{X}$ is simply connected, $\pi_i^F(\underline{X}) = 0$ for $i=0,1$. By the Hurewicz theorem 6.1, then $H_1^F(\underline{X}) = 0$ and certainly $\varphi\colon \pi_1^F(\underline{X}) \longrightarrow \rho^! H_1^F(\underline{X})$ is an isomorphism. Thus the inductive hypothesis holds for k=1.

Assume the inductive hypothesis holds for $k\leq n-1$. Then $\pi_i^F(\underline{X}) = 0$ for $i\leq k-1$ and $\varphi\colon \pi_k^F(\underline{X}) \longrightarrow \rho^! H_k^F(\underline{X})$ is an isomorphism. By hypothesis $H_j^F(\underline{X}) = 0$ for $j\leq n-1$. In particular, $H_k^F(\underline{X}) = 0$. Thus $\pi_k^F(\underline{X}) = 0$ and by the Hurewicz theorem 6.1, $\psi\colon \rho_! \pi_{k+1}^F(\underline{X}) \longrightarrow H_{k+1}^F(\underline{X})$ is an isomorphism.

Since $\rho_!\colon \mathfrak{D} \longrightarrow \mathcal{E}$ is a left adjoint for $\rho^!$ by I.5.4, where $\mathfrak{D} = \mathcal{A}b^{\mathfrak{B}G_1(\underline{X})}(\Sigma^{-1})$ and $\mathcal{E} = \mathcal{A}b^{\mathfrak{B}}(\Sigma^{-1})$, for every $R \in |\mathfrak{D}|$ and $S \in |\mathcal{E}|$ there is a commutative diagram

$$\begin{array}{ccc}
\mathcal{E}(\rho_! R, S) & \xrightarrow{\quad\sigma\quad} & \mathfrak{D}(R, \rho^! S) \\
\Big\downarrow{\scriptstyle \rho^!} & & \Big\| \\
\mathcal{E}(\rho^!\rho_! R, \rho^! S) & \xrightarrow{\quad\eta^*\quad} & \mathcal{E}(R, \rho^! S)
\end{array}$$

where σ (respectively $\eta\colon I \longrightarrow \rho^!\rho_!$) is the isomorphism (respectively, the unit) of the adjunction between $\rho^!$ and $\rho_!$ and η^* is composition on the right with η (cf. [Ma, Theorem 1(i), p.80]). We note, in particular, that since $\psi = \sigma^{-1}(\varphi)$, the diagram

$$\begin{array}{ccc} \pi^F_{k+1}(\underline{X}) & \xrightarrow{\ \varphi\ } & \rho^! H^F_{k+1}(\underline{X}) \\ \eta' \big\downarrow & \nearrow{\scriptstyle \rho^!(\psi)} & \\ \rho^!\rho_!\pi^F_{k+1}(\underline{X}) & & \end{array}$$

commutes where η' is η evaluated on $\pi^F_{k+1}(\underline{X})$. Since ψ is an isomorphism, so is $\rho^!(\psi)$ and it now suffices to prove that η' is an isomorphism.

Since $\underline{X}$ is coextensive with $\mathcal{B}$ and simply connected and $\mathcal{A}b$ is cocomplete, $\rho_!$ is an equivalence of categories by 5.2 and I.6.1. It now follows immediately from [Ma, pp. 90-93] that $\eta\colon I \longrightarrow \rho^!\rho_!$ is a natural equivalence. In particular, η' is an isomorphism and the proof of case 1 of 8.1 is complete.

Case 2. Suppose $\underline{X}$ is not necessarily coextensive with $\mathcal{B}$. We let $\mathcal{B}^0$ be the full subcategory whose objects are those $K \in |\mathcal{B}|$ with $X_K \neq \emptyset$ and let $\underline{X}^0$ be the restriction of $\underline{X}$ to $\mathcal{B}^0$. Clearly $\underline{X}^0$ is coextensive with $\mathcal{B}^0$. Since the inclusion functor $j\colon \mathcal{B}^0 G_1(\underline{X}^0) \longrightarrow \mathcal{B}G_1(\underline{X})$ is an isomorphism of categories and $\underline{X}$ is simply connected, $\underline{X}^0$ is simply connected. Finally, since $H^F_i(\underline{X}^0) = k^! H^F_i(\underline{X})$ where $k\colon \mathcal{B}^0 \longrightarrow \mathcal{B}$ is the inclusion functor, and $H^F_i(\underline{X}) = 0$ for $i \leq n-1$, $H^F_i(\underline{X}^0) = 0$ for $i \leq n-1$. Hence $\varphi\colon \pi^F_n(\underline{X}^0) \longrightarrow \rho^{0!} H^F_n(\underline{X}^0)$ is an isomorphism by case 1. It is now easy to see that there are isomorphisms making the following diagram commute

$$\begin{array}{ccc} j_!\pi^F_n(\underline{X}^0) & \xrightarrow{\ j_!(\varphi)\ } & j_!\rho^{0!}H^F_n(\underline{X}^0) \\ \simeq \big\downarrow & & \big\downarrow \simeq \\ \pi^F_n(\underline{X}) & \xrightarrow{\ \varphi\ } & \rho^! H^F_n(\underline{X}) \end{array}$$

This completes the proof of case 2 and of 8.1.

Proof of 8.2: To prove (i), we note that since $\underline{X}$ and $\underline{Y}$ are simply connected, $\pi_1^F(\underline{X},\underline{Y}) = 0$. Hence $\psi^{ab}: (\rho_!\pi_2^F(\underline{X},\underline{Y}))^{ab} \longrightarrow H_2^F(\underline{X},\underline{Y})$ is an isomorphism by 7.1 where $\psi: \rho_!\pi_2^F(X,Y) \longrightarrow U_!H_2^F(\underline{X},\underline{Y})$ is the adjoint of φ. Since the diagram

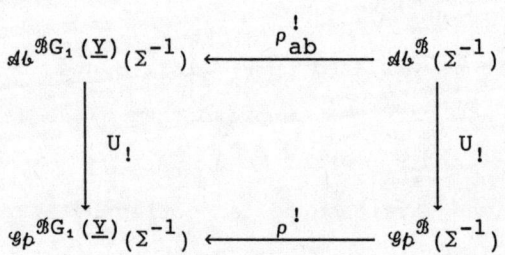

$$\begin{array}{ccc}
\mathcal{A}b^{\mathcal{B}G_1(\underline{Y})}(\Sigma^{-1}) & \xleftarrow{\rho^!_{ab}} & \mathcal{A}b^{\mathcal{B}}(\Sigma^{-1}) \\
\Big\downarrow U_! & & \Big\downarrow U_! \\
\mathcal{G}p^{\mathcal{B}G_1(\underline{Y})}(\Sigma^{-1}) & \xleftarrow{\rho^!} & \mathcal{G}p^{\mathcal{B}}(\Sigma^{-1})
\end{array}$$

commutes, we can regard φ as a morphism $\varphi: \pi_2^F(\underline{X},\underline{Y}) \longrightarrow U_!\rho^!_{ab}H_2^F(\underline{X},\underline{Y})$. Since $(\ \ )^{ab}: \mathcal{G}p \longrightarrow \mathcal{A}b$ is a left adjoint for U, it preserves colimits (cf. the discussion in I.5 preceeding I.5.7). It now follows from I.5.9 that ψ^{ab} is the adjoint of $\varphi^{ab}: (\pi_2^F(\underline{X},\underline{Y})) \xrightarrow{ab} \rho^!_{ab}H_2(\underline{X},\underline{Y})$. Since $\underline{Y}$ is coextensive with $\underline{X}$, an argument similar to one in the proof of 8.1 shows that φ^{ab} is an isomorphism. We now note that there is a commutative diagram

$$\begin{array}{ccc}
\pi_2^F(\underline{X},\underline{Y}) & \xrightarrow{\varphi} & U_!\rho^!_{ab}H_2^F(\underline{X},\underline{Y}) \\
\Big\downarrow \eta & \nearrow U_!(\varphi^{ab}) & \\
U_!(\pi_2^F(\underline{X},\underline{Y}))^{ab} & &
\end{array}$$

where $\eta: I \longrightarrow U_!(\ \ )^{ab}$ is the unit of adjunction between $U_!$ and $(\ \ )^{ab}$ (cf. [Ma, Theorem 1, p. 90]). Part (i) of 8.2 is now a consequence of the following lemma:

Lemma 8.3 *If* $\partial_*: \pi_2^F(\underline{X},\underline{Y}) \longrightarrow \pi_1^F(\underline{Y})$ *is the trivial morphism, then* $\eta: \pi_2^F(\underline{X},\underline{Y}) \longrightarrow U_!(\pi_2^F(\underline{X},\underline{Y}))^{ab}$ *is an isomorphism.*

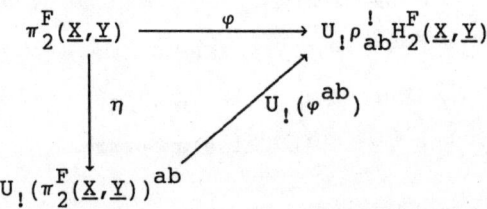

Proof: Let $(y,K) \in |\mathcal{B}G_1(\underline{Y})|$ and

$$\eta_{(y,K)}: \pi_2(X_K,Y_K,y) \longrightarrow U(\pi_2(X_K,Y_K,y)^{ab})$$

be the obvious quotient map with kernel the commutator subgroup of $\pi_2(X_K,Y_K,y)$. Then η is represented by $\{\eta_{(y,K)}\}$ and since every $\eta_{(y,K)}$ is onto, η is eventually epimorphic. On the other hand, let $x = [\alpha,\beta] \in \ker \eta_{(y,K)}$. Then $\alpha\beta\alpha^{-1} = \partial_{(y,K)}\alpha \cdot \beta$ where

$$\partial_{(y,K)}: \pi_2(X_K,Y_K,y) \longrightarrow \pi_1(Y_K,y)$$

and $\cdot$ denotes the usual action of π_1 on π_2. Since ∂ is trivial, there exists an integer $r \geq 0$ such that for all (y,K), $\tau_*^r\partial_{(y,K)} = 1$. Since the action of π_1 on π_2 is natural under maps of spaces, it is easy to see that $\tau_*^r[\alpha,\beta] = 1$. It follows that $\tau_*^r(\ker \eta_{(y,K)}) = 1$ and that η is eventually monomorphic. Lemma 8.3 follows from I.4.7.

The proof of (ii) of 8.2 is similar to the proof of 8.1 and is left to the reader.

9. Algebraic invariants for boundedly controlled spaces.

In this section we define the homology and homotopy of a bc space (or pair) and record their main properties.

Let (Z,P,C) be a boundedness control space. We define $H_n^c: \mathcal{T}op^c/Z \longrightarrow \mathcal{A}b^{\mathcal{F}}(\Sigma^{-1})$ to be the composite

$$\mathcal{T}op^c/Z \xrightarrow{Fr_1} \mathcal{F}\mathcal{T}op/\mathcal{F} \xrightarrow{H_n^F} \mathcal{A}b^{\mathcal{F}}(\Sigma^{-1})$$

of the functors Fr_1 of section 2 and H_n^F of section 3. If $(X,p) \in |\mathcal{T}op^c/Z|$, we call $H_n^c(X,p)$ the *boundedly controlled* (or simply *bc*) *homology* of (X,p). $H_n^c(X,p)$ may be described more concretely by remembering that for any $K \in |\mathcal{F}|$, $Fr_1(X,p)(K) = p^{-1}(K)$ (cf. 2.1). Thus for any $K \in |\mathcal{F}|$, $H_n^c(X,p)(K) = H_n(p^{-1}K)$. The bc homology of a bc pair or triple is defined similarly.

With these definitions all the results of section 3 carry over to bc homology without change. Thus, there are exact sequences for the bc homology of pairs and triples, bc homotopic maps induce the same morphisms on bc homology, and the bc homology of a space is the homology of its bc singular chains $\Delta_*^C = \{\Delta_n^C, \partial_n^C\}$ where Δ_n^C: $\mathcal{T}op^C/Z \to \mathcal{A}b^{\mathcal{P}}(\Sigma^{-1})$ is the composite

$$\mathcal{T}op^C/Z \xrightarrow{Fr_1} \mathcal{F}\mathcal{T}op/\mathcal{P} \xrightarrow{\Delta_n} \mathcal{A}b^{\mathcal{P}}(\Sigma^{-1})$$

and Δ_n is the functor of 3.6.

If $(X,p) \in |\mathcal{T}op^C/Z|$, we define the n-th *boundedly controlled* (or simply *bc*) *homotopy* of (X,p) by setting $\pi_n^C(X,p) = \pi_n^F(Fr_1(X,p))$. In this case, we simplify notation and let $\mathcal{P}G_1(X,p)$ denote $\mathcal{P}(G_1 Fr_1(X,p))$. Then for any $(x,K) \in |\mathcal{P}G_1(X,p)|$, $\pi_n^C(X,p)(x,K) = \pi_n(p^{-1}(K),x)$. The bc homotopy of a bc pair or triple is defined similarly. As in the case of homology, all the results about homotopy of fragmented spaces in section 4 carry over to bc spaces essentially without change. The details are left to the reader.

Let (X,p) be a bc space over Z. It follows from the discussion above and in section 6 that there are morphisms $\varphi: \pi_1^C(X,p) \to \rho^! U_! H_1^C(X,p)$ and $\varphi: \pi_n^C(X,p) \to \rho^! H_n^C(X,p)$ for $n \geq 2$ induced by the Hurewicz homomorphisms. Let $\psi: \rho_! \pi_1^C(X,p) \to U_! H_1^C(X,p)$ and $\psi: \rho_! \pi_n^C(X,p) \to H_n^C(X,p)$ for $n \geq 2$ be the adjoints of the morphisms φ. (Here $\rho: \mathcal{P}G_1(X,p) \to \mathcal{P}$ is the forgetful functor and for any category with endomorphism $\mathcal{B}$, $U_!: \mathcal{A}b^{\mathcal{B}}(\Sigma^{-1}) \to \mathcal{G}p^{\mathcal{B}}(\Sigma^{-1})$ is induced by the forgetful functor $U: \mathcal{A}b \to \mathcal{G}p$.)

Theorem 9.1 *Let (X,p) be a bc space over the boundedness control space (Z,P,C). Then $\psi^{ab}: (\rho_! \pi_1^C(X,p))^{ab} \to H_1^C(X,p)$ is an isomorphism. In addition, if for some $n \geq 2$, $\pi_i^C(X,p) = 0$ for $1 \leq i \leq n-1$, then $\psi: \rho_! \pi_n^C(X,p) \to H_n^C(X,p)$ is an isomorphism.*

If (X,p) *is simply connected and, for some* $n \geq 2$, $H_i^c(X,p) = 0$ *for* $1 \leq i \leq n-1$, *then* $\pi_i^c(X,p) = 0$ *for* $1 \leq i \leq n-1$ *and* $\varphi\colon \pi_n^c(X,p) \longrightarrow \rho^! H_n^c(X,p)$ *is an isomorphism.*

In this theorem ψ^{ab} is the image of ψ under the abelianization functor $\mathcal{G}p^{\mathcal{P}}(\Sigma^{-1}) \longrightarrow \mathcal{A}b^{\mathcal{P}}(\Sigma^{-1})$. Furthermore, to say (X,p) is *simply connected* means that $\pi_i^c(X,p) = 0$ for $i=0,1$.

Let (X,p) and (Y,q) be bc spaces over Z. We say (Y,q) *is coextensive with* (X,p) if there exists an integer $m \geq 0$ such that for every $K \in |\mathcal{P}|$ if $p^{-1}(K) \neq \emptyset$, then $q^{-1}(C^m K) \neq 0$ and if $q^{-1}(K) \neq \emptyset$, then $p^{-1}(C^m K)$.

Let (X,Y,p) be a bc pair over Z. Then there are morphisms $\varphi\colon \pi_2^c(X,Y,p) \longrightarrow \rho^! U_! H_2^c(X,Y)$ and $\varphi\colon \pi_n^c(X,Y,p) \longrightarrow \rho^! H_n^c(X,Y)$ for $n \geq 3$. Let $\psi\colon \rho_! \pi_2^c(X,Y) \longrightarrow U_! H_2^c(X,Y)$ and $\psi\colon \rho_! \pi_n^c(X,Y) \longrightarrow H_n^c(X,Y)$ be the adjoints of the morphisms φ.

Theorem 9.2 *Let* (X,Y,p) *be a bc pair over the boundedness control space* (Z,P,C) *and suppose that* $\pi_i^c(X,Y,p) = 0$ *for* $i=0,1$. *Then* $\psi^{ab}\colon (\rho_! \pi_2^c(X,Y,p))^{ab} \longrightarrow H_2^c(X,Y,p)$ *is an isomorphism. In addition if, for some* $n \geq 3$, $\pi_i^c(X,Y,p) = 0$ *for* $i \leq n-1$, *then* $\psi\colon \rho_! \pi_n^c(X,Y,p) \longrightarrow H_n^c(X,Y,p)$ *is an isomorphism.*

If $(Y,p|Y)$ *is coextensive with* (X,p) *and both bc spaces are simply connected, then* $\varphi\colon \pi_2^c(X,Y,p) \longrightarrow \rho^! U_! H_2^c(X,Y,p)$ *is an isomorphism. Furthermore, if for some* $n \geq 3$, $H_i^c(X,Y,p) = 0$ *for* $i \leq n-1$, *then* $\pi_i^c(X,Y,p) = 0$ *for* $i \leq n-1$ *and* $\varphi\colon \pi_n^c(X,Y,p) \longrightarrow \rho^! H_n^c(X,Y,p)$ *is an isomorphism.*

Theorems 9.1 and 9.2 are immediate corollaries of Theorems 6.1 and 8.1 and 7.1 and 8.2, respectively.

The reader will note that we have consistently used the inverse image fragmentation in defining algebraic invariants for bc spaces, pairs, and triples. We have done so in order to establish easily the usual properties of these invariants. In practice, to calculate these invariants we often replace the inverse image fragmentation by some other equivalent fragmentation. For example, in dealing with a bc CW complex (X,p) it is usually more convenient to use the smallest subcomplex fragmentation $Fr_2(X,p)$ (or some variation on it) than the inverse image fragmentation.

Remark 9.3 If $Fr_2(X,p)(K) = X_K$ is the smallest subcomplex of X containing $p^{-1}(K)$, then the inclusions $p^{-1}(K) \subset X_K$ define a map of fragmented spaces $i\colon Fr_1(X,p) \longrightarrow Fr_2(X,p)$ and induce a strong eventual equivalence of categories

$$i_*\colon \mathcal{F}G_1(X,p) = \mathcal{F}G_1(Fr_1(X,p)) \longrightarrow \mathcal{F}G_1(Fr_2(X,p)).$$

Hence we may regard $\pi_n^F(X,p)$ as being represented by the functor on $\mathcal{F}G_1(X,p)$ whose value on (x,K) is $\pi_n(X_K,x)$. Similarly, we may regard the universal cover $\tilde{X}$ of X (cf. 2.3) as being represented by the functor on $\mathcal{F}G_1(X,p)$ whose value on (x,K) is $P(X_K,x)/\sim$. In particular $\tilde{X}(x,K)$ is the universal cover of the component of X_K containing x as was observed in 2.3.

10. The Whitehead Theorem.

It is the object of this section to prove the following results:

Theorem 10.1 *Let* (Z,P,C) *be a boundedness control space. Let* $f\colon (X,A,p) \longrightarrow (Y,B,q)$ *be a bc map in* $\mathcal{CW}^c/Z$. *If* $(B,q|B)$ *is coextensive with* (Y,q) *and for some integer* $r \geq 0$, $\pi_n^c(Y,B,q) = 0$ *for* $n \leq r$, *then there exists a map* $g\colon (X, A \cup X^{(r)}, p) \longrightarrow (Y,B,q)$ *bc homotopic to* f *relative to* A *such that* $g|A = f|A$.

In this theorem, $X^{(r)}$ denotes the r-skeleton of X.

Corollary 10.2 *Let* (Y,B,q) *be a pair in* $\mathcal{CW}^c/Z$ *such that* $(B,q|B)$ *is coextensive with* (Y,q) *and* $\pi_n^c(Y,B,q) = 0$ *for all* $n \leq \dim(Y-B)$. *Then there exists a strong deformation retraction of* (Y,q) *onto* $(B,q|B)$.

Corollary 10.3 *(Cellular Approximation Theorem)* *For any map* $f\colon (X,A,p) \longrightarrow (Y,B,q)$ *in* $\mathcal{CW}^c/Z$, *there exists a map* $g\colon (X,A,p) \longrightarrow (Y,B,q)$ *homotopic to* f *relative to* A *such that for every* $r \geq 0$, $g(A \cup X^{(r)}) \subseteq B \cup Y^{(r)}$.

Corollary 10.4 *(Whitehead Theorem)* *Let* $f\colon (X,p) \longrightarrow (Y,q)$ *be a map in* $\mathcal{CW}^c/Z$. *Then* f *is a homotopy equivalence if and only if* (Y,q) *is coextensive with* (X,p) *and for all* $n \geq 0$, $f_*\colon \pi_n^c(X,p) \longrightarrow f^!\pi_n^c(Y,q)$ *is an isomorphism.*

Corollary 10.5 *(Whitehead Theorem)* *Let* (X,p) *and* (Y,q) *be coextensive, simply connected bc CW-complexes over* Z *and* $f\colon (X,p) \longrightarrow (Y,q)$ *be a bc map. Then* f *is a homotopy equivalence if and only if* $f_*\colon H_n^c(X,p) \longrightarrow H_n^c(Y,q)$ *is an isomorphism for all* $n \geq 0$.

The following lemma is used in proving 10.1:

Lemma 10.6 *Let* (X,A,p) *and* (Y,B,q) *be pairs in* $\mathcal{CW}^c/Z$ *and let* $n \geq 0$ *and* $h\colon ((X^{(n)} \cup A)\times 0 \cup (X^{(n-1)} \cup A)\times I,\ (X^{(n-1)} \cup A)\times 1,\ p\pi) \longrightarrow (Y,B,q)$ *be bc where* $\pi\colon X\times I \longrightarrow X$ *is the projection. Suppose* $\pi_n^c(Y,B,q) = 0$ *and if* $n=0$ *suppose also that* $(B,q|B)$ *is coextensive with* (Y,q). *Then there exists a bc map* $H\colon ((X^{(n)} \cup A)\times I,\ (X^{(n)} \cup A)\times 1,\ p\pi) \longrightarrow (Y,B,q)$ *extending* h.

Proof: Let $\{e_\lambda \mid \lambda\in\Lambda\}$ be the set of n-cells of $X^{(n)}-A$. Since $(X,p) \in |\mathscr{CW}^c/Z|$, there exists an integer $a\geq 0$ such that for every $\lambda\in\Lambda$, there exists a minimal element $K_\lambda \in |\mathscr{F}|$ with $p(e_\lambda) \subseteq C^a K_\lambda$. Let $\varphi_\lambda\colon (D^n,S^{n-1}) \longrightarrow (X^{(n)},X^{(n-1)})$ be a characteristic map for e_λ ($\lambda\in\Lambda$) and let ψ_λ be the composite

$$(D^n\times 0 \cup S^{n-1}\times I,\ S^{n-1}\times 1) \xrightarrow{\varphi\cup(\varphi|)\times 1}$$
$$((X^{(n)} \cup A)\times 0 \cup (X^{(n-1)} \cup A)\times I,\ (X^{(n-1)} \cup A)\times 1) \xrightarrow{h} (Y,b).$$

Since h is bc, there exists an integer $b\geq 0$ such that $q(\mathrm{Im}\ \psi_\lambda) \subseteq C^{a+b}K_\lambda$. There are now two cases to consider: n=0 and $n\geq 1$.

Suppose n=0. Then $(D^n,S^{n-1}) = (\mathrm{pt.},\emptyset)$ and $\mathrm{Im}\ \psi_\lambda = \psi_\lambda(\mathrm{pt})$ is a single point $y_\lambda\in Y$. Since $(B,q|B)$ is coextensive with (Y,q) there exists an integer $m\geq 0$ such that for every $L\in|\mathscr{F}|$ if $q^{-1}(L) \neq \emptyset$, then $q^{-1}(C^m L) \cap B \neq \emptyset$. In particular, $q^{-1}(C^u K_\lambda) \cap B \neq \emptyset$ ($u = a+b+m$) and we may choose a point $b_\lambda \in q^{-1}(C^u K_\lambda) \cap B$. In this case, we regard ψ_λ as determining the element $[y_\lambda] \in \pi_0(q^{-1}(C^u K_\lambda),\ q^{-1}(C^u K_\lambda) \cap B,\ b_\lambda) = \pi_o^c(Y,B,q)(b_\lambda,C^u K_\lambda)$.

If $n\geq 1$, we let $* \in S^{n-1}$ be a basepoint and set $\psi_\lambda(*) = b_\lambda$. Since we can identify $(D^n\times 0 \cup S^{n-1}\times I,\ S^{n-1}\times 1,\ *\times 1)$ with $(D^n,\ S^{n-1},\ *)$, ψ_λ represents an element

$$\xi_\lambda \in \pi_n(q^{-1}(C^u K_\lambda),\ q^{-1}(C^u K_\lambda) \cap B,\ b_\lambda) = \pi_n^c(Y,B,q)(b_\lambda,\ C^u k_\lambda)$$

($u = a+b$ in this case).

In either case, since $\pi_n^c(Y,B,q) = 0$, by I.4.5 there exists an integer $d\geq 0$ such that for all $(b,K) \in \mathscr{F}G_1(B,q|B)$,

$$\pi_n(q^{-1}(K),q^{-1}(K) \cap B,\ b) \longrightarrow \pi_n(q^{-1}(C^d K),\ q^{-1}(C^d K) \cap B,b)$$

is the zero map. In particular, this holds for $(b,K) = (b_\lambda,C^u K_\lambda)$. If n=0, this implies that there is a map $f_\lambda\colon D^0\times I \longrightarrow q^{-1}(C^v K_\lambda)$ ($v = u+d$) with $f_\lambda(0) = y_\lambda$ and $f_\lambda(1) = b_\lambda \in B$. If $n\geq 1$, this implies there is a map $f_\lambda\colon (D^n\times I,D^n\times 1) \longrightarrow (q^{-1}(C^v K_\lambda),q^{-1}(C^v K_\lambda) \cap B)$ extending ψ_λ. Define an extension $H\colon (X^{(n)} \cup A)\times I \longrightarrow Y$ of $h\colon (X^{(n)} \cup A)\times 0 \cup (X^{(n-1)} \cup A)\times I \longrightarrow Y$ by setting $h|e_\lambda\times I = f_\lambda(\varphi_\lambda\times 1)^{-1}$. Clearly H is continuous. That H is bc follows easily from the following lemmas the first of which is obvious:

Lemma 10.7 *Let* (V,p) *and* (W,q) *be in* $\mathcal{T}op^c/Z$ *and* $f\colon V \longrightarrow W$. *Suppose* $V = V_1 \cup V_2$ *and that* $f_i = f|V_i\colon (V_i,p_i) \longrightarrow (W,q)$ *is bc, where* $p_i = p|V_i$ $(i=1,2)$. *Then* f *is bc.*

Lemma 10.8 *Let* (V,p) *and* (W,q) *be in* $\mathcal{T}op^c/Z$, $\{V_\lambda \mid \lambda\in\Lambda\}$ *be a bounded family of subspaces of* V, *and* $f_\lambda\colon V_\lambda \longrightarrow W$ $(\lambda\in\Lambda)$ *be a family of maps such that* $f_\lambda|V_\lambda\cap V_\mu = f_\mu|V_\ell\cap V_\mu$ *for all* $\lambda,\mu\in\Lambda$. *If* $\{f(V_\lambda) \mid \lambda\in\Lambda\}$ *is a bounded family of subsets of* (W,q), *then* $f = \cup\, f_\lambda\colon (\cup_\lambda V_\lambda, p|) \longrightarrow (W,q)$ *is bc.*

Proof: Since $\{V_\lambda \mid \lambda\in\Lambda\}$ and $\{f(V_\lambda) \mid \lambda\in\Lambda\}$ are bounded, there exist integers $a,b\geq 0$ such that for every $\lambda\in\Lambda$ there exists a minimal element $K_\lambda\in|\mathcal{S}|$ with $V_\lambda \subset p^{-1}(C^a V_\lambda)$ and $f(V_\lambda) \subset q^{-1}(C^{a+b}V_\lambda)$. Let $K \in |\mathcal{S}|$ and suppose $(p|)^{-1}(K) \neq \emptyset$. Then there exists a $\lambda' \in \Lambda$ and a point $v \in V_{\lambda'}$ with $p(v) \in K$. Since $p(v) \in p(V_{\lambda'}) \subseteq C^a K_{\lambda'}$, $K \cap C^a K_{\lambda'} \neq \emptyset$. Hence, by 1.1.(iv), $K_{\lambda'} \subseteq C^{\theta(a)}K$ and $C^{a+b}K_{\lambda'} \subseteq C^{a+b+\theta(a)}K$. Since $f(v) = f_\lambda(v) \in f_\lambda(V_\lambda) \subseteq q^{-1}(C^{a+b+\theta(b)}K)$, $f(p^{-1}(K)) \subseteq q^{-1}(C^r K)$ for $r = a+b+\theta(b)$ and f is bc.

Proof of 10.1: A simple argument proceeding by induction up the skeleta of $X-A$ and using 10.6 shows that there exist bc maps

$$h_n\colon (X^{(n)} \cup A)\times I,\ (X^{(n)} \cup A)\times 1, p\pi) \longrightarrow (Y,B,q)$$

$(0\leq n\leq r)$ such that $h_n|(X^{(n-1)} \cup A)\times I = h_{n-1}, h_n|A\times\{t\} = f$ $(0\leq t\leq 1)$, and $h_n|X^{(n)}\times 0 = f$. If we now replace (Y,B,q) with (Y,Y,q) in 10.6, the same lemma shows that h_r may be extended to a bc map $H\colon (X\times I, p\pi) \longrightarrow (Y,q)$ with $h|X\times 0 = f$. Then $g = h|X\times 1$ satisfies the conditions of 10.1.

Proof of 10.2: Apply 10.1 to the identity map $1\colon (Y,B,q) \longrightarrow (Y,B,q)$.

Proof of 10.3: We note first that for every $r \geq 0$, the pair $(Y, B \cup Y^{(r)}, q)$ satisfies the hypothesis of 10.1. For certainly these bc spaces are coextensive and consider the fragmented spaces $(\underline{Y}, \underline{B} \cup \underline{Y}^{(r)})$ and $(\underline{Z}, \underline{B} \cup \underline{Z})$ that send $K \in |\mathcal{F}|$ to $(q^{-1}(K), q^{-1}(K) \cap (B \cup Y^{(r)}))$ and $(Y_K, Y_K \cap (B \cup Y^{(r)}))$ where Y_K is the smallest subcomplex of Y containing $q^{-1}(K)$. Since the obvious inclusion of fragmented spaces $i\colon (\underline{Y}, \underline{B} \cup \underline{Y}^{(r)}) \longrightarrow (\underline{Z}, \underline{B} \cup \underline{Z}^{(r)})$ is an equivalence by 2.5, $i_*\colon \pi_n^F(\underline{Y}, \underline{B} \cup \underline{Y}^{(r)}) \longrightarrow i^! \pi_n^F(\underline{Z}, \underline{B} \cup \underline{Z})$ is an isomorphism by 4.10. Since, for $n \leq r$, $\pi_n^F(\underline{Z}, \underline{B} \cup \underline{Z}^{(r)})(y,K) = \pi_n(Y_K, Y_K \cap (B \cup Y_K^{(r)}), y) = 0$ for all $(y,K) \in |\mathcal{F}G_1(\underline{Z})|$, $0 = \pi_n^F(\underline{Y}, \underline{B} \cup \underline{Y}^{(r)}) = \pi_n^c(Y, B \cup Y^{(r)}, q)$.

The map of g of 10.3 is now constructed by induction. Suppose $g_{r-1}\colon (X, A \cup X^{(r-1)}, p) \longrightarrow (Y, B \cup Y^{(r-1)}, q)$ has been constructed such that g_{r-1} is homotopic to f relative to A and $g_{r-1}(A \cup X^{(n)}) \subseteq B \cup Y^{(n)}$ for all $n \leq r-1$. Apply 10.1 to g_{r-1} thought of as a map into $(Y, B \cup Y^{(r)}, q)$ with $A \cup X^{(r-1)}$ playing the role of A. The result is a map $g_r\colon (X, A \cup X^{(r)}, p) \longrightarrow (Y, B \cup Y^{(r)}, q)$ homotopic to g_{r-1} relative to $A \cup X^{(r-1)}$. It follows easily that g_r satisfies the inductive hypothesis. Since Y is assumed to be finite dimensional, 10.3 now follows easily.

Proof of 10.4: Suppose f is a homotopy equivalence. That (Y,q) is coextensive with (X,p) is obvious. That f induces an isomorphism $f_*\colon \pi_n^c(X,p) \longrightarrow f^! \pi_n^c(Y,q)$ follows from the discussion in section 9 and from 4.10.

Now suppose that (Y,q) is coextensive with (X,p) and that $f\colon (X,p) \longrightarrow (Y,q)$ induces an isomorphism $f_*\colon \pi_n^c(X,p) \longrightarrow f^! \pi_n^c(Y,q)$ for all $n \geq 0$. By 10.3 and the results of section 4, we may assume f is cellular. Let M_f be the usual mapping cylinder of f (the disjoint union $X \times I \amalg Y$ with $(x,1)$ identified with $f(x)$) and r be the obvious deformation retraction $(r(x,t) = f(x)$ for $0\leq t\leq 1$, $r(y) = y)$. Then (M_f,qr) is in $\mathcal{CW}^c/Z$ and $r\colon (M_f,qr) \longrightarrow (Y,q)$ is a bc homotopy equivalence with inverse the inclusion $j\colon (Y,q) \longrightarrow (M_f,qr)$. Now consider the homotopy commutative diagram

$$\begin{array}{ccc}
(X,p) & \xrightarrow{\quad f \quad} & (Y,q) \\
{\scriptstyle 1}\downarrow & & \downarrow{\scriptstyle j} \\
(X,qf) = (X,qr|) & \xrightarrow{i} & (M_f,qr)
\end{array}$$

where $i(x) = (x,0) \in M_f$ and 1 is the identity map. Since 1 is a bc homeomorphism by 1.9, it now suffices to prove i is a bc homotopy equivalence.

To show this we note first that (Y,qf) is coextensive with (M_f,qr). Also by 4.4 there is a natural equivalence η making the diagram

$$\begin{array}{ccc}
\pi_n^c(X,qf) & \xrightarrow{\quad i_* \quad} & i^!\pi_n^c(M_f,qr) \\
 & & \uparrow{\scriptstyle \eta} \\
{\scriptstyle 1_*}\downarrow & & 1^!f^!j^!\pi_n^c(M_f,qr) \\
 & & \uparrow{\scriptstyle (1^!f^!)(j_*)} \\
1^!\pi_n^c(X,p) & \xrightarrow{\quad 1^!(f_*) \quad} & 1^!f^!\pi_n^c(Y,q)
\end{array}$$

commute. It now follows easily that i_* is an isomorphism for every $n\geq 0$. Since the objects in the exact sequence

$$\pi_n^c(X,qf) \xrightarrow{i_*} i^!\pi_n^c(M_f,qr) \xrightarrow{j_*} \pi_n^c(M_f,X,qr)$$

$$\xrightarrow{\partial} \pi_{n-1}^c(X,qf) \xrightarrow{i_*} i^!\pi_{n-1}^c(M_f,qr)$$

lie in an abelian category when $n\geq 3$, standard exactness arguments show $\pi_n^c(M_f,X,qr) = 0$ for $n\geq 3$. If $n\leq 2$, the same result follows by using I.4.7. Then 10.4 follows from 10.2.

Proof of 10.5: If f is a homotopy equivalence, then f_* is an isomorphism for every n by 3.3. If f_* is an isomorphism for every n, form the mapping cylinder (M_f,qr) of f as in the proof of 10.4 and observe that $H_n^c(M_f,X,qf) = 0$ for every n. Since (X,p) and (Y,q) are coextensive and simply connected, so are (X,qf) and (M_f,qr). Then by the variation on the relative Hurewicz theorem 8.2, $\pi_n^c(M_f,X,qr) = 0$ for all n. That f is a homotopy equivalence now follows directly from the proof of 10.4.

CHAPTER III
THE GEOMETRIC, BOUNDEDLY CONTROLLED WHITEHEAD GROUP

In this chapter we develop the foundations of simple homotopy theory for the category $\mathcal{CW}_{\ell}^{c}/Z$. In particular, we introduce appropriate notions of elementary expansions and collapses and use these to define the boundedly controlled geometric Whitehead group $Wh^{c}(X,p)$ for (X,p) in $\mathcal{CW}_{\ell}^{c}/Z$. We prove that $Wh^{c}: \mathcal{CW}_{\ell}^{c}/Z \longrightarrow \mathcal{Ab}$ is a homotopy functor, and we establish a 2-Index Theorem by representing each element of $Wh^{c}(X,p)$ in a simplified form.

Section 1 deals with the notions of boundedly controlled n-cells and bc CW complexes. Elementary expansions and collapses are defined in section 2 and are used in section 3 to define $Wh^{c}(X,p)$. Section 3 also contains the proof that Wh^{c} is a homotopy functor. Section 4 contains the statement, as well as the proof, of the 2-Index Theorem. Finally, in section 5 we define the (geometric) torsion of a bc homotopy equivalence and establish some of its standard properties.

The discussion in this chapter differs from that of our announcement, [AM2], in two respects: the reparametrizations included there have been eliminated since they are just the isomorphisms in $\mathcal{CW}_{\ell}^{c}/Z$ and are not needed, see Proposition 3.4 below; the complicated definition of elementary expansions given there has been simplified since a sharper understanding of the algebra developed in Chapters IV and V has shown that the complications are unnecessary.

The whole chapter owes much to Marshall Cohen's treatment of the compact case, [Co], although we do feel that we have been able to introduce a couple of simplifications.

1. Boundedly controlled cells and CW complexes.

Throughout this chapter Z denotes a space with a boundedness control structure (P,C) in the sense of II.1.1. We denote the corresponding category with endomorphism by $\mathcal{F}$, see I.1.2. We refer the reader to II.1 for the notions of bc CW complexes and finite, bc CW complexes.

We denote by I^n the n-dimensional unit cube; I^{n-1} is embedded in I^n as $I^{n-1}\times 0$ and $J^n = cl(\partial I^n - I^{n-1})$ where ∂I^n is the boundary of I^n.

If S is a discrete topological space and $q: S\times I^n \longrightarrow Z$ is a map such that $\{q(s\times I^n) \mid s\in S\}$ is bounded, then $(S\times I^n,q)$ is called a *bc n-cell* (over Z). If, in addition $\{s\in S \mid q(s\times I^n) \subseteq K\}$ is finite for each $K\in|\mathcal{F}|$ then we speak of a *finite bc n-cell* (over Z).

Let $(S\times I^n,q)$ be a bc n-cell over Z, let $(X,p) \in |\mathcal{T}op^c/Z|$, and let $f: (S\times\partial I^n,q|) \longrightarrow (X,p)$ be a bc map. We say that f is an *attaching map for a bc n-cell* if there exists a bc n-cell $(S\times I^n,q')$ with $q'|S\times\partial I^n = pf$. In that case we define a new space in $\mathcal{T}op^c/Z$ by setting

$$(X,p)\cup_f(S\times I^n,q) = (X\cup_f S\times I^n, p\cup q'),$$

and we say this space *is obtained from* (X,p) *by attaching the bc n-cell* $(S\times I^n,q)$ *along the attaching map f*. The identification map

$$\psi: (S\times I^n,q) \longrightarrow (X\cup_f S\times I^n, p\cup q')$$

will be called a *characteristic map* for the bc n-cell in question.

Lemma 1.1 *In the situation described above, let* $q',q'': S\times\partial I^n \longrightarrow Z$ *be maps such that* $(S\times I^n,q')$ *and* $(S\times I^n,q'')$ *are bc n-cells and such that* $q'|S\times\partial I^n = pf = q''|S\times\partial I^n$. *Then* $1: (S\times I^n,q') \longrightarrow (S\times I^n,q'')$ *and* $1: (X\cup_f S\times I^n, p\cup q') \longrightarrow (X\cup_f S\times I^n, p\cup q'')$ *are bc homeomorphisms.*

Proof: This follows immediately from II.1.8 and II.10.7.

Proposition 1.2 *Finite dimensional* bc CW *complexes over* Z *are precisely those* bc *spaces over* Z *which can be obtained from the empty space by attaching a finite number of* bc n-*cells with* n *increasing in the order of attaching. A* bc CW *complex obtained in this way will be finite if and only if each of the finitely many* bc *cells is finite.*

We leave the easy proof to the reader.

In chapter II the *cells* of a CW complex, X, refer to the *open* cells, i.e. X is the disjoint union of its cells. We explicitly notice that if $\psi: (S\times I^n, q_n) \longrightarrow (X,q)$ is a characteristic map for the bc n-cell of (X,q), then the open n-cells of X are $\psi(s\times \mathrm{int}(I^n))$, $s\in S$. We also record the following useful observation concerning the closed cells which follows directly from II.2.5:

Lemma 1.3 *Let* $(X,p) \in |\mathcal{CW}^c_{\mathcal{E}}/Z|$. *Then* $\{p(\mathrm{Cl}(e)) \mid e\in X\}$ *is a bounded family of subsets of* Z

2. Elementary expansions and collapses.

Let (Y,X,q) be a bc CW pair and set $p=q|X$. Since we want to fix (X,p) and let (Y,q) vary, in the sequel we shall often denote this pair by ((Y,q),(X,p)) and shall write $(X,p) \le (Y,q)$.

Let ((Y,q),(X,p)) and ((W,u),(X,p)) be two pairs of finite bc CW complexes. We say that (W,u) is an *elementary expansion* of (Y,q) and we write $(Y,q)\overset{\nearrow}{_e}(W,u)$ if the following conditions hold:

(2.1) ((W,u),(Y,q)) is a pair of bc CW complexes;

(2.2) $(W,u) = (Y,q)\cup_f(S\times I^r, q_r)\cup_g(S\times I^{r+1}, q_{r+1})$ for some $r \geq 0$, some finite bc (r+i)-cell $(S\times I^{r+i}, q_{r+i})$ (i=0,1), and suitable attaching maps f and g; and

(2.3) There is a characteristic map $\psi_{r+1}: (S\times I^{r+1},q_{r+1}) \longrightarrow (W,u)$ for the bc r+1-cell such that $\psi_{r+1}|S\times I^r: (S\times I^r,q_r) \longrightarrow (W,u)$ is characteristic for the bc r-cell.

For fixed $s\in S$ the two cells $\psi_{r+1}(s\times I^{r+1})$, $\psi_{r+1}(s\times I^r)$ "generate" an elementary expansion of Y in the classical sense. Note that ψ_{r+1} maps $S\times J^r$ into the r-skeleton of W and ψ_{r+1} maps $S\times\partial I^r$ into the (r-1)-skeleton of W.

We also express the above relationship between (Y,q) and (W,u) by saying that (W,u) *collapses* to (Y,q) *by an elementary collapse*, written $(W,u)\overset{e}{\searrow}(Y,q)$.

More generally we say that (Y,q) *expands to* (W,u), written $(Y,q)\nearrow(W,q)$, or that (W,u) *collapses to* (Y,q), written $(W,u)\searrow(Y,q)$ if there is a string of elementary expansions

$$(Y,q) = (Y_1,q_1)\underset{e}{\nearrow}(Y_2,q_2)\underset{e}{\nearrow}\cdots\underset{e}{\nearrow}(Y_n,q_n) = (W,u).$$

Finally we say that there is a *formal deformation from* (Y,q) to (W,u) written $(Y,q)\curvearrowright(W,u)$, if $(Y,q)\nearrow(\bar{Y},\bar{q})\searrow(W,u)$ for some $(\bar{Y},\bar{q})$.

Note that, in the above, we are always dealing with finite bc CW complexes containing (X,p). When we want to emphasize (X,p) we shall write $(Y,q)\curvearrowright(W,u)$ rel(X,p).

The following trivial remark is used repeatedly in the sequel.

Remark Assume that $(Y,q)\leq(W,u)$ and that W has the form

$$W = Y\cup(\bigcup_{s\in S} e_s^r)\cup(\bigcup_{s\in S} e_s^{r+1})$$

for some indexing set S, some $r\in\mathbb{Z}_+$, and some collection of open cells e_s^r, e_s^{r+1} in W. Then $(Y,q)\underset{e}{\nearrow}(W,u)$ if and only if there is a permutation $\alpha: S \longrightarrow S$ such that for each $s\in S$, $W_s = Y\cup e_s^r\cup e_{\alpha(s)}^{r+1}$ is a subcomplex of W and $Y \nearrow W_S$ in the classical sense. Of course, in most cases the cells will be indexed such that $\alpha = 1_S$.

Proposition 2.4 *If* $(Y,q) \nearrow (W,u)$, *then there is a* bc *strong deformation retraction* $r\colon (W,u) \longrightarrow (Y,q)$. *If* $(Y,q) \nearrow\!\!\searrow (W,u)$ $\mathrm{rel}(X,p)$, *then there is a* bc *homotopy equivalence of pairs* $f\colon ((Y,q),(X,p)) \longrightarrow ((W,u),(X,p))$ *which is the identity on* (X,p).

Proof: For the first part we may assume that $(Y,q) \nearrow_e (W,u)$. Thus let

$$(W,u) = (Y,q) \cup_f (S \times I^r, u_r) \cup_g (S \times I^{r+1}, u_{r+1})$$

as above. The standard deformation of I^{r+1} onto J^r now gives the desired bc deformation. The second part of the proposition is an immediate consequence of the first part.

A homotopy equivalence of pairs $f'\colon ((Y,q),(X,p)) \longrightarrow ((W,u),(X,p))$ which is the identity on (X,p) is said to *correspond to a formal deformation rel* (X,p) if there exists a formal deformation $(Y,q) \nearrow\!\!\searrow (W,u)$ rel (X,p) for which f' is homotopic to the map f of 2.4.

Proposition 2.5 *The relation* $\nearrow\!\!\searrow$ $\mathrm{rel}(X,p)$ *is an equivalence relation.*

Proof: Only transitivity needs comments. It suffices to show that if $(Y,q) \searrow^e (W,u) \nearrow_e (V,v)$ $\mathrm{rel}(X,p)$ then there exists $(\bar{W},\bar{u})$ such that $(Y,q) \nearrow_e (\bar{W},\bar{u}) \searrow^e (V,v)$. For $(\bar{W},\bar{u})$ one simply takes the pushout of the diagram

$$\begin{array}{ccc} (W,u) & \leq & (V,v) \\ \wedge & & \\ (Y,q) & & \end{array}$$

Then $(\bar{W},\bar{u})$ arises from (Y,q) by attaching those cells which create (V,v) from (W,u). Hence $(Y,q)\nearrow(\bar{W},\bar{u})$. Similarly $(V,v)\nearrow(\bar{W},\bar{u})$.

3. The boundedly controlled geometric Whitehead group.

For a fixed $(X,p) \in |\mathscr{CW}^c_\ell/Z|$, we let $DR^c(X,p)$ be the collection of all pairs $((Y,q),(X,p))$ in $\mathscr{CW}^c_\ell/Z$ for which (X,p) is a bc strong deformation retract of (Y,q).

The equivalence class of such a pair under the relation $\curvearrowright$ rel(X,p) is denoted by $[Y,X,q]$ (or by $[(Y,q),(X,p)]$ if we want to emphasize p). The collection of all such equivalence classes is denoted $Wh^c(X,p)$.

Given two such equivalence classes, say $[Y,X,q]$ and $[W,X,s]$, we define their sum by

(3.1) $\qquad [Y,X,q]+[W,X,s] = [Y\cup_X W,X,q\cup s].$

Furthermore, given a cellular bc map $g\colon (X,p) \longrightarrow (X_1,p_1)$, we define the induced map $g_*\colon Wh^c(X,p) \longrightarrow Wh^c(X_1,p_1)$ by

(3.2) $\qquad g_*([Y,X,q]) = [X_1\cup_g Y,X_1,p_1\cup p_1 g\rho],$

where $\rho\colon (Y,q) \longrightarrow (X,p)$ is a deformation retraction.

The rest of this section is devoted to proving

Theorem 3.3 *The constructions described above define a homotopy functor* $Wh^c\colon \mathscr{CW}^c_\ell/Z \longrightarrow \mathscr{Ab}$.

Another proof of this theorem is given in our paper [AM3].

To see that $Wh^c(X,p)$ is a set we need the following proposition:

Proposition 3.4 *Let* (Y,X,q), $(W,X,s) \in DR^c(X,p)$. *If* $f: (Y,q) \longrightarrow (W,s)$ *is a* CW *isomorphism in* $\mathcal{CW}^c_{\ell}/Z$ *and* $f|X = 1_X$, *then* $[Y,X,q] = [W,X,s]$ *in* $Wh^c(X,p)$.

Proof: Let $M(f)$ be the mapping cylinder of f with its natural CW structure. We note that since f is a homeomorphism, $M(f)$ is homeomorphic to $Y\times I$. Also let $M(f)_X$ be the relative mapping cylinder, i.e. $M(f)_X = M(f)/\sim$ where $\sim$ identifies (x,t) to x for all $t\in I$ and for each $x\in X$. There is a natural CW structure on $M(f)_X$ where each cell e of $Y-X$ gives rise to three cells in $M(f)_X-X$, viz. e itself, $e\times(0,1)$, and $f(e) \in W-X$. We note that f establishes a 1-1 correspondence between the cells of $Y-X$ and those of $W-X$.

We finish the proof of 3.4 by showing that there exists a map $m: M(f)_X \longrightarrow Z$ such that $(M(f)_X,m) \in |\mathcal{CW}^c_{\ell}/Z|$, $m|X=p$, and

$$(3.5) \qquad (Y,q)\nearrow(M(f)_X,m)\searrow(W,s).$$

To do this let $\pi: Y\times I \longrightarrow Y$ be the projection and let $h: (Y\times I,q\pi) \longrightarrow (Y,q)$ be a bc strong deformation with $h|Y\times 0 = 1$, $h(y,1) \in X$ for every $y \in Y$, and $h(x,t) = x$ for every $x \in X$. We define $\bar{m}: M(f) \longrightarrow Z$ by

$$\bar{m}([y,t]) = \begin{cases} qh(y,2t) & , 0 \le t \le 1/2 \\ sfh(y,2-2t) & , 1/2 \le t \le 1 \end{cases}.$$

Then $\bar{m}$ is well defined because $qh(y,1) = ph(y,1) = sh(y,1) = sfh(y,1)$. For $x\in X$ one has $qh(x,2t) = p(x) = sfh(x,2-2t)$ so $\bar{m}$ induces a map $m: M(f)_X \longrightarrow Z$ with $m|X = p$. Also $m|Y = q$ and $m|W = s$ (write $w\in W$ as $f(y)$ so that $w \in M(f)_X$ equals $[y,1]$ and $m(w) = sfh(y,0) = sf(y) = s(w)$).

To show that $(M(f)_X,m)$ is a bc CW complex, we note that by 1.3 since (Y,q) and (W,s) are in $\mathcal{CW}^c_{\ell}/Z$ and f is a CW isomorphism, there exists a $d \ge 0$ such that for every cell $e \in Y$ there are minimal elements K_e and $K'_{f(e)}$ in $|\mathcal{F}|$ with $q(Cl(e)) \subseteq C^dK_e$ and $s(Cl(f(e))) \subseteq C^dK'_{f(e)}$. If d' is a common bound for h and for f then it easily follows that

$$\bar{m}(Cl(e\times(0,1))) \subseteq C^{d+d'}K_e \cup C^{d+2d'}K'_{f(e)}.$$

Moreover, for any point y of e, $\bar{m}([y,1/2]) \in C^{d+d'}K_e \cap C^{d+2d'}K'_{f(e)}$.

Therefore, by II.1.1.(iv), $K'_{f(e)} \subseteq C^{d+d'+\theta(d+d')}K_e$. Hence

$$\bar{m}(Cl(e\times(0,1))) \subseteq C^{2d+3d'+\theta(d+d')}K_e.$$

Since $\bar{m}$ = qUs on YUW, it follows that $(M(f)_X,m)$ is a bc CW complex.

To see that $(M(f)_X,m)$ is finite let $K\in|\mathcal{F}|$. It suffices to show that $\bar{m}^{-1}(K)$ is contained in some finite subcomplex of M(f). We put $D = 2d+3d'+\theta(d+d')$ and we let $L = C^{d+\theta(D)}K$. Since (Y,q) is finite there is a finite subcomplex Y_L of Y which contains $q^{-1}(L)$. We claim that $\bar{m}^{-1}(K)$ is contained in the finite subcomplex $Y_L \cup Y_L\times(0,1) \cup f(Y_L)$. Indeed, let $x\in\bar{m}^{-1}(K)$ and pick a cell e of Y such that $x\in Cl(e\times(0,1))$. By the above, $\bar{m}(x) \in C^D(K_e)\cap K$. By II.1.1.(iv), it follows that K_e is contained in $C^{\theta(D)}(K)$. But then $e \in Y_L$ so that x belongs to the subcomplex as stated. This completes the proof that $(M(f)_X,m) \in|\mathcal{CW}^c_{\ell}/Z|$.

Next, to prove that $(Y,q)\nearrow(M(f)_X,m)$ one just has to note that the cells of $M(f)_X$-Y come in pairs $(e\times(0,1),f(e))$ where e ranges over the cells of Y-X. Using descente finie according to the dimension of e one sees that the desired collapse can be obtained as a composition of dimY+1 elementary collapses.

Similarly, one shows that $(M(f)_X,m)\searrow(W,s)$.

Let us then show that $Wh^c(X,p)$ is a set. Let $[Y,X,q] \in Wh^c(X,p)$. Since $q^{-1}K$ is contained in a finite complex for each $K \in |\mathcal{F}|$ and since $Z = \cup_n C^n K$ for an arbitrary K, the CW complex Y has at most countably many cells. Up to CW isomorphism relative to X this leaves only a set of possibilities for Y. Moreover each Y only supports a set of maps $q\colon Y \longrightarrow Z$ extending p. Hence, the desired conclusion follows.

To prove that the sum is well defined we need nothing but the following lemma:

Lemma 3.6 *Assume that* (Y,X,q) *and* (W,X,s) $\in DR^c(X,p)$. *If* $(Y,q)\nearrow(Y_1,q_1)$ *then* $(Y\cup_X W, q\cup s)\nearrow(Y_1\cup_X W, q_1\cup s)$.

Proof: Using a simple induction on the number of elementary expansions, we only have to treat the case where $(Y,q)\nearrow_e(Y_1,q_1)$. That case is trivial since $Y_1\cup_X W$ is obtained from $Y\cup_X W$ by attaching the cells of Y_1-Y.

Next, from Proposition 3.4 it is clear that the addition defined in 3.1 is commutative and associative and has [X,X,p] as a zero element.

To show that each [Y,X,q] has an inverse and to establish the functoriality of $Wh^c(X,p)$, we need to study bc mapping cylinders.

We assume that g: (X,p) ⟶ (Y,q) is a cellular bc map between bc CW complexes over Z. We denote by M(g) the mapping cylinder of g with its preferred CW structure (X and Y are subcomplexes; all (open) cells of W-X-Y have the form $e\times(0,1)$ with e varying over the cells of X). Also, let ρ: M(g) ⟶ Y be the standard retraction given by $\rho([x,t]) = g(x)$, and consider $q\rho$: M(g) ⟶ Z. The following proposition is obvious:

Proposition 3.7 *If* $g \in \mathcal{CW}^c/Z((X,p),(Y,q))$, *then* $(M(g),q\rho) \in |\mathcal{CW}^c/Z|$. *If* (X,p) and (Y,q) *are finite, then so is* $(M(g),q\rho)$.

Remark Since $q\rho|Y = q$ and $q\rho|X\times 0 = qg$, (Y,q) and (X,qg) are bc sub complexes of $(M(g),q\rho)$. Although $qg \neq p$ in general, 1: (X,p) ⟶ (X,qg) is a bc homeomorphism by II.1.9.

The mapping cylinders above are special cases of *bc adjunction spaces*. Assume given a diagram in $\mathcal{CW}^c/Z$

$$
(3.8)\qquad \begin{array}{ccc} (X,p) & \le & (Y,q) \\ \downarrow{\scriptstyle g} & & \\ (X_1,p_1) & & \end{array}
$$

where g is assumed to be cellular and bc.

Proposition 3.9 *The diagram* (3.8) *fits into a pushout diagram*

$$
\begin{array}{ccc} (X,p) & \le & (Y,q) \\ {\scriptstyle g}\downarrow & & \downarrow{\scriptstyle g_1} \\ (X_1,p_1) & \le & (Y_1,q_1) \end{array}
$$

in $\mathcal{CW}^c/Z$ *if and only if there exists* $q' : Y \longrightarrow Z$ *such that* $p_1g = q'|X$ *and* $1_Y \in \mathcal{CW}^c/Z((Y,q),(Y,q'))$.

Proof: First assume that such a map q' exists. One then lets $(Y_1,q_1) = (X_1 \cup_g Y, p_1 \cup q')$ and one defines g_1 in the obvious way. The pushout property is easily verified using II.1.9 and the hypothesis that 1_Y is bc.

Conversely if the pushout exists, one takes $q' = q_1g_1$ and invokes II.1.9.

Notation: In the above situation we write (Y_1,q_1) as $(X_1,p_1)\cup_g(Y,q)$.

We need 3 lemmas concerning pushouts and mapping cylinders.

Lemma 3.10 *(The relativity principle)* *Suppose given a diagram in* $\mathcal{CW}^c_{\ell}/Z$

$$
\begin{array}{ccccc} (X,p) & \le & (Y,q) & \curvearrowright & (W,s)\ \mathrm{rel}(X,p) \\ \downarrow{\scriptstyle g} & & & & \\ (X_1,p_1) & & & & \end{array}
$$

where g *is cellular. If* $(X_1,p_1)\cup_g(Y,q)$ *exists, then so does*

$(X_1,p_1)\cup_g(W,s)$; *furthermore*

$$(X_1,p_1)\cup_g(Y,q) \nearrow\!\searrow (X_1,p_1) \cup_g(W,s) \text{ rel } (X_1,p_1).$$

Proof: We may assume that $(Y,q)\underset{e}{\nearrow}(W,s)$ and we let $q'\colon Y \longrightarrow Z$ satisfy the hypotheses of 3.9 so that $(X_1,p_1)\cup_g(Y,q) = (X_1\cup_g Y, p_1\cup q')$. By 2.4 there is a strong deformation retraction $r\colon (W,s) \longrightarrow (Y,q)$. We put $s' = qr$. By II.1.9 $1_W\colon (W,s) \longrightarrow (W,s')$ is a bc homeomorphism. Since $s'|X = qr|X = q|X = p_1 g$, $(X_1,p_1)\cup_g(W,s) = (X_1\cup_g W, p_1\cup s')$ exists. Moreover, the cells of $X_1\cup_g W - X_1\cup_g Y$ are precisely those of $W-Y$ so $(X_1\cup_g Y, p_1\cup q')\underset{e}{\nearrow}(X_1\cup_g W, p_1\cup s')$ as claimed.

Lemma 3.11 *(Formal deformations of mapping cylinders)* *Let* $g \in \mathcal{CW}^c_{\mathcal{L}}((X,p),(Y,q))$ *and let* $(X_0,p_0) \le (X,p)$. *Write* $g_0 = g|X_0$. *Then*

(i) $(M(g_0),q\rho)\nearrow(M(g),q\rho)$

(ii) $(Y,q)\nearrow(M(g),q\rho)$

(iii) $(X_0\times I \cup X\times\{i\}, p\pi|)\nearrow(X\times I, p\pi)$ *for* $i=0,1$

(iv) *If* $(X_0,p_0)\nearrow(X,p)$, *then* $(X\times\{0\}\cup M(g_0), q\rho|)\nearrow(M(g),q\rho)$.

Proof: Parts (ii) and (iii) are special cases of (i) taking $X_0=\emptyset$ and $g = 1_{(X,p)}$, respectively (as usual $\pi\colon X\times I \longrightarrow X$ is the projection).

The proof of (i) uses induction on $\dim(X-X_0)$ and the observation that the (open) cells of $M(g)-M(g_0)$ come in pairs $(e\times(0,1),e)$, indexed on the cells e of $X-X_0$.

Finally to prove (iv) one first assumes that $(X_0,p_0)\underset{e}{\nearrow}(X,p)$. Thus there is a family of pairs of open cells $(e^{r+1}_\alpha, e^r_\alpha)$ such that $X = X_0 \cup (\cup_\alpha e^r_\alpha) \cup (\cup_\alpha e^{r+1}_\alpha)$, $X_\alpha = X_0 \cup e^r_\alpha \cup e^{r+1}_\alpha$ is a subcomplex of X for each α, and X_0 expands to X_α (in the classical sense) for each α.

But then $M(g) = X\times\{0\} \cup M(g_0) \cup (\cup_\alpha e^r_\alpha\times(0,1)) \cup_\alpha(\cup e^{r+1}_\alpha\times(0,1))$ has $M(g)_\alpha = X\times\{0\} \cup M(g_0) \cup e^r_\alpha\times(0,1) \cup e^{r+1}_\alpha\times(0,1)$ as a subcomplex for each α, and $X\times\{0\} \cup M(g_0)$ expands to $M(g)_\alpha$ (in the classical sense) for each α. And this means that $(X\times\{0\}\cup M(g_0), q\rho|)\underset{e}{\nearrow}(M(g),q\rho)$ as desired.

The general case follows by a straightforward induction.

Lemma 3.12 *Let* $h: f\simeq g: (X,p) \longrightarrow (Y,q)$ *be a cellular homotopy in* $\mathscr{CW}^c_{\mathscr{L}}/Z$. *Then there exists* $q': M(g) \longrightarrow Z$ *such that*

(i) $(M(f),q\rho) \nearrow\!\!\!\searrow (M(g),q')$ rel$(X,qf) \cup (Y,q)$; *and*

(ii) $1_{M(g)}: (M(g),q') \longrightarrow (M(g),q\rho)$ *is a bc homeomorphism.*

Proof: Define $m: M(h) \longrightarrow Z$ by

$$m([x,t,s]) = qh(x,ts), \quad (x,t) \in X\times I \ , \ s\in I$$
$$m(y) = q(y), \quad , \ y\in Y.$$

One easily checks that $(M(h),m) \in |\mathscr{CW}^c_{\mathscr{L}}/Z|$ and that $1_{M(h)}: (M(h),q\rho) \longrightarrow (M(h),m)$ is bc. From 3.11 (iv) we see that

$$(X\times I\times\{0\}\cup Y,m|) \leq (X\times I\times\{0\}\cup M(h|X\times\{0\}),m|) \nearrow\!\!\!\searrow$$
$$(X\times I\times\{0\}\cup M(h|X\times\{1\}),m|) \ \ \text{rel}(X\times I\times\{0\}\cup Y,m|).$$

The relativity principle applies to this formal deformation and the map $\pi\cup 1_Y: (X\times I\times\{0\}\cup Y,m|) \longrightarrow (X\cup Y,qf\cup q)$ to give

$$(X\cup Y,qf\cup q) \leq (M(f),q\rho) \nearrow\!\!\!\searrow (M(g),q') \ \text{rel}(X\cup Y,qf\cup q)$$

where $q': M(g) \longrightarrow Z$ has

$$q'([x,s]) = qh(x,s) \ , \quad (x,s) \in X\times I$$
$$q'(y) = q(y) \ , \quad y \in Y.$$

Finally with this explicit description of q' one easily checks that the map $1_{M(g)}: (M(g),q\rho) \longrightarrow (M(g),q')$ is bc.

With the above lemmas at our disposal we show that each $[Y,X,q] \in Wh^c(X,p)$ has an inverse. By 3.4 and II.1.9 we can assume that $q=pr$ for some strong deformation retraction $r: (Y,q) \longrightarrow (X,p)$. Let $i: (X,p) \leq (Y,q)$ be the inclusion and let $h: (Y\times I,q\pi) \longrightarrow (Y,q)$ be a homotopy from ir to 1_Y. Consider the diagram

$$(X_-,p_-) \xleftarrow{r_-} (Y_0,q_0) \xrightarrow{r} (X,p)$$

where (X_-,p_-) is a copy of (X,p), (Y_0,q_0) is a copy of (Y,q), and r_- a copy of r. We have the mapping cylinders $(M(r_-),p_-\rho)$, $(M(r),p\rho)$ (where

ρ is the generic name for the standard retraction in a mapping cylinder), and one easily checks that

$$(M(r\_)\cup_{Y_0} M(r), X, p\_\rho \cup p\rho) \in DR(X,p).$$

To see that this pair represents an inverse for $[Y,X,q]$, we form the sum

$$[M(r\_)\cup_{Y_0} M(r)\cup_X Y, X,\ p\_\rho\cup p\rho\cup q] = [M(r\_)\cup M(ir), X,\ p\_\rho\cup q\rho],$$

and we proceed to check that this element vanishes in $Wh^c(X,p)$.

Since $ir \simeq 1\colon (Y,q) \longrightarrow (Y,q)$, 3.12 implies that

$$(M(ir), q\rho) \curvearrowright (Y_0\times I, q') \quad rel(Y_0,q_0)\cup(Y,q)$$

where $Y \equiv Y_0\times\{1\} \subseteq Y_0\times I$, $Y_0 \equiv Y_0\times\{0\} \subseteq Y_0\times I$ and $1\colon (Y_0\times I, q\rho) \longrightarrow (Y_0\times I, q')$ is bc.

Since $q'|Y_0\times\{0,1\} = q\rho|Y_0\times\{0,1\}$, it follows from 3.4 that

$$(M(ir), q\rho) \curvearrowright (Y_0\times I, q\rho) \quad rel(Y_0,q_0)\cup(Y,q).$$

Using 3.10, it follows that

$$(M(r\_)\cup_{Y_0} M(ir), p\_\rho\cup q\rho) \curvearrowright (M(r\_)\cup_{Y_0} Y_0\times I, p\_\rho\_\cup q\rho)$$

$rel(M(r\_), p\_\rho)\cup(Y,q)$. Thus we just have to see that

$$(M(r\_)\cup_{Y_0} Y_0\times I, p\_\rho\cup q\rho) \curvearrowright (X,p)\ rel(X,p),$$

and this follows in two steps. First (by 3.11 (iii))

$$(Y_0\times I, q\rho)\searrow(X_0\times I\cup Y_0\times\{0\}, p\rho\cup q_0)$$

so (by 3.10)

$$(M(r\_)\cup Y_0\times I, p\_\rho\cup q\rho)\searrow(M(r\_)\cup_{X_0} X_0\times I, p\_\rho\cup p\rho).$$

Next (by 3.11 (i))

$$(M(r\_), p\_\rho)\searrow(M(r\_|X_0) = X_0\times[-1,0], p\rho)$$

(here $X_0 \subseteq Y_0$ is the copy of X in Y_0, and $X\_ = X_0\times\{-1\}$), so (by 3.10)

$$(M(r\_)\cup_{X_0} X_0\times I, p\_\rho\cup p\rho)\searrow(X_0\times[-1,1], p\rho).$$

Finally, $(X_0\times[-1,1], p\rho)$ obvious formally deforms to (X,p) $rel(X,p)$. (Recall that $X = X_0\times\{0\}$.)

This finishes the proof that $Wh^c(X,p)$ is an abelian group.

Next let $g: (X,p) \longrightarrow (X_1,p_1)$ be a bc map. If g is cellular, then 3.10 suffices to guarantee that (3.2) gives a well defined function $g_*: Wh^c(X,p) \longrightarrow Wh^c(X_1,p_1)$. A straightforward argument shows that g_* is a homomorphism.

Moreover, if one also has a cellular bc map $g_1: (X_1,p_1) \longrightarrow (X_2,p_2)$, then, obviously, $(g_1g)_* = g_{1*}g_*$.

To check that cellular maps $g_0,g_1: (X,p) \longrightarrow (X_1,p_1)$ which are cellularly homotopic induce identical homomorphisms $g_{0*} = g_{1*}: Wh^c(X,p) \longrightarrow Wh^c(X_1,p_1)$ one only needs to verify that $i_\nu: (X,p) \longrightarrow (X\times I,p\pi)$ $(\nu=0,1)$ have $i_{0*} = i_{1*}$. And this follows from 3.11 (iii).

Finally, if $g: (X,p) \longrightarrow (X_1,p_1)$ is bc but not necessarily cellular, one defines g_* to be g_{1*} where g_1 is some bc cellular approximation to g (cf. II.10.3). This finishes the proof of Theorem 3.3.

4. The 2 Index theorem.

A representative (Y,X,q) for an element of $Wh^c(X,p)$ is said to be in *simplified form* if there exist an integer r, two finite, bc cells, $(S_r\times I^r,q_r)$ and $(S_{r+1}\times I^{r+1},q_{r+1})$, and attaching maps, ϕ_r and ϕ_{r+1}, such that

(4.1) $\quad (Y,q) = (X,p)\cup_{\phi_r}(S_r\times I^r,q_r)\cup_{\phi_{r+1}}(S_{r+1}\times I^{r+1},q_{r+1}),$

(4.2) $\quad \phi_r|s\times\partial I^r$ is constant for every $s\in S_r$,

(4.3) $\quad \phi_{r+1}|s\times J^r$ is constant for every $s\in S_{r+1}$.

This section is totally devoted to the proof of the following theorem:

Theorem 4.4 *(The 2 Index Theorem) Each element of* $Wh^c(X,p)$ *has a representative in simplified form.*

For the proof we need two lemmas whose proofs are deferred:

Lemma 4.5 *Let* $(Y,q) = (X,p)\cup_{\phi_0}(S\times I^r,q_r)$ *for some finite, bc r-cell* $(S\times I^r,q_r)$ *and some attaching map* $\phi_0\colon (S\times\partial I^r,q_r|) \longrightarrow (X,p)$. *If* $h\colon \phi_0 \simeq \phi_1$ *is a bc cellular homotopy, then* ϕ_1 *is also an attaching map and*

$$(Y,q) = (X,p)\cup_{\phi_0}(S\times I^r,q_r) \nearrow\!\!\!\searrow (X,p)\cup_{\phi_1}(S\times I^r,q_r).$$

Lemma 4.6 *(The formal deformation extension theorem) Suppose given a commutative diagram in* $\mathscr{CW}^c_{\ell}/Z$

$$\begin{array}{ccccc} (X,p) & \leq & (Y,q) & \leq & (U,t) \\ \downarrow 1 & & \downarrow f & & \\ (X,p) & \leq & (W,s) & & \end{array}$$

in which f *is a homotopy equivalence corresponding to some formal deformation* $(Y,q)\nearrow\!\!\!\searrow(W,s)$ rel(X,p). *If* $(W,s)\cup_f(U,t)$ *exists, then* $(U,t)\nearrow\!\!\!\searrow(W,s)\cup_f(U,t)$ rel(X,p).

Proof of 4.4: By 1.2,

$$(Y,q) = (X,p)\cup_{\phi_0}(S_0\times I^0,q_0)\cup_{\phi_1}\cdots\cup_{\phi_r}(S_r\times I^r,q_r)$$

for some $r \geq 0$ and some finite bc i-cells $(S_i\times I^i,q_i)$ with characteristic maps $\psi_i\colon (S_i\times I^i,q_i) \longrightarrow (Y,q)$ and attaching maps $\varphi_i = \psi_i|(S_i\times\partial I^i,q_i|)$. We may further assume that $q\psi_i = q_i$. Let $h\colon (Y\times I,q\pi) \longrightarrow (Y,q)$ be a bc cellular deformation with $h(y,1) \in X$, $h(x,t) = x$, and $h(y,0) = y$ for all $x\in X$, all $y\in Y$ and all $t\in I$.

Assume inductively that $S_0 = S_1 = \cdots = S_{u-1} = \emptyset$ for some u with $0 \leq u < r-1$. We want to modify (Y,q), keeping $[Y,X,q] \in Wh^c(X,p)$ unchanged, so that also $S_u = \emptyset$. Let Φ be the map

$$\Phi = h(\phi_u\times 1)\colon (S_u\times I^u\times I,q_u\pi) \longrightarrow (Y,q)$$

and notice that Φ maps into the (u+1) skeleton of Y and takes $S_u \times \partial I^{u+1}$ into the u-skeleton. Therefore, we can think of the adjunction space (cf. 3.9)

$$(U,t) = (Y,q) \cup_{\Phi} (S_u \times I^{u+2}, q'_{u+2}),$$

where

$$q'_{u+2}(s,v,t) = q\Phi(s,v), \quad (v \in I^{u+1}, \ t \in I),$$

as a finite bc CW complex obtained from (Y,q) by attaching a bc u+1 cell $(S_u \times J^{u+1}, q'_{u+2}|)$ and a bc u+2 cell $(S_u \times I^{u+2}, q'_{u+2})$. We use some standard homeomorphism (rel $\partial I^{u+1} = \partial J^{u+1}$) to identify J^{u+1} with I^{u+1}. We then have characteristic maps

$$\psi'_{u+1}: (S_u \times I^{u+1}, q'_{u+1}) \longrightarrow (U,t) \text{ and } \psi'_{u+2}: (S_u \times I^{u+2}, q'_{u+2}) \longrightarrow (U,t)$$

which satisfy (2.3). Therefore,

(4.7) $$(Y,q) \nearrow (U,t)$$

Let ϕ'_{u+1} and ϕ'_{u+2} be the attaching maps corresponding to the charateristic maps ψ'_u and ψ'_{u+1}, respectively. Then

$$(V,t|V) = (X,p) \cup_{\phi_u} (S_u \times I^u, q_u) \cup_{\phi'_{u+1}} (S_u \times I^{u+1}, q'_{u+1})$$

is a subobject of (U,t). Moreover, by the remark preceding 2.4,

(4.8) $$(X,p) \underset{e}{\nearrow} (V,t|V).$$

We can now apply 4.6 to the diagram

$$\begin{array}{ccccc} (X,p) & \le & (V,t|V) & \le & (U,t) \\ \downarrow 1 & & \downarrow r & & \\ (X,p) & = & (X,p) & & \end{array}$$

where r is "the" retraction of (V,t|V) onto (X,p), to conclude that

(4.9) $$(U,t) \curvearrowright (X,p) \cup_r (U,t) \text{ rel}(X,p).$$

Since (U,t)-(V,t|V) has bc cells only in dimensions u+1,...,r so does

$(X,p)\cup_r(U,t)$ - (X,p), and the inductive step is finished by referring also to (4.7).

We can now assume that our given element $[Y,X,q] \in Wh^c(X,p)$ satisfies (4.1).

To establish (4.2) we note that $\phi_r\colon (S_r\times\partial I^r, q_r|) \longrightarrow (X,p)$ is bc homotopic to a map which is constant on each $s\times\partial I^r$ ($s\in S_r$). In fact the desired homotopy obviously exists in the bigger space (Y,q); now use a retraction of (Y,q) onto (X,p) to press the homotopy down into (X,p).

Using such a homotopy and (4.5) we see that

$$(X,p)\cup_{\phi_r}(S_r\times I^r,q_r) \simeq (X,p)\cup_{\phi_r''}(S_r\times I^r,q_r)$$

where ϕ_r'' satisfies (4.2).

Finally apply 4.6 to the diagram

$$\begin{array}{ccccc}(X,p) & \le & (X,p)\cup_{\phi_r}(S_r\times I^r,q_r) & \le & (Y,q)\\ \downarrow 1 & & \downarrow & & \\ (X,p) & \le & (X,p)\cup_{\phi_r''}(S_r\times I^r,q_r) & & \end{array}$$

to get the desired representative.

The proof that (4.3) can be obtained, is similar starting from the fact that there is a deformation of ∂I^{r+1} which deforms J^r to a point.

We finish this section with the proofs of 4.5 and 4.6.

Proof of 4.5: Let $\psi\colon (S\times I^r,q_r) \longrightarrow (Y,q)$ be a characteristic map for the given bc r-cell. By remark 1.1 and II.1.9, we can assume that $q_r=q\psi$. Since h is cellular, one has $\phi_1(S\times\partial I^r) \subseteq X^{r-1}$. Let $\rho\colon I^r\times I \longrightarrow I^r\times\{0\}\cup\partial I^r\times I$ be some standard retraction and put

(4.10) $\quad \bar{b} = (q_r\cup ph)(1_S\times\rho)\colon S\times I^r\times I \longrightarrow Z$, and $b = \bar{b}(-,-,1)\colon S\times I^r \longrightarrow Z$.

Then $(S\times I^r,b)$ is a bc r-cell, the map $\phi_1\colon (S\times\partial I^r,b|) \longrightarrow (X,p)$ is bc, and $b|S\times\partial I^r = p\phi_1$. Thus ϕ_1 is an attaching map as claimed.

Let $(Y_1,q_1) = (X,p)\cup_{\phi_1}(S\times I^r,q_r)$ and note that $Y\cap Y_1 = X$. Clearly $(W,w) = (Y\cup_X Y_1, q\cup q_1)$ is in $\mathcal{CW}^c_{\ell}/Z$. We can attach the bc $(r+1)$-cell $(S\times I^{r+1},\bar{b})$ (where $\bar{b}$ is defined in (4.10)) to (W,w) along the bc map $\Phi\colon (S\times\partial I^{r+1},\bar{b}|) \longrightarrow (W,w)$ given by requiring that

$$\Phi|S\times\partial I^r\times I = h\colon S\times\partial I^r\times I \longrightarrow X\ (\subseteq W)$$
$$\Phi|S\times I^r\times\{0\} = \psi\colon S\times I^r \longrightarrow Y\ (\subseteq W)$$
$$\Phi|S\times I^r\times\{1\} = \psi_1\colon S\times I^r \longrightarrow Y_1\ (\subseteq W)$$

where ψ_1 is characteristic for the r-cell of $(Y_1,q_1)-(X,p)$. The result

$$(U,t) = (W,w)\cup_\Phi(S\times I^{r+1},\bar{b}) = (X,p)\cup_\phi(S\times I^r,q_r)\cup_{\phi_1}(S\times I^r,b)\cup_\Phi(S\times I^{r+1},\bar{b})$$

and it is clear that $(Y,q)\nearrow(U,t)\searrow(Y_1,q_1)$.

Proof of 4.6: In 4.6 $f\colon (Y,q) \longrightarrow (W,s)$ is a composition of maps of two types, viz. inclusions $i\colon (Y,q) \longrightarrow (W,s)$ where $(Y,q)\nearrow_e(W,s)$, and retractions $\rho\colon (Y,q) \longrightarrow (W,s)$ where $(W,s)\nearrow_e(Y,q)$. If $f = i$ is of the first type, then the desired result follows by applying 3.10 to the dia gram

$$\begin{array}{ccc} (Y,q) & = & (Y,q)\nearrow(W,s)\ \mathrm{rel}(Y,q) \\ \cap & & \\ (U,t) & & \end{array}$$

Using an obvious induction we may now assume that $(W,s)\nearrow(Y,q)$ and that $f=\rho$ is a retraction as above. A subsidiary induction on the number of bc cells in $(U,t)-(Y,q)$ allows us to assume that (U,t) has the form $(U,t) = (Y,q)\cup_{\phi_r}(S\times I^r,q_r)$ for some finite bc r-cell $(S\times I^r,q_r)$ with $q\phi_r = q_r|S\times\partial I^r$. Now $\rho\phi_r$ is also an attaching map for a bc r-cell $(S\times I^r,q_r)$ and $(W,s)\cup_\rho(U,t) = (W,s)\cup_{\rho\phi_r}(S\times I^r,q_r)$. If $j\colon (W,s) \longrightarrow (Y,q)$ denotes the inclusion then

$$(Y,q)\cup_j((W,s)\cup_\rho(U,t)) = (Y,q)\cup_{j\rho\phi_r}(S\times I^r,q_r).$$

But $j\rho\phi_r$ is cellularly homotopic to ϕ_r; so by 4.5 we have

(4.11) $$(Y,q)\cup_j((W,s)\cup_\rho(U,t))\nearrow\!\!\searrow(U,t).$$

Next apply the relativity principle, 3.10, to the diagram

$$\begin{array}{ccc} (W,s) & = & (W,s)\nearrow_e(Y,q) \\ \cap| & & \\ (W,s)\cup_\rho(U,t) & & \end{array}$$

to conclude that

(4.12) $$(W,s)\cup_\rho(U,t)\nearrow(Y,q)\cup_j((W,s)\cup_\rho(U,t)).$$

It follows from (4.11) and (4.12) that $(W,s)\cup_\rho(U,t)\nearrow\!\!\searrow(U,t)$ (actually rel(W,s) and not just rel(X,p)).

Remark: One cannot apply the relativity principle to the diagram

$$\begin{array}{ccc} (X,p) & \leq & (Y,q)\nearrow\!\!\searrow(W,s) \text{ rel}(X,p) \\ & & \cap \\ & & (U,t) \end{array}$$

to get the conclusion of 4.6. The reason is that the formal deformation leading from (Y,q) to (W,s) may change part of the domain of the inclusion $(Y,q)\subset(U,t)$.

5. The Whitehead torsion of a bc homotopy equivalence.

Let f_0: (X,p) $\longrightarrow$ (Y,q) be a homotopy equivalence in $\mathscr{CW}^c_\ell/Z$. By II.10.3 there is a cellular map f: (X,p) $\longrightarrow$ (Y,q) which is homotopic to f_0 in $\mathscr{CW}^c_\ell/Z$. We then have the pair $(M(f),X,q\rho) \in DR^c(X,qf)$ and the identity map i: (X,qf) $\longrightarrow$ (X,p) is bc (cf. II.1.9).

Definition 5.1 *The Whitehead torsion of the homotopy equivalence* f_0: (X,p) $\longrightarrow$ (Y,q) *is defined to be*

$$\tau(f_0) = i_*[M(f),X,q\rho] \in Wh^c(X,p).$$

To see that $\tau(f_0)$ is well defined, we need the first part of the following lemma whose proof is deferred:

Lemma 5.2 *If* $h\colon f \simeq g\colon (X,p) \longrightarrow (Y,q)$ *is a cellular homotopy in* $\mathcal{CW}^c_{\ell}/Z$ *then* $i_*[M(f),X,q\rho] = j_*[M(g),X,q\rho]$ *where* $j = 1\colon (X,qg) \longrightarrow (X,p)$.

If (Y,X,q) *is a pair of finite bc CW complexes and* $f\colon (X,p) \longrightarrow (Y,q)$ *is the inclusion, then* $[M(f),X,q\rho] = [Y,X,q]$.

A homotopy equivalence f_0 is called *simple* if $\tau(f_0)=0$. There is the following description of the simple homotopy equivalences (cf. [Co, 22.2]).

Proposition 5.3 *The bc homotopy equivalence* f_0 *is simple if and only if there exist expansions* $(X,p)\nearrow(U,t)$ *and* $(Y,q)\nearrow(U,t')$ *such that* $1_U\colon (U,t) \longrightarrow (U,t')$ *is* bc *and the diagram*

$$\begin{array}{ccc} (X,p) & \leq & (U,t) \\ {\scriptstyle f_0}\big\downarrow & & \big\downarrow{\scriptstyle 1} \\ (Y,q) & \leq & (U,t') \end{array}$$

homotopy commutes.

Again the proof is deferred.

One has the usual formula for compositions of homotopy equivalences (cf. [Co, 22.4]).

Proposition 5.4 *Let* $(X,p) \xrightarrow{f} (Y,q) \xrightarrow{g} (W,s)$ *be homotopy equivalences in* $\mathcal{CW}^c_{\ell}/Z$. *Then* $f_*\tau(gf) = \tau(g)+f_*\tau(f)$.

The proof is given later in this section.

The classical sum theorem (see e.g. [Co, 23.1]) has the following counterpart.

Proposition 5.5 *Let* $f_i\colon (X_i,p_i) \longrightarrow (Y_i,q_i)$ *be homotopy equivalences in* $\mathscr{CW}^c_{\mathscr{L}}/Z$ *for* $i=1,2$. *Assume that* $(X,p) = (X_1,p_1)\cup(X_2,p_2)$, $(Y,q) = (Y_1,q_1)\cup(Y_2,q_2)$ *with* $(X_0,p_0) = (X_1,p_1)\cap(X_2,p_2)$ *and* $(Y_0,q_0) = (Y_1,q_1)\cap(Y_2,q_2)$ *and that* $f_1|X_0 = f_2|X_0 = f_0\colon (X_0,p_0) \longrightarrow (Y_0,q_0)$ *is a homotopy equivalence. Then* $f = f_1\cup f_2\colon (X,p) \longrightarrow (Y,q)$ *is a homotopy equivalence and*

$$\tau(f) = j_{1*}\tau(f_1)+j_{2*}\tau(f_2)-j_{0*}\tau(f_0)$$

where $j_\nu\colon (X_\nu,p_\nu) \leq (X,p)$.

The rest of this section gives proofs of the above results.

Proof of 5.2: Let $k = i^{-1}j\colon (X,qg) \longrightarrow (X,qf)$. We must show that $k_*[M(g),X,q\rho] = [M(f),X,q\rho]$. By 3.12 one has

$$(5.6)\qquad (M(f),q\rho)\nearrow\!\searrow(M(g),q') \quad \text{rel } (X,qf)\cup(Y,q)$$

where $1\colon (M(g),q') \longrightarrow (M(g),q\rho)$ is a bc homeomorphism. Moreover, from the proof of 3.12, we see that $q'|X\times\{0\} = qf$. Therefore, the diagram

$$\begin{array}{ccc} (X,qg) & \leq & (M(g),q\rho)) \\ \big\downarrow k=1_X & & \\ (X,qg) & & \end{array}$$

has $(X,qf)\cup_k(M(g),q') = (M(g),q')$ as its pushout. Hence

$$(5.7)\qquad k_*[M(g),X,q\rho] = [M(g),X,q'].$$

But by 3.4, $[M(g),X,q'] = [M(f),X,q\rho]$. This finishes the proof of the first part.

Next let f denote an inclusion $(X,p) \leq (Y,q)$. Then, by 3.11 (iii), $(M(f),q\rho) = (X\times I\cup Y\times\{1\},p\pi\cup q)\nearrow(Y\times I,q\pi)\searrow(Y,q)$ rel(X,p) where $Y = Y\times 0$. This proves the latter part of the lemma.

Proof of 5.3: For the "if part" let $i: (X,p) \longrightarrow (U,t)$, and $j: (Y,q) \longrightarrow (U,t')$ be the inclusions and let $r: (U,t') \longrightarrow (Y,q)$ be a cellular retraction. By II.1.9 $1_U: (U,t') \longrightarrow (U,qr)$ is bc, so we may assume that $t'=qr$. Then $f = ri: (X,p) \longrightarrow (Y,q)$ is cellular and homotopic to f_0. Thus we only need to show that $[M(ri),X,q\rho] = 0$ in $Wh^c(X,qri)$. Since $(X,qri) \cong (X,p)$ and $(U,t') \cong (U,qr)$, there is the homotopy equivalence $i: (X,qri) \longrightarrow (U,qr)$ with $i_*[M(ri),X,q\rho] = [U\cup_X M(ri),U,qr\cup q\rho]$, and it suffices to show that this element vanishes. This will be done in 4 steps. First $(U\cup_X M(ri),qr\cup q\rho) \nearrow (M(r),q\rho)$ by 3.11.(iv). Next, 3.10 applies to the diagram

$$\begin{array}{ccc} (Y,q) & = & (Y,q)\nearrow(U,qr) \\ \cap & & \\ (M(r),q\rho) & & \end{array}$$

to give $(M(r),q\rho) \curvearrowright (M(r)\cup_Y U,q\rho\cup qr) = (M(jr),qr\rho)$.

For the third step use the fact that $jr \simeq 1: (U,qr) \longrightarrow (U,qr)$ to conclude from 3.12 that

$$(M(jr),qr\rho) \curvearrowright (M(1),q') = (U\times I,q') \ \mathrm{rel}(U\times\{0,1\},qr\cup qr)$$

where $1_{U\times I}: (U\times I,qr\pi) \longrightarrow (U\times I,q')$ is an isomorphism in $\mathscr{CW}^c_{\ell}/Z$.

It follows from by 3.4, that $(U\times I,qr\pi) \curvearrowright (U\times I,q') \searrow (U,qr)$ rel(U,qr), completing the proof of the "if part" of 5.3.

The "only if" part is proved as follows. With the notation above, if $\tau(f_0) = 0$, then there is a cellular map $f \simeq f_0$ such that

$$(X,qf) \curvearrowright (M(f),q\rho) \ \mathrm{rel}(X,qf).$$

Thus we have expansions $(X,qf) \nearrow (U,t') \searrow (M(f),q\rho)$ such that

$$\begin{array}{ccc} (X,qf) & \leq & (M(f),q\rho) \\ \downarrow f & & \cap \\ (Y,q) & \leq & (U,t') \end{array}$$

homotopy commutes, and all the inclusions are expansions.

Now pick a retraction $r': (U,t') \longrightarrow (X,qf)$. The composition $(U,t') \xrightarrow{r'} (X,qf) \xrightarrow{1} (X,p)$ is bc so II.1.9 implies that $1_U: (U,t') \longrightarrow$

(U,t) is a bc homeomorphism where $t = pr'$. Also $t|X=p$ so $(X,p) \nearrow (U,t)$ and the proof is finished.

Proof of 5.4: If $f\colon (X,p) \longrightarrow (Y,q)$ is an inclusion, we write $[f]$ for $[(Y,q),(X,p)] \in Wh^c(X,p)$. We first prove 5.4 when both f and j are inclusions. In this case, there is the pushout diagram

$$\begin{array}{ccccc} (X,p) & \xrightarrow{f} & (Y,q) & \xrightarrow{g} & (W,s) \\ \downarrow{\scriptstyle f} & & \downarrow{\scriptstyle f_2} & & \downarrow{\scriptstyle f_3} \\ (Y,q) & \xrightarrow[f_1]{} & (U,u) & \xrightarrow[g_1]{} & (V,v) \end{array}$$

and

$$f_*\tau(f)+\tau(f) = f_*[f]+[g] = [f_2]+[g] = [g_1f_2];$$

while

$$f_*\tau(gf) = f_*[gf] = [g_1f_1].$$

But $g_1f_1f = g_1f_2f$ and f is a homotopy equivalence. Hence $g_1f_2 \simeq g_1f_2$ so $[g_1f_1] = \tau(g_1f_1) = \tau(g_1f_2) = [g_1f_2]$. This finishes the proof in the special case.

To treat the general case one may assume that qf=p and sg=q. One has the homotopy commutative diagram

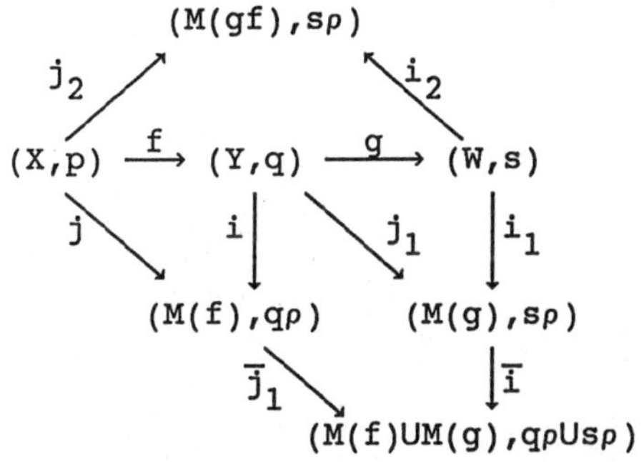

where the diamond is a pushout. Now i_* is an isomorphism and (using the special case)

$$\begin{aligned} i_*(f_*\tau(f)+\tau(g)) &= j_*([j])+i_*([j_1]) = j_*([j])+[\bar{j}_1] \\ &= j_*([\bar{j}_1j]) = i_*f_*[\bar{j}_1j] \end{aligned}$$

so $\tau(f)+f_*^{-1}\tau(g) = [\bar{j}_1 j]$. Since one also has $\tau(gf) = [j_2]$, one needs only to prove that

$$(M(f)\cup_Y M(g), q\rho \cup s\rho) \curvearrowright (M(gf), s\rho).$$

To do so, define $G: (M(f),q\rho) \longrightarrow (W,s)$ by

$$G([x,t]) = gf(x)\ ,\quad x\in X\ ,\ t\in I$$

$$G(y) = g(y)\ ,\quad y\in Y.$$

Then G is a bc map and we may form $(M(G),s\rho)$. From 3.11 (iv), we see that $(M(f)\cup_Y M(g), q\rho\cup s\rho) \nearrow (M(G),s\rho)$ and from 3.11 (i), we see that $(M(gf),s\rho) \nearrow (M(G),s\rho)$. This finishes the proof of 5.4.

Proof of 5.5: We leave it to the reader to check that f is a bc homotopy equivalence. In the proof of the addition formula we may assume that $qf=p$ (so that also $q_i f_i = p_i$ for $i=0,1,2$, since $q_i = q|Y_i$ and $p_i = p|X_i$). There are the inclusions ($\upsilon=1,2$)

$$\begin{array}{ccccc}
(X,p) & \xrightarrow{k_0} & (M(f_0)\cup_{X_0} X, q_0\rho\cup p) & \xrightarrow{k_\upsilon} & (M(f_\upsilon)\cup_{X_\upsilon} X, q_\upsilon\rho\cup p) \\
 & & & \searrow^{\ell_0} & \downarrow \\
 & & & & (M(f),q\rho)
\end{array}$$

and by definition

$$(5.8)\qquad j_{0*}\tau(f_0) = [k_0];\ j_{\upsilon *}\tau(f_\upsilon) = [k_\upsilon k_0],\ (\upsilon=1,2);\ \tau(f) = [\ell_0 k_0].$$

Also by definition

$$(5.9)\qquad [k_1]+[k_2] = [\ell_0].$$

Finally, from 5.4 one has

$$(5.10)\qquad k_{0*}[k_0]+[\ell_0] = k_{0*}[\ell_0 k_0];$$

$$k_{0*}[k_0]+[k_\upsilon] = k_{0*}[k_\upsilon k_0]\ ,\ (\upsilon=1,2).$$

By combining (5.8)-(5.10), one gets the desired formula.

CHAPTER IV

FREE AND PROJECTIVE RPG MODULES

THE ALGEBRAIC WHITEHEAD GROUPS OF RPG

It is the main purpose of this chapter to introduce an algebraic Whitehead group, Wh($R\mathcal{P}G$), derived from the category of $R\mathcal{P}G$ modules in essentially the same way that the classical Whitehead group, Wh(π), is obtained from the category of $\mathbb{Z}[\pi]$ modules.

In section 1 we work in the generality of $R\mathcal{B}$ modules where $\mathcal{B}$ is an arbitrary category with endomorphism. We introduce the notion of a *free* $R\mathcal{B}$ module and a *basis* for a free $R\mathcal{B}$ module. The projective objects of $R\mathcal{B}$-*mod* admit the usual characterization as summands of free $R\mathcal{B}$ modules (see Corollary 1.4). Also free (and hence projective) objects behave well under functors of the type $\phi_!$: $R\mathcal{B}$-*mod* $\longrightarrow$ $R\mathcal{B}'$-*mod* (see Propositions 1.9 and 1.11).

In section 2 we specialize to $R\mathcal{B}$ modules where $\mathcal{B}$ has the form $\mathcal{P}G$ for some space Z with a boundedness control structure (P,C) and some functor G: $\mathcal{P} \longrightarrow \mathcal{G}poid$. Generalizing the notion of a finitely generated $\mathbb{Z}[\pi]$ module we introduce the notion of an $R\mathcal{P}G$ module being *boundedly finitely generated* (*bfg*, for short) (see Definitions 2.2 and 2.4). The main result is Theorem 2.6 which states that two bfg based $R\mathcal{P}G$ modules are isomorphic if and only if their bases are isomorphic. This result is crucial later on when we want to turn isomorphisms of bfg based modules (which the geometry is willing to deliver) into automorphisms (which the algebra is willing to accept). We also study the behaviour of bfg modules viz-a-viz functors of the form $\phi_!$: $R\mathcal{P}G$-*mod* $\longrightarrow$ $R\mathcal{P}'G'$-*mod* (see Proposition 2.7).

In section 3 we define the groups $K_1(R\mathcal{P}G)$ and Wh($R\mathcal{P}G$). For the first one we use Bass's definition. The second one is given as a quotient of the first (see Definition 3.7). In the very short section 3^*, we point out that Quillen's Q-construction, as well as his $\mathcal{S}^{-1}\mathcal{S}$ construction, permit one to define $K_i(R\mathcal{P}G)$ for any $i \geq 0$.

The functoriality of $K_1(R\mathcal{P}G)$ and $Wh(R\mathcal{P}G)$ is studied in section 4. Proposition 4.1 summarizes the properties of a functor ϕ: $\mathcal{P}G \longrightarrow \mathcal{P}'G'$ which are needed to guarantee that there are induced homeomorphisms on $K_1(R(\ ))$ and on $Wh(R(\ ))$. The general results from chapter I concerning equivalences of categories of the form $\phi_!$: $R\mathcal{B}$-*mod* $\longrightarrow$ $R\mathcal{B}'$-*mod* are used to give, in Theorem 4.4, sufficient conditions on a bc map $f: (X,p) \longrightarrow (Y,q)$ to ensure that the categories $R\mathcal{P}G_1(X,p)$-*mod* and $R\mathcal{P}G_1(Y,q)$-*mod* are equivalent. Essentially the conditions state that f must induce an isomorphism on π_0^c and π_1^c. In Theorem 4.7 we show that somewhat weaker conditions on ϕ: $\mathcal{P}G \longrightarrow \mathcal{P}'G'$ suffice to guarantee that the categories of bfg free modules over $R\mathcal{P}G$ and $R\mathcal{P}'G'$ are equivalent (and hence that $K_1(R\mathcal{P}G) \cong K_1(R\mathcal{P}'G')$ and $Wh(R\mathcal{P}G) \cong Wh(R\mathcal{P}'G')$). In Example 4.11 two corollaries of this result (see 4.8 and 4.10) are applied to the case where $Z = \mathbb{R}^k$. The outcome is that there is a wide range of choices for $\mathcal{P}$ which leaves the category of bfg free $R\mathcal{P}G_1(X,p)$ modules essentially unchanged.

Finally, in section 5, we generalize the classical involutions on $K_1(\mathbb{Z}[\pi])$ and $Wh(\mathbb{Z}[\pi])$ to *involutions on* $K_1(R\mathcal{P}G)$ *and* $Wh(R\mathcal{P}G)$, see Theorem 5.2. For this one also needs a *generalized Stiefel-Whitney class*, which is defined in 5.1. The involution will be needed in our proof of the Realization Theorem in section VI.7.

1. Free and projective $R\mathscr{B}$ modules.

Let R be a ring and let $\mathscr{B}$ be a category with endomorphism. We shall introduce and study free (and projective) $R\mathscr{B}$ modules. Any free $R\mathscr{B}$ module has a *basis* and we first must explain in which category the basis is to live.

We start by defining a category with endomorphism $\mathscr{B}as_*(\mathscr{B})$. An object is a pair (S,σ) where S is a set and $\sigma\colon S \longrightarrow |\mathscr{B}|$ is a function. To define the morphisms we shall think of S as the object set of a category (also called S) whose morphism set is $\{1_s \mid s\in S\}$. With this interpretation $\sigma\colon S \longrightarrow \mathscr{B}$ is a functor. A morphism $(\alpha,\nu,n)\colon (S,\sigma) \longrightarrow (T,\rho)$ consists of a functor (i.e. a function) $\alpha\colon S \longrightarrow T$, an integer n and a natural transformation $\nu\colon \rho\alpha \longrightarrow C^n\sigma$. Note that ν simply is a collection of morphisms in $\mathscr{B}$, $\nu(s)\colon \rho\alpha(s) \longrightarrow C^n\sigma(s)$, with no compatibility conditions. The composition is given by the formula

$$(\beta,\mu,m)(\alpha,\nu,n) = (\beta\alpha,(C^m\nu)(\mu\alpha),m+n)$$

$\mathscr{B}as_*(\mathscr{B})$ becomes a category with endomorphism when one lets C be the identity functor and defines $\tau\colon \mathrm{Id} \longrightarrow C = \mathrm{Id}$ by the formula

$$\tau(S,\sigma) = (1_S,\tau\sigma,1)\colon (S,\sigma) \longrightarrow (S,\sigma).$$

We define $\mathscr{B}as(\mathscr{B})$ to be the category $\mathscr{B}as_*(\mathscr{B})(\Sigma^{-1})$. Thus a basis (over $\mathscr{B}$) is a pair (S,σ) as above and a morphism $(\alpha,\nu)\colon (S,\sigma) \longrightarrow (T,\rho)$ is represented by a map $\alpha\colon S \longrightarrow T$ and a natural transformation $\nu_n\colon \rho\alpha \longrightarrow C^n\sigma$; the equivalence relation is generated by declaring (α,ν_n) equivalent to $(\alpha,\tau\nu_n)$ where $\tau\colon C^n\sigma \longrightarrow C^{n+1}\sigma$.

Let (S,σ) be a basis over $\mathscr{B}$. The *free* $R\mathscr{B}$ *module*, $F(\sigma)$, with basis (S,σ) is defined as follows (recall that $F(\sigma)$ is to be a functor $F(\sigma)\colon \mathscr{B} \longrightarrow R\text{-}\mathit{mod}$).

(1.1) For any $b \in |\mathscr{B}|$, $F(\sigma)(b)$ is the free R module on $\{(\beta,s) \mid s\in S,\ \beta\in\mathscr{B}(\sigma(s),b)\}$.

(1.2) For any $\gamma \in \mathcal{B}(b,c)$, $\gamma_*: F(\sigma)(b) \longrightarrow F(\sigma)(c)$ has $\gamma_*(\beta,s) = (\gamma\beta,s)$.

Clearly $F(\sigma)$ does belong to $R\text{-}mod^{\mathcal{B}}$. It is free in the following sense:

Proposition 1.3 *Let* $M \in |R\mathcal{B}\text{-}mod|$. *For any integer* $d \geq 0$ *and any collection of elements* $m(s) \in M(C^d\sigma(s))$, *there is a unique morphism* $\varphi: F(\sigma) \longrightarrow M$ *represented by a natural transformation* $\varphi_d: F(\sigma) \longrightarrow MC^d$ *with*

$$\varphi_d(\sigma(s))(1,s) = m(s) \in MC^d(\sigma(s)) \ , \ s \in S.$$

Another integer $d' \geq 0$ *and another collection* $m'(s) \in M(C^{d'}\sigma(s))$, $s \in S$, *will define the same morphism precisely if there is some* $n \geq \max\{d,d'\}$ *such that* $m(s)$ *and* $m'(s)$ *meet under the maps*

$$M(C^d\sigma(s)) \xrightarrow[\tau_*^{n-d}]{} M(C^n\sigma(s)) \xleftarrow[\tau_*^{n-d'}]{} M(C^{d'}\sigma(s))$$

for all $s \in S$. *Finally, any morphism* $\varphi: F(\sigma) \longrightarrow M$ *arises in this way.*

The proof is obvious and is left to the reader.

Abuse of notation When using the above proposition to define $\varphi: F(\sigma) \longrightarrow M$ we shall often write $\varphi(1,s) = m(s)$ instead of $\varphi_d(\sigma(s))(1,s) = m(s)$.

Corollary 1.4 *Any* $F(\sigma)$ *is a projective object in* $R\mathcal{B}G\text{-}mod$. *For any* $R\mathcal{B}G$ *module* M, *there exist a basis* (S,σ) *and an epimorphism* $\varphi: F(\sigma) \longrightarrow M$. *Finally,* M *is projective if and only if* M *is a direct summand of some* $F(\sigma)$.

Proof: Consider a diagram

$$\begin{array}{ccccc} & & F(\sigma) & & \\ & & \big\downarrow \phi & & \\ M & \xrightarrow{\ \mu\ } & N & \longrightarrow & 0 \end{array}$$

(in $R\mathscr{B}$-*mod*) with exact row. It is represented by a diagram in R-$mod^{\mathscr{B}}$

$$\begin{array}{ccc} & & F(\sigma) \\ & & \big\downarrow \phi_n \\ M & \xrightarrow{\ \mu_n\ } & NC^n \end{array}$$

Since μ is epic, there is some $p \in \mathbb{Z}_+$ such that for each $s \in S$ one can find $m(s) \in MC^p(\sigma(s))$ with $m(s)$ and $(1,s)$ meeting under the maps

$$MC^p(\sigma(s)) \xrightarrow[\mu_n]{} NC^{n+p}(\sigma(s)) \ ,$$

respectively

$$F(\sigma)(\sigma(s)) \xrightarrow[\phi_n]{} NC^n(\sigma(s)) \xrightarrow[\tau^p]{} NC^{n+p}(\sigma(s)).$$

By 1.3, the integer p and the collection m(s) define a morphism $\hat{\phi}$: $F(\sigma) \longrightarrow M$. One readily checks that $\mu\hat{\phi} = \phi$. Thus $F(\sigma)$ is projective.

Next let M be an arbitrary $R\mathscr{B}$ module. For each $b \in |\mathscr{B}|$, let S_b be a set of R module generators for M(b), S be the disjoint union of all S_b, and define σ: $S \longrightarrow |\mathscr{B}|$ to have $\sigma(S_b) = b$. The collection of elements $m(s) = s \in M(\sigma(s))$ defines a morphism ϕ: $F(\sigma) \longrightarrow M$ which is obviously epic.

The last part of 1.4 is now trivial (and well known).

Now let $(\alpha,\upsilon,n) \in \mathscr{B}as_*(\mathscr{B})((S,\sigma),(T,\rho))$. For each $s \in S$, one gets the element $m(s) = (\upsilon(s),\alpha(s)) \in F(\rho)(C^n\sigma(s))$. By 1.3, there results a morphism of $R\mathscr{B}$ modules $F(\alpha,\upsilon,n)$: $F(\sigma) \longrightarrow F(\rho)$ and a straightforward naturality argument shows that this definition makes F: $\mathscr{B}as_*(\mathscr{B}) \longrightarrow R\mathscr{B}$-*mod* into a functor. Also, for any (S,σ)

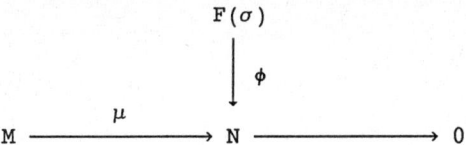

$$F(\tau_{(S,\sigma)}) = F(1,\tau\sigma,1)\colon F(\sigma) \longrightarrow F(\sigma)$$

is given by the elements $(\tau\sigma(s),s) \in F(\sigma)(C\sigma(s))$. Thus $F(\tau_{(S,\sigma)}) = 1_{F(\sigma)}$. It follows that F induces a functor (which we also call) $F\colon \mathscr{B}as(\mathscr{B}) \longrightarrow R\mathscr{B}\text{-}mod$ such that the following diagram commutes:

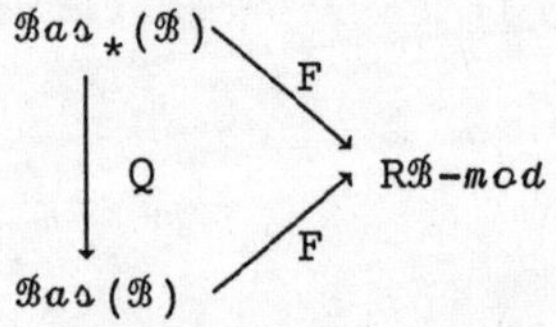

Remark On objects Q is the identity so there is no confusion as to the meaning of $F(\sigma)$. For morphisms confusion is excluded because morphisms in $\mathscr{B}as_*(\mathscr{B})$ carry 3 components; those in $\mathscr{B}as(\mathscr{B})$, only 2. We explicitly mention that when $(\alpha,\nu)\colon (S,\sigma) \longrightarrow (T,\rho)$ has $\nu_n\colon \rho\alpha \longrightarrow C^n\sigma$ as a representative, then $F(\alpha,\nu)\colon F(\sigma) \longrightarrow F(\rho)$ is given by the formula

(1.5) $\quad F(\alpha,\nu)(1,s) = (\nu(s),\alpha(s)) \in F(\rho)(C^n\sigma(s))\ ,\ s\in S.$

Example 1.6 Let (X,p) be a bc CW complex over Z, $X^{(n)}$ be the n-skeleton, and S_n be the set of n-cells. For each $e\in S_n$, pick a point $x_e \in \bar{e}-e$ and a $K_e \in |\mathscr{P}|$ such that, for some $d \geq 0$ and for all $e\in S_n$

(1.7) $\quad \{K_e \mid e\in S_n\}$ is bounded, $p(x_e) \in K_e$, and $p(\bar{e}) \subseteq C^dK_e$.

Letting $\sigma_n(e) = (x_e,K_e) \in \mathscr{P}G_1(X^{(n-1)},p)$ gives us a basis over $\mathscr{P}G_1(X^{(n-1)},p)$. In V.2.5.(ii) we shall show that the corresponding free $\mathscr{P}G_1(X^{(n-1)},p)$ module is isomorphic to $\pi_n^c(X^{(n)},X^{(n-1)},p)$ when $n\geq 3$. We remark that it is easy to prove that (S_n,σ_n) is well defined up to an isomorphism in $\mathscr{B}as(\mathscr{P}G_1(X^{(n-1)},p))$, at least if $n\geq 2$.

We next study the behaviour of free (and projective) $R\mathscr{B}$ modules viz-a-viz functors of the type $\phi_!\colon R\mathscr{B}\text{-}mod \longrightarrow R\mathscr{B}'\text{-}mod$ where $\phi\colon \mathscr{B} \longrightarrow \mathscr{B}'$ is assumed to be endomorphism preserving.

Recall from I.5 that for such a ϕ there are natural transformations α and β such that

$$\begin{array}{ccc} \phi & \xrightarrow{\phi\tau} & \phi C \\ & {\scriptstyle \tau^n\phi}\searrow & \downarrow{\scriptstyle \alpha} \\ & & C^n\phi \end{array} \qquad \text{and} \qquad \begin{array}{ccc} \phi & \xrightarrow{\tau\phi} & C\phi \\ & {\scriptstyle \phi\tau^m}\searrow & \downarrow{\scriptstyle \beta} \\ & & \phi C^m \end{array}$$

commute. This defines a morphism

$$(1.8) \qquad A_{(S,\sigma)} = (1,\alpha\sigma,n)\colon (S,\phi\sigma) \longrightarrow (S,\phi C\sigma).$$

Proposition 1.9 *Any endomorphism preserving functor* $\phi\colon \mathcal{B} \longrightarrow \mathcal{B}'$ *induces functors* ϕ_* *and* $\phi_!$ *such that*

$$\begin{array}{ccc} \mathcal{B}as_*(\mathcal{B}) & \xrightarrow{\phi_*} & \mathcal{B}as(\mathcal{B}') \\ \downarrow{\scriptstyle Q} & & \downarrow{\scriptstyle Q} \\ \mathcal{B}as(\mathcal{B}) & \xrightarrow[\phi_!]{} & \mathcal{B}as(\mathcal{B}') \end{array}$$

commutes. In fact ϕ_* *is given by the formula*

$$\phi_*[(\gamma,\upsilon,d)\colon (S,\sigma) \longrightarrow (T,\rho)] = [(\gamma,\phi(\upsilon),0)A^{(d)}_{(S,\sigma)}\colon (S,\phi\sigma) \longrightarrow (T,\phi\rho)]$$

where $A^{(d)}_{(S,\sigma)}$ is defined inductively by

$$A^{(d)}_{(S,\sigma)} = A_{(S,C^{d-1}\sigma)} A^{(d-1)}_{(S,\sigma)}\colon (S,\phi\sigma) \longrightarrow (S,\phi C^d\sigma)$$

and $(\gamma,\phi(\upsilon),0)\colon (S,\phi C^d\sigma) \longrightarrow (T,\phi\rho)$.

Proof: Naturality of α is all that is needed to verify that ϕ_* is functorial. Since $\phi_*(\tau_{(S,\sigma)}) = \tau^{(n)}_{\phi_*(S,\sigma)}$, ϕ_* covers a unique functor $\phi_!$.

Remark 1.10 In more pedestrian terms $\phi_!$ is given as follows: On objects $\phi_!(S,\sigma) = (S,\phi\sigma)$. If $(\gamma,\nu)\colon (S,\sigma) \longrightarrow (T,\rho)$ is represented by $\nu_d\colon \rho\gamma \longrightarrow C^d\sigma$ then $\phi_!(\gamma,\nu) = (\gamma,\lambda)$ with λ represented by the natural transformation

$$\phi\rho\gamma \xrightarrow{\phi(\nu_d)} \phi C^d\sigma \xrightarrow{\alpha^{(d)}\sigma} C^{nd}\phi\sigma$$

Here $\alpha^{(d)}$ is the obvious iteration of α. Although it seems as if $\phi_!$ might depend not only on ϕ but also on the choice of α, this is not so. In fact, $A_{(S,\sigma)}$ fits into a commutative diagram

$$\begin{array}{ccccc}
(S,\phi C\sigma) & \xrightarrow{(1,C^n\phi\tau\sigma,0)} & (S,\phi\sigma) & & \\
& \searrow^{\tau^{(n)}_{(S,\phi C\sigma)}} & \downarrow{\scriptstyle A_{(S,\sigma)}} & \searrow^{\tau^{(n)}_{(S,\phi\sigma)}} & \\
& & (S,\phi\sigma) & \xrightarrow[(1,\phi\tau\sigma,0)]{} & (S,\phi\sigma)
\end{array}$$

(cf. the proof of I.3.2). Since the two diagonal maps become isomorphisms under Q, so does $A_{(S,\sigma)}$, and $Q(A_{(S,\sigma)}) = Q(\tau^{(n)}_{(S,\phi\sigma)})^{-1}Q(1,\phi\tau\sigma,0)$ is independent of the choice of α.

Proposition 1.11 *For any endomorphism preserving functor* $\phi\colon \mathfrak{B} \longrightarrow \mathfrak{B}'$, *the diagram of functors*

$$\begin{array}{ccc}
\mathfrak{B}\mathit{as}(\mathfrak{B}) & \xrightarrow{F} & R\mathfrak{B}\text{-}\mathit{mod} \\
\downarrow{\scriptstyle \phi_!} & & \downarrow{\scriptstyle \phi_!} \\
\mathfrak{B}\mathit{as}(\mathfrak{B}') & \xrightarrow[F]{} & R\mathfrak{B}'\text{-}\mathit{mod}
\end{array}$$

commutes up to a natural isomorphism. Consequently, $\phi_!$ *preserves projectives.*

Proof: In view of the commutative diagram

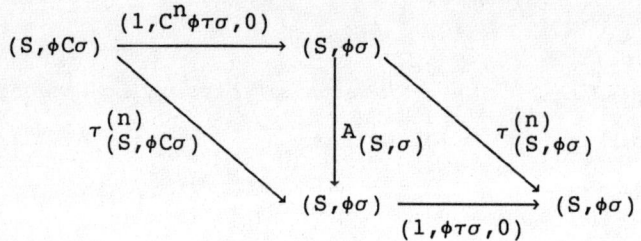

$$\begin{array}{ccccccc}
\mathcal{B}as_*(\mathcal{B}) & & & \xrightarrow{\phi_*} & & & \mathcal{B}as_*(\mathcal{B}') \\
\Big\downarrow F & \searrow Q & & & & Q \swarrow & \Big\downarrow F \\
 & & \mathcal{B}as(\mathcal{B}) & \xrightarrow{\phi_!} & \mathcal{B}as(\mathcal{B}') & & \\
 & \swarrow F & & & & F \searrow & \\
R\mathcal{B}\text{-}mod & & & \xrightarrow{\phi_!} & & & R\mathcal{B}'\text{-}mod
\end{array}$$

and the universal property of Q, it suffices to show that $F\phi_*$ and $\phi_!F$ are naturally isomorphic functors from $\mathcal{B}as_*(\mathcal{B})$ to $R\mathcal{B}'\text{-}mod$. Let (S,σ) be any basis over $\mathcal{B}$. We must compare $\phi_!F(\sigma)$ and $F(\phi_!\sigma) = F(\phi\sigma)$. Let us write $M = F(\sigma)$, $N = F(\phi\sigma)$, and let

$$\eta = \eta(M)\colon M \longrightarrow \phi^*\phi_*M \quad \text{and} \quad \varepsilon = \varepsilon(N)\colon \phi_*\phi^*N \longrightarrow N$$

be the unit and counit for the adjoint pair (ϕ_*,ϕ^*). There are natural transformations (i.e. morphisms in $R\text{-}mod^{\mathcal{B}}$, respectively, in $R\text{-}mod^{\mathcal{B}'}$)

(1.12) $\mu\colon M \longrightarrow \phi^*N = N\phi$, and $\nu\colon N \longrightarrow \phi_*M$

given by (cf. Proposition 1.3 and the abuse of notation following it)

(1.13) $\mu(1,s) = (1,s) \in N\phi(\sigma(s)) = F(\phi\sigma)(\phi\sigma(s))$, $s\in S$

(1.14) $\nu(1,s) = \eta(1,s) \in \phi^*\phi_*M(\sigma(s)) = \phi_*M(\phi\sigma(s))$, $s\in S$.

(Here η is short for $\eta(\sigma(s))\colon M(\sigma(s)) \longrightarrow \phi^*\phi_*M(\sigma(s))$.)

We define $\hat{\mu}\colon \phi_*M \longrightarrow N$ to be the adjoint of μ and claim that $\hat{\mu}$ and ν are inverses. We note that $\hat{\mu}$ is given as the composition

$$\phi_*M \xrightarrow[\phi_*\mu]{} \phi_*\phi^*N \xrightarrow[\varepsilon]{} N$$

To check that $\hat{\mu}\nu = 1$ we consider the diagram

$$\begin{array}{ccccccc}
N(\phi\sigma(s)) & \xrightarrow{\nu} & \phi_*M(\phi\sigma(s)) & \xrightarrow{\phi_*\mu} & \phi_*\phi^*N(\phi\sigma(s)) & \xrightarrow{\varepsilon} & N(\phi\sigma(s)) \\
 & & \| & & \| & & \\
 & & \phi^*\phi_*M(\sigma(s)) & \xrightarrow[\phi^*\phi_*\mu]{} & \phi^*\phi_*\phi^*N(\sigma(s)) & & \\
 & & \Big\uparrow \eta & & \Big\uparrow \eta(\phi^*N) & & \\
 & & M(\sigma(s)) & \xrightarrow[\mu_0]{} & \phi^*N(\sigma(s)) & &
\end{array}$$

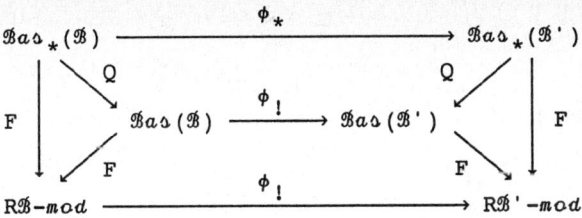

where all arrows are natural transformations to be evaluated at $\sigma(s)$ or $\phi\sigma(s)$. Now natural transformations $N \to N$ are determined by the action on $(1,s) \in N(\phi\sigma(s))$, $s\in S$. Hence we must check that the upper row maps $(1,s)$ to itself. By (1.13) and (1.14), this reduces to showing that $(1,s) \in \phi^*N(\sigma(s))$ maps to itself under $\varepsilon\eta(\phi^*N)$. The full name of this ε is $\varepsilon(N)(\phi\sigma(s)) = \phi^*(\varepsilon(N))(\sigma(s))$ and the composite natural transformation $\phi^*(\varepsilon(N))\cdot\eta(\phi^*N)$ is the identity natural transformation by general facts about adjunctions (cf. [Ma, p. 80]).

We proceed to show that $\nu\hat{\mu} = 1$. Fix some $b' \in |\mathcal{B}'|$. We must prove that the endomorphism of $\phi_*M(b')$ in the diagram

$$\begin{array}{ccccc}
\phi_*M(b') & \xrightarrow{\phi_*(\mu)(b')} & \phi_*\phi^*N(b') & \xrightarrow{\varepsilon(b')} & N(b') \\
\uparrow{\scriptstyle c(b,\beta')} & & & & \downarrow{\scriptstyle \nu(b')} \\
M(b) & & \xrightarrow{\qquad c(b,\beta') \qquad} & & \phi_*M(b')
\end{array}$$

is the identity map (of R-modules). We recall (from I.5.3) the construction of $\phi_*M(b')$. One has the category $\phi\backslash b'$ with objects (b,β') where $b \in |\mathcal{B}|$ and $\beta' \in \mathcal{B}'(\phi(b),b')$; a morphism $\gamma\colon (b_1,\beta_1') \longrightarrow (b_2,\beta_2')$ is an element $\gamma \in \mathcal{B}(b_1,b_2)$ with $\beta_2'\phi(\gamma) = \beta_1'$. One further has a functor $M_{b'}\colon \phi\backslash b' \longrightarrow R\text{-}mod$ with $M_{b'}(\gamma\colon(b_1,\beta_1') \longrightarrow (b_2,\beta_2')) = (M(\gamma)\colon M(b_1) \longrightarrow M(b_2))$. By definition $\phi_*(M)(b') = \operatorname{colim} M_{b'}$. In the above diagram $c(b,\beta')$ is the universal arrow from $M_{b'}(b,\beta') = M(b)$ to the colimit. By properties of colim it suffices to check that the diagram commutes for each $(b,\beta') \in |\phi\backslash b'|$.

Now each R-module generator for $M(b) = F(\sigma)(b)$ has the form $(\beta,s) = M(\beta)(1,s)$, $(1,s) \in M(\sigma(s))$. Thus by naturality we only have to consider the diagram when $b = \sigma(s)$ for some $s\in S$. In that case (most of) the diagram fits into the following diagram:

$$\begin{array}{ccccccc}
\phi_*M(b') & \xrightarrow{\phi_*(\mu)(b')} & \phi_*\phi^*N(b') & \xrightarrow{\varepsilon(b')} & N(b') & \xrightarrow{\upsilon(b')} & \phi_*M(b') \\
\uparrow c(\sigma(s),\beta') & & \uparrow c(\sigma(s),\beta') & \nearrow N(\beta') & & \nearrow \phi_*M(\beta') & \\
M(\sigma(s)) & \xrightarrow[\mu(\sigma(s))]{} & \phi^*N(\sigma(s)) & & & & \\
& & \| & & & & \\
& & N(\phi\sigma(s)) & & \xrightarrow[\upsilon(\phi\sigma(s))]{} & & \phi_*M(\phi\sigma(s))
\end{array}$$

The definition of $\phi_*(\mu)$ shows that the rectangle commutes. The triangle commutes by the definition of the counit ε, and the rhombus by naturality of υ.

The final step consists in showing that $c(\sigma(s),\beta') = \phi_*M(\beta')\circ\upsilon(\phi\sigma(s))\circ\mu(\sigma(s))$. By (1.13) and (1.14), this amounts to proving that

$$\begin{array}{ccc}
M(\sigma(s)) & \xrightarrow{\eta(\sigma(s))} & \phi^*\phi_*M(\sigma(s)) \\
c(\sigma(s),\beta') \downarrow & & \| \\
\phi_*M(b') & \xleftarrow[\phi_*M(\beta')]{} & \phi_*M(\phi\sigma(s))
\end{array}$$

commutes. But the front adjunction η has $\eta(\sigma(s)) = c(\sigma(s),1)$, so commutativity of the diagram is part of the definition of $\phi_*M(\beta')$. This finishes the proof that $\hat{\mu}\colon \phi_*F(\sigma) \longrightarrow F(\phi\sigma)$ is an isomorphism. Hence so is $Q(\hat{\mu})\colon \phi_!F(\sigma) \longrightarrow F(\phi\sigma)$.

It follows that $\phi_!$ preserves free $R\mathcal{B}$ modules. It clearly also preserves direct sums. Therefore, it preserves projectives.

The next few pages are devoted to proving that $Q(\hat{\mu})\colon \phi_!F(\sigma) \longrightarrow F(\phi\sigma)$ is natural viz-a-viz morphisms $(\gamma,\upsilon,d)\colon (S,\sigma) \longrightarrow (T,\rho)$. We need to show that

$$\begin{array}{ccc}
\phi_!F(\sigma) & \xrightarrow{Q(\hat{\mu})} & F(\phi\sigma) \\
\downarrow \phi_!F(\gamma,\upsilon,d) & & \downarrow F(\phi_!(\gamma,\upsilon,d)) \\
\phi_!F(\rho) & \xrightarrow[Q(\hat{\mu})]{} & F(\phi\rho)
\end{array}$$

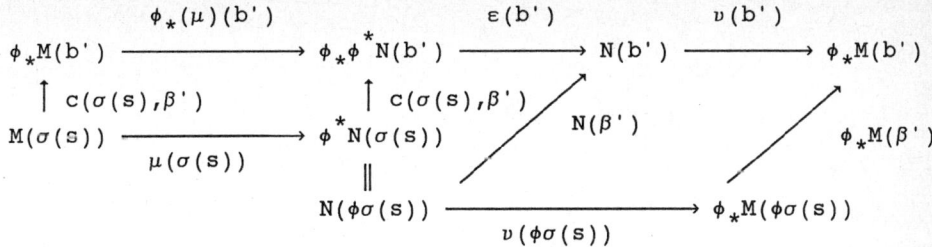

commutes. We apply $\phi^!$ to the diagram, precompose with $\eta\colon I \longrightarrow \phi^!\phi_!$ in the left hand column, and use the universal property of η to conclude that it suffices to show that

$$(1.15)\qquad \begin{array}{ccc} F(\sigma) & \xrightarrow{\;Q(\mu)\;} & \phi^! F(\phi\sigma) \\ \downarrow{\scriptstyle F(\gamma,\upsilon,d)} & & \downarrow{\scriptstyle \phi^! F(\phi_!(\gamma,\upsilon,d))} \\ F(\rho) & \xrightarrow[\;Q(\mu)\;]{} & \phi^! F(\phi\rho) \end{array}$$

commutes.

In this diagram both instances of $Q(\mu)$ are represented by natural transformations given by (1.13). Also, $F(\gamma,\upsilon,d)$ is represented by $\psi_d\colon F(\sigma) \longrightarrow F(\rho)C^d$ having

$$(1.16)\qquad \psi_d(1,s) = (\upsilon(s),\gamma(s)) \in F(\rho)(C^d\sigma(s)) \ , \ s\in S,$$

and $F(\phi_!(\gamma,\upsilon,d))$ is represented by $\xi_{nd}\colon F(\phi\sigma) \longrightarrow F(\phi\rho)C^{nd}$ having

$$(1.17)\qquad \xi_{nd}(1,s) = (\alpha^{(d)}(\sigma(s))\phi(\upsilon(s)),\gamma(s)) \in F(\phi\rho)(C^{nd}\sigma(s)).$$

Hence $\phi^! F(\phi_!(\gamma,\upsilon,d))$ is represented by the composition $\zeta_{mnd}\colon F(\phi\sigma)\phi \longrightarrow F(\phi\rho)\phi C^{mnd}$ given by

$$(1.18)\qquad F(\phi\sigma)\phi \xrightarrow[\;\xi_{nd}\phi\;]{} F(\phi\rho)C^{nd}\phi \xrightarrow[\;F(\phi\rho)\beta^{(nd)}\;]{} F(\phi\rho)\phi C^{mnd}$$

where $\beta^{(nd)}$ is an iteration of the natural transformation $\beta\colon C\phi \longrightarrow \phi C^m$ entering into the definition of ϕ being endomorphism preserving (cf. section I.5). In view of the above descriptions of representatives for the various maps entering into (1.15) it now suffices to check that the following diagram commutes for some sufficiently large N

$$\begin{array}{ccc}
F(\sigma)(\sigma(s)) & \xrightarrow{\quad\mu\quad} & F(\phi\sigma)(\phi\sigma(s)) \\
\Big\downarrow \psi_d & & \Big\downarrow \xi_{nd}\phi \\
F(\rho)(C^d\sigma(s)) & & F(\phi\rho)(C^{nd}\phi\sigma(s)) \\
\Big\downarrow \mu C^d & & \Big\downarrow F(\phi\rho)\beta^{(nd)} \\
F(\phi\rho)(\phi C^d\sigma(s)) & & F(\phi\rho)(\phi C^{mnd}\sigma(s)\ ) \\
F(\phi\rho)\phi C^d\tau^{N-d} \searrow & & \swarrow F(\phi\rho)\phi C^{mnd}\tau^{N-mnd} \\
& F(\phi\rho)(\phi C^N\sigma(s)) &
\end{array} \qquad (1.19)$$

Here all the natural transformations are evaluated at $\sigma(s)$. In (1.19) it suffices to check what happens to the element $(1,s) \in F(\sigma)(\sigma(s))$. It maps to (γ_1,s), respectively (γ_2,s), under the two maps $F(\sigma)(\sigma(s)) \longrightarrow F(\phi\rho)(\phi C^N\sigma(s))$. Here γ_1 and γ_2 are the two ways around in the diagram

$$\begin{array}{ccc}
\phi\rho\gamma(s) & \xrightarrow{\quad\phi(\upsilon(s))\quad} & \phi C^d\sigma(s) \\
\Big\downarrow \phi(\upsilon(s)) & & \Big\downarrow \alpha^{(d)} \\
\phi C^d\sigma(s) & & C^{nd}\phi\sigma(s) \\
\Big\downarrow \phi C^d\tau^{N-d} & & \Big\downarrow \beta^{(nd)} \\
\phi C^N\sigma(s) & \xleftarrow{\quad\phi C^{mnd}\tau^{N-mnd}\quad} & \phi C^{mnd}\sigma(s)
\end{array}$$

where the transformations are evaluated at $\sigma(s)$. Hence it suffices to show that

$$\begin{array}{ccc}
\phi C^d & \xrightarrow{\quad\alpha^{(d)}\quad} & C^{nd}\phi \\
\phi C^d\tau^{N-d} \Big\downarrow & & \Big\downarrow \beta^{(nd)} \\
\phi C^N & \xleftarrow[\quad\phi C^{mnd}\tau^{N-mnd}\quad]{} & \phi C^{mnd}
\end{array}$$

commutes for N sufficiently large. This question for general d quickly reduces to the same question for d=1. And in that case $N = mn+1$ works as is shown by the diagram

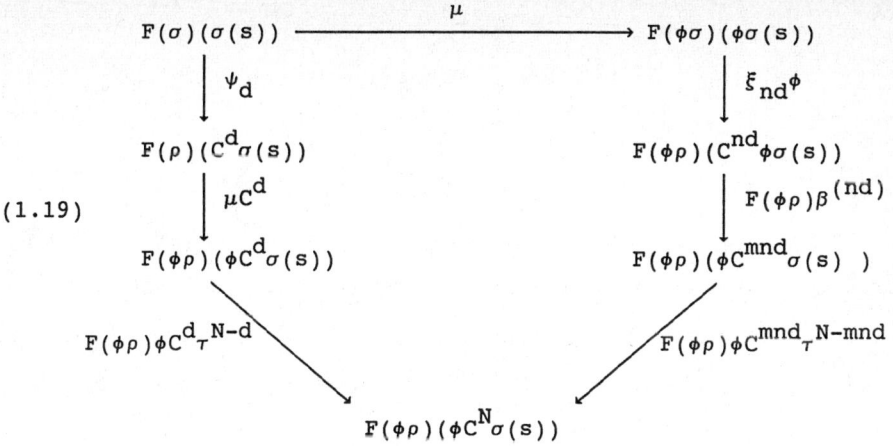

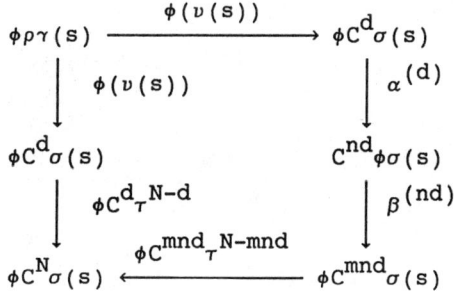

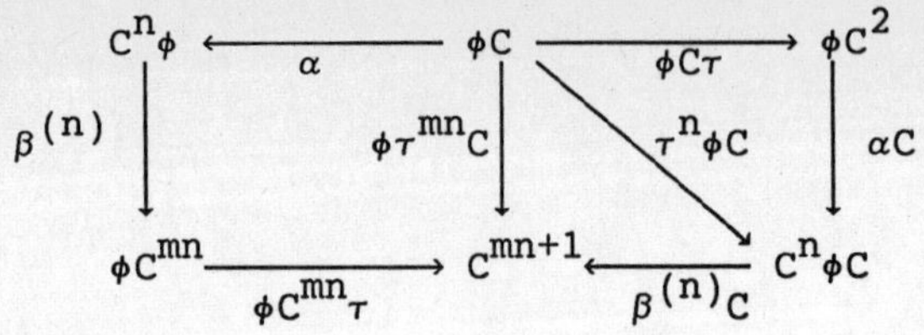

in which the big rectangle commutes because $\beta^{(n)}\alpha$ is a natural transformation and the two triangles commute because of the characteristic property of α, respectively $\beta^{(n)}$. It follows that the two squares commute. (For general d it follows that N = mnd+d will do.)

Remark 1.20 Let $\phi\colon \mathcal{B} \longrightarrow \mathcal{B}'$ be a functor which is not necessarily endomorphism preserving but where the endomorphism structure in $\mathcal{B}$ is assumed trivial, i.e. $C = 1_{\mathcal{B}}$, and $\tau=1$. In this case there are still induced functors

$$\phi_!\colon \mathcal{B}\mathit{as}(\mathcal{B}) \longrightarrow \mathcal{B}\mathit{as}(\mathcal{B}') \text{ and } \phi_!\colon R\mathcal{B}\text{-}\mathit{mod} \longrightarrow R\mathcal{B}'\text{-}\mathit{mod}$$

and the conclusion of 1.11 holds, i.e. $\phi_!(F(\sigma))$ is naturally isomorphic o $F(\phi\sigma)$ for any basis (S,σ) over $\mathcal{B}$.

We proceed with a simple proof of these claims. It is important to notice that $\alpha=1\colon \varphi C = \varphi \longrightarrow \varphi = C^0\varphi$ makes

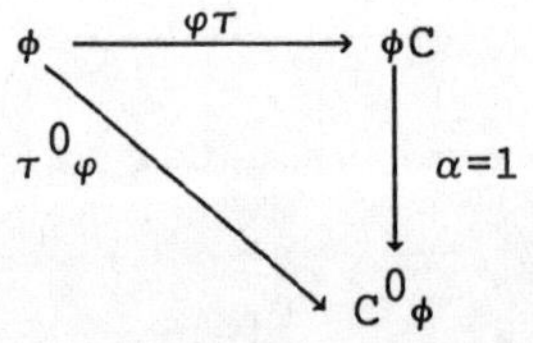

commute. Hence ϕ is almost endomorphism preserving.

Let (S,σ) be a basis over $\mathcal{B}$. The morphism $A_{(S,\sigma)}$ of 1.8 becomes $1_{(S,\sigma)}$, so the definition of ϕ_* in 1.9 takes the form

$$\phi_*[(\alpha,\upsilon,n)\colon (S,\sigma) \longrightarrow (T,\rho)] = [(\alpha,\phi(\upsilon),0)\colon (S,\phi\tau) \longrightarrow (T,\phi\rho)].$$

Since $\phi_*(\tau_{(S,\sigma)}) = 1_{(S,\sigma)}$ is an isomorphism, ϕ_* induces a functor

$\phi_!: \mathcal{B}as(\mathcal{B}) \longrightarrow \mathcal{B}as(\mathcal{B})$. In the pedestrian terms of 1.10, any morphism $(\gamma,\nu): (S,\sigma) \longrightarrow (T,\rho)$ is represented by a natural transformation $\nu_0: \rho\gamma \longrightarrow \sigma$ and $\phi_!(\gamma,\nu) = (\gamma,\lambda)$ where λ is represented by $\phi(\nu_0): \phi\rho\gamma \longrightarrow \phi\sigma$.

Next we turn to the functor $\phi_!$ at the module level. We have the commutative diagram

$$\begin{array}{ccc} \phi^* & \xrightarrow{\tau\varphi^*} & C\phi^*(=\phi) \\ & {\scriptstyle 1=\phi^*\tau^0}\searrow & \Big\downarrow {\scriptstyle \alpha^*=1} \\ & & \phi^*C^0 \end{array}$$

Since the functor ϕ_* of I.5.3 is a left adjoint of ϕ^*, the proof of I.3.6 shows that ϕ_* is almost endomorphism preserving. Therefore, by I.3.2, ϕ_* induces a functor $\phi_!: R\mathcal{B}\text{-}mod \longrightarrow R\mathcal{B}'\text{-}mod$ as claimed.

Finally, to check that $\phi_!F(\sigma) \cong F(\phi\sigma)$ for any basis (S,σ) over $\mathcal{B}$, let $\mu = \mu_{(S,\sigma)}: F(\sigma) \longrightarrow \phi^*F\phi_*(\sigma)$ be the morphism of (1.13). A tedious calculation shows that, since $\alpha=1$, μ is a natural transformation. Hence, so is its adjoint $\hat{\mu}$ in the proof of 1.11. Moreover, the proof that $\hat{\mu}_{(S,\sigma)}$ is an isomorphism carries over without change to the present setting. Hence

$$\begin{array}{ccc} \mathcal{B}as_*(\mathcal{B}) = \mathcal{B}as(\mathcal{B}) & \xrightarrow{F} & R\mathcal{B}\text{-}mod \\ \Big\downarrow \phi_! & & \Big\downarrow \phi_! \\ \mathcal{B}as(\mathcal{B}') & \xrightarrow[F]{} & R\mathcal{B}'\text{-}mod \end{array}$$

commutes up to the natural equivalence $\hat{\mu}$, as claimed. In particular, if $(\alpha,\nu): (S,\sigma) \longrightarrow (T,\rho)$ in $\mathcal{B}as(\mathcal{B})$ then $\phi_!F(\alpha,\nu) = F(\alpha,\phi\nu)$.

We finish this section with some considerations of (infinite) direct sums of $R\mathcal{B}$ modules. Let M_i, $i\in I$, be a collection of $R\mathcal{B}$ modules. There is the $R\mathcal{B}$ module $\oplus_i M_i$ defined by

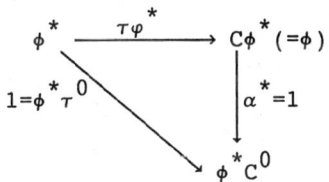

$$(\oplus_i M_i)(\beta\colon b \longrightarrow c) \;=\; \oplus_i M_i(\beta)\colon \oplus_i M_i(b) \longrightarrow \oplus_i M(c).$$

However, if I is infinite this module is not a direct sum of the modules M_i in the categorical sense. In fact, suppose given representatives $\varphi_{i,d_i}\colon M_i \longrightarrow NC^{d_i}$, $i\in I$, for morphisms $\varphi_i\colon M_i \longrightarrow N$ for some $R\mathcal{B}$ module N. If d_i is minimal for a representative for φ_i and $\{d_i \mid i\in I\}$ is unbounded, then there is no way to define a morphism $\varphi\colon \oplus_i M_i \longrightarrow N$ such that the following diagram commutes for all j:

$$\begin{array}{ccc} M_j & \longrightarrow & \oplus_i M_i \\ & {\scriptstyle\varphi_j}\searrow & \downarrow{\scriptstyle\varphi} \\ & & N \end{array}$$

On the positive side we note the following obvious proposition:

Proposition 1.21 *If the morphisms* $\varphi_i\colon M_i \longrightarrow N$, $i\in I$, *have representatives* $\varphi_{i,d_i}\colon M_i \longrightarrow NC^{d_i}$ *with* $\{d_i \mid i\in I\}$ *bounded, then there is a unique* $\varphi\colon \oplus_i M_i \longrightarrow N$ *which makes the above diagram commute for all* $j\in I$.

If (S,σ) is a basis, $S = \amalg_i S_i$, and $\sigma_i = \sigma|S_i$, then there is an obvious isomorphism $F(\sigma) \cong \oplus_i F(\sigma_i)$ which we shall always treat as an identification.

2. Boundedly finitely generated $R\mathcal{P}G$ modules.

In this section we continue the study of free and projective $R\mathcal{B}$ modules specializing $\mathcal{B}$ to be of the form $\mathcal{P}G$ where $\mathcal{P}$ is the category with endomorphism arising from some space Z with a boundedness control structure (thus each object of $\mathcal{P}$ is a subset of Z and each morphism of $\mathcal{P}$ is an inclusion among subsets) and $G\colon \mathcal{P} \longrightarrow \mathcal{G}poid$ is a functor.

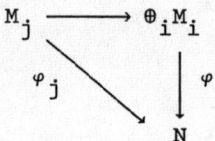

General assumption Throughout the remainder of this book we assume that R has the following property

(2.1) If for some $n,m \in \mathbb{Z}_+$, R^n is a direct summand in R^m then $n \leq m$.

We note that (2.1) holds for any ring R which admits a homomorphism into some field. If (2.1) holds, then R has the invariant basis property. We do not know whether the converse is true.

Definition 2.2 *Let* $(S,\sigma) \in \mathcal{B}as(\mathcal{F}G)$ *have* $\sigma(s) = (x_s,K_s)$. *We call* (S,σ) bounded *if* $\{K_s \mid s \in S\}$ *is a bounded family of subsets of* Z. *We call* (S,σ) locally finite *if* $\{s \mid K_s \subseteq K\}$ *is finite for each* $K \in |\mathcal{F}|$.

Example 2.3 Any one of the bases (S_n,σ_n) introduced in Example 1.6 is bounded because of the requirement (1.7). If X is a *finite* bc CW complex then (S_n,σ_n) is also locally finite. In fact, if $K_e \subseteq K$ then $p(Cl(e)) \subseteq C^d K$. Therefore the union of all cells e with $K_e \subseteq K$ is contained in $p^{-1}(C^d K)$ which in turn is contained in some finite subcomplex of X. This example is the main reason for introducing and studying the notions of bounded (respectively, locally finite) bases.

Definition 2.4 *Let* M *be an* $R\mathcal{F}G$*-module. We call* M boundedly generated *(respectively,* locally finitely generated*) if there exists an epimorphism* $\varphi: F(\sigma) \longrightarrow M$ *with* (S,σ) *bounded (respectively, locally finite). If* M *is both boundedly and locally finitely generated, then we call* M bfg *(*boundedly finitely generated*).*

The main results of the present section are the following proposition and theorem whose proofs are postponed:

Proposition 2.5 *A free* R$\mathcal{P}$G *module* F(σ) *is boundedly generated (respectively, locally finitely generated) if and only if* (S,σ) *is bounded (respectively, locally finite).*

Theorem 2.6 *Two bfg free* R$\mathcal{P}$G *modules,* F(σ) *and* F(σ'), *are isomorphic if and only if their bases,* (S,σ) *and* (S',σ'), *are isomorphic in* $\mathcal{B}as(\mathcal{P}G)$.

We need to know when functors of the form $\phi_!$: R$\mathcal{P}$G-*mod* $\longrightarrow$ R$\mathcal{P}$'G'-*mod* will preserve bfg modules. Let ρ: $\mathcal{P}G \longrightarrow \mathcal{P}$ and ρ': $\mathcal{P}'G' \longrightarrow \mathcal{P}'$ be the obvious "forgetful" functors [$\rho((\omega,i): (x,K) \longrightarrow (y,L)) = (i: K \subseteq L)$].

Proposition 2.7 *Let* ϕ: $\mathcal{P}G \longrightarrow \mathcal{P}'G'$ *be an endomorphism preserving functor. If* ϕ *satisfies*

(2.8) *If* $\{b_s \mid s \in S\}$ *is any collection of objects in* $\mathcal{P}$G *such that* $\{\rho(b_s) \mid s \in S\}$ *is bounded in* $|\mathcal{P}|$, *then* $\{\rho'\phi(b_s) \mid s \in S\}$ *is bounded in* $|\mathcal{P}'|$,

then $\phi_!$ *preserves boundedly generated modules. If* ϕ *satisfies*

(2.9) *For each* $K' \in |\mathcal{P}'|$ *there exists* $K \in |\mathcal{P}|$ *such that whenever an object* $b \in |\mathcal{P}G|$ has $\rho'\phi b \subseteq K'$, *it also has* $\rho b \subseteq K$,

then $\phi_!$ *preserves locally finitely generated modules.*

The following lemma and its corollary are essential for the proofs of the above mentioned results.

Lemma 2.10 *Let* γ_s: $(y_s,L_s) \longrightarrow (x_s,K_s)$, $s \in S$, *be a collection of morphisms in* $\mathcal{P}$G. *Assume that* $\{K_s \mid s \in S\}$ *is bounded. Then there exists a* $d \geq 0$ *and morphisms* γ'_s: $(x_s,K_s) \longrightarrow C^d(y_s,L_s)$ *such that*

$$\begin{array}{ccccc}
(Y_s,L_s) & \xrightarrow{\gamma_s} & (X_s,K_s) & & \\
& \searrow^{\tau^d} & \downarrow{\scriptstyle \gamma_s'} & \searrow^{\tau^d} & \\
& & C^d(Y_s,L_s) & \xrightarrow{C^d\gamma_s} & C^d(X_s,K_s)
\end{array}$$

commutes.

Proof: Write $\gamma_s = (\omega_s, i_s)$ where $i_s\colon L_s \subseteq K_s$ and ω_s is in $G(K_s)(i_{s*}(Y_s),X_s)$. Pick an integer n and for each $s\in S$ a minimal $K_{0,s}$ such that $K_s \subseteq C^n K_{0,s}$. By II.1.1.(iv) there is an inclusion $j_s\colon K_s \subseteq C^{n+\theta(n)}L_s$. We take $d = n+\theta(n)$ and define γ_s' to be $\gamma_s' = (\bar{\omega}_s^{-1}, j_s)$, where $\bar{\omega}_s \in G(C^d L_s)(j_{s*}i_{s*}(Y_s), j_{s*}(X_s))$ is the image of ω_s under $G(j_s)$.

Corollary 2.11 *Let $(\alpha,\nu)\colon (S,\sigma) \longrightarrow (T,\rho)$ be a morphism of bases over $\mathcal{P}G$. If (S,σ) is bounded and $\alpha\colon S \longrightarrow T$ is bijective, then (α,ν) is an isomorphism of bases.*

Proof: Let ν be represented by $\nu_n\colon \rho\alpha(s) \longrightarrow C^n\sigma(s)$ in $\mathcal{P}G$. By 2.10, there are morphisms $\nu_{n,s}'$ such that the diagrams

$$\begin{array}{ccc}
\rho\alpha(s) & \xrightarrow{\nu_{n,s}} & C^n\sigma(s) \\
& \searrow^{\tau^d} & \downarrow{\scriptstyle \nu_{n,s}'} \\
& & C^d\rho\alpha(s)
\end{array}$$

commute, $(s\in S)$. Define $(\beta,\mu)\colon (T,\rho) \longrightarrow (S,\sigma)$ by letting $\beta = \alpha^{-1}$ and μ be represented by

$$\sigma\beta(t) \xrightarrow{\ \tau^n\ } C^n\sigma\beta(t) \xrightarrow{\ \nu_{n,\beta(t)}'\ } C^d\rho\alpha\beta(t) = C^d\rho(t) .$$

A straightforward calculation shows that (β,μ) is an inverse to (α,ν).

For the proofs of 2.5 and 2.6 we also need the following lemma:

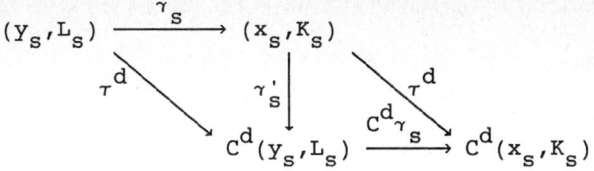

Lemma 2.12 *Let ϕ: $F(\sigma') \longrightarrow F(\sigma)$ be an epimorphism of free* $R\mathcal{P}G$ *modules. Then there exists a morphism* (α,ν): $(S,\sigma) \longrightarrow (S',\sigma')$ *of bases with* α: $S \longrightarrow S'$ *monic.*

Proof: Choose a splitting ψ: $F(\sigma) \longrightarrow F(\sigma')$ for ϕ and an integer $m\geq 0$ such that ϕ and ψ are represented by natural transformations

$$\psi_m: F(\sigma) \longrightarrow F(\sigma')C^m \;,\; \phi_m: F(\sigma') \longrightarrow F(\sigma)C^m \;.$$

For each $s\in S$ let $A(s) = \{s'\in S' \mid (2.13) \text{ below holds}\} \subseteq S'$

(2.13) For some $\beta \in \mathcal{P}(G)(\sigma'(s'),C^m\sigma(s))$ the R-module generator (β,s') of $F(\sigma')(C^m(\sigma(s)))$ appears non-trivially in $\psi_m(\sigma(s))(1,s) \in F(\sigma')(C^m(\sigma(s)))$.

For each pair (s,s') with $s' \in A(s)$, choose a particular $\beta_{s,s'} \in \mathcal{P}(G)(\sigma'(s'),C^m(\sigma(s)))$ satisfying (2.13).

We claim that there is a monomorphism α: $S \longrightarrow S'$ with each $\alpha(s) \in A(s)$, i.e. that the collection $\{A(s) \mid s\in S\}$ admits a transversal in the sense of combinatorics (cf. [Br, pp. 25-26). By Theorem 4.2 of [Br], we only have to show that

(2.14) For each finite subset J of S one has $|\cup_{j\in J}A(j)| \geq |J|$.

Thus we consider some finite $J\subseteq S$. Write $L = \cup_j A(j)$, $\iota=\sigma|J$, $\lambda'=\sigma'|L$. Clearly the restriction of ψ to $F(\iota)$ is a split monomorphism ξ: $F(\iota) \longrightarrow F(\lambda')$. There is a unique functor π: $(\mathcal{P}G,C,\tau) \longrightarrow (\mathcal{P}_0,C_0,\tau_0)$ where P_0 is the poset with one element (so that C_0,τ_0 are unique). Clearly π is endomorphism preserving so we get $\pi_!$: $R\mathcal{P}G$-*mod* $\longrightarrow R\mathcal{P}_0$-*mod* and by (1.11) a split monomorphism

$$F(\pi\iota) \cong \pi_!F(\iota) \longrightarrow \pi_!F(\lambda') \cong F(\pi\lambda').$$

But $F(\pi\iota)$ and $F(\pi\lambda')$ are simply the free R modules on the sets J and L, respectively. Thus, by (2.1), one has $|J| \leq |L|$.

This gives us the monomorphism α. We let ν be represented by $\nu_m: \sigma'\alpha \longrightarrow C^m\sigma$ given by

$$\nu_m(s) = \beta_{s,\alpha(s)} \in \mathcal{G}G(\sigma'\alpha(s), C^m\sigma(s)) , \ s \in S.$$

Proof of 2.5: Only the "only if" part needs proof. Thus let $\phi: F(\sigma') \longrightarrow F(\sigma)$ be an epimorphism. We have to show that if (S',σ') is bounded (or locally finite), then so is (S,σ). By 2.12, we have $(\alpha,\nu): (S,\sigma) \longrightarrow (S',\sigma')$ with α monic. Write $\sigma(s) = (x_s, K_s)$, $\sigma'(s') = (x'_{s'}, K'_{s'})$. If ν is represented by $\nu_m: \sigma'\alpha \longrightarrow C^m\sigma$, then α maps $\{s \in S \mid K_s \subseteq K\}$ monomorphically into $\{s' \in S' \mid K'_{s'} \subseteq C^m K\}$. Now if (S',σ') is locally finite, the latter set is finite for all K. Hence so is the former, so (S,σ) is locally finite.

Next assume that (S',σ') is bounded. For each $\ell \in \mathbb{Z}_+$, let $S_\ell \subseteq S$ be the set of all $s \in S$ such that rad $K_s \leq \ell$, and put $S_{(\ell)} = S - S_\ell$. Let $\sigma_\ell = \sigma|S_\ell$, $\sigma_{(\ell)} = \sigma|S_{(\ell)}$, and notice that $F(\sigma) = F(\sigma_\ell) \oplus F(\sigma_{(\ell)})$ for any ℓ. If ϕ is represented by $\phi_n: F(\sigma') \longrightarrow F(\sigma)C^n$ and rad $K'_{s'} \leq d$ for all s', then ϕ maps $F(\sigma')$ into the summand $F(\sigma_\ell)$ whenever $\ell \geq k+n$. Since ϕ is epic, $F(\sigma_{(\ell)})$ must vanish for $\ell \geq k+n$. Since this is possible only if $S_{(k+n)} = \emptyset$, $S = S_{k+n}$ and (S,σ) is bounded.

Proof of 2.6: Since F is a functor, only the "only if" needs proof. Thus let $\phi: F(\sigma') \longrightarrow F(\sigma)$ be an isomorphism. Lemma 2.12 supplies morphisms of bases

$$(\alpha,\nu): (S,\sigma) \longrightarrow (S',\sigma') \text{ and } (\beta,\mu): (S',\sigma') \longrightarrow (S,\sigma)$$

where $\alpha: S \longrightarrow S'$ and $\beta: S' \longrightarrow S$ are monic. We can assume $m \geq 0$ chosen so that we have representing natural transformations

$$\nu_m: \sigma'\alpha \longrightarrow C^m\sigma \ , \text{ and } \mu_m: \sigma\beta \longrightarrow C^m\sigma' .$$

We can now manufacture a bijection $\gamma: S \longrightarrow S'$ by using a version of the Schröder-Bernstein theorem due to Banach. For a convenient formulation we refer to [Br, Corollary 4.5]. We can assume that S and S' are disjoint, and we can form the bipartite graph (S,Λ,S') where Λ is the set of all (s,s') with $\alpha(s) = s'$ or $s = \beta(s')$ or both. In the terminology of the corollary mentioned above, the monomorphisms α and β are matchings of S with a subset of S', respectively, of S' with a subset of S. Therefore the corollary guarantees a matching of S with S', i.e. a bijection $\gamma\colon S \longrightarrow S'$ such that

(2.15) For each $s \in S$ either $\gamma(s) = \alpha(s)$ or $s = \beta\gamma(s)$ (or both).

Let $S_1 = \{s \in S \mid \gamma(s) = \alpha(s)\}$, $S_2 = S - S_1$, $S_1' = \gamma(S_1)$, and $S_2' = \gamma(S_2)$. Then $(\alpha,\nu)\colon (S,\sigma) \longrightarrow (S',\sigma')$ restricts to a morphism $(\alpha_1,\nu_1)\colon (S_1,\sigma_1) \longrightarrow (S_1',\sigma_1')$ which is an isomorphism by 2.11. Similarly (β,μ) restricts to an isomorphism $(\beta_2,\mu_2)\colon (S_2',\sigma_2') \longrightarrow (S_2,\sigma_2)$. The disjoint union (in the obvious sense) is the desired isomorphism $(S,\sigma) \longrightarrow (S',\sigma')$.

Proof of 2.7: Let $\psi\colon F(\sigma) \longrightarrow M$ be an epimorphism. Using 1.11 we get

$$F(\phi\sigma) \cong \phi_! F(\sigma) \xrightarrow{\phi_!(\psi)} \phi_! M$$

This is an epimorphism (because any left adjoint functor preserves epimorphisms). The rest of the proof is now obvious.

3. $K_1(R\mathcal{F}G)$ and $Wh(R\mathcal{F}G)$.

Let R be a ring satisfying (2.1). Let $\mathcal{F}$ be the category with endomorphism arising from a boundedness control structure (P,C) on a space Z and let $G\colon \mathcal{F} \longrightarrow \mathcal{G}poid$ be any functor. In this section, we define abelian groups $K_1(R\mathcal{F}G)$ and $Wh(R\mathcal{F}G)$ and establish their main properties. These groups are the bc analogues of the classical groups $K_1(R\pi)$ and $Wh(\pi)$. We begin with the definition of the K_1-group and go on to define the Wh-group as a suitable quotient of it.

Let bfgfree$R\mathcal{G}G$-*mod* be the full subcategory of $R\mathcal{G}G$-*mod* with objects the bfg free $R\mathcal{G}G$ modules. Since this is a category with $\oplus$ as a product in the sense of VII.5 of Bass, [Ba], we follow [Ba] and define $K_1(R\mathcal{G}G)$ to be

$$K_1(R\mathcal{G}G) = K_1(\text{bfgfree}R\mathcal{G}G\text{-}mod).$$

We recall Bass's definition: Every automorphism $\varphi\colon F \longrightarrow F$ of any bfg free $R\mathcal{G}G$ module represents an element $[F,\varphi]$ of $K_1(R\mathcal{G}G)$. The collection of such elements generates $K_1(R\mathcal{G}G)$, and the relations are

(3.1) $[F,\varphi] = [F_1,\varphi_1]$ if there is an isomorphism $\phi\colon F \longrightarrow F_1$ with $\phi\varphi = \varphi_1\phi$

(3.2) $[F,\varphi] + [F',\varphi'] = [F\oplus F',\varphi\oplus\varphi']$

(3.3) $[F,\varphi\psi] = [F,\varphi] + [F,\psi]$

for all possible choices of bfg free $R\mathcal{G}G$ modules F,F_1,F' and automorphisms φ,ψ,φ_1, and φ'.

Clearly every element of $K_1(R\mathcal{G}G)$ has the form $[F(\sigma),\varphi]$ for some bfg basis (S,σ). An automorphism $\varepsilon\colon F(\sigma) \longrightarrow F(\sigma)$ is called *elementary* if there is a splitting, $S = S_1 \amalg S_2$, of S into disjoint subsets such that

$$\varepsilon = \begin{bmatrix} 1 & \gamma \\ 0 & 1 \end{bmatrix}\colon F(\sigma_1)\oplus F(\sigma_2) \longrightarrow F(\sigma_1)\oplus F(\sigma_2)$$

for some $\gamma\colon F(\sigma_2) \longrightarrow F(\sigma_1)$. Here $\sigma_i = \sigma|S_i$ and we identify $F(\sigma)$ with $F(\sigma_1)\oplus F(\sigma_2)$. Note that in the classical case the corresponding notion is not an elementary matrix in the sense that only *one element* off the diagonal is allowed to be non zero. Algebraically it might, therefore, be better to use the term block-elementary. Geometrically, though, the terminology chosen is justified.

Proposition 3.4 *Let* (S_i,σ_i) *be a bfg basis and let* $\varphi_i\colon F(\sigma_i) \longrightarrow F(\sigma_i)$ *be an automorphism* $(i=1,2)$. *Then* $[F(\sigma_1),\varphi_1] = [F(\sigma_2),\varphi_2]$ *in* $K_1(R\mathcal{G}G)$ *if and only if there exist a bfg basis* (T,ρ) *and elementary* *automorphisms* $\varepsilon_j\colon F(\sigma_1 \amalg \sigma_2 \amalg \rho) \longrightarrow F(\sigma_1 \amalg \sigma_2 \amalg \rho)$ $(1 \le j \le k \le 16)$ *such that*

$$\varphi_1 \oplus \varphi_2^{-1} \oplus 1_{F(\rho)} = \varepsilon_1 \varepsilon_2 \cdots \varepsilon_k\colon F(\sigma_1 \amalg \sigma_2 \amalg \rho) \longrightarrow F(\sigma_1 \amalg \sigma_2 \amalg \rho)$$

Proof: By (3.3) $[F, 1_F] = 0$ and $[F, \varphi^{-1}] = -[F, \varphi]$. Hence $[F(\sigma_1), \varphi_1] = [F(\sigma_2), \varphi_2]$ if and only if $[F(\sigma_1 \amalg \sigma_2), \varphi_1 \oplus \varphi_2^{-1}]$ vanishes in $K_1(R\mathcal{G}G)$. Let us write $(S, \sigma) = (S_1 \amalg S_2, \sigma_1 \amalg \sigma_2)$ and $\varphi = \varphi_1 \oplus \varphi_2^{-1}$.

If $[F(\sigma), \varphi] = 0$ then by VII.1.9 of Bass, [Ba], there is a bfg basis (T, ρ) and an automorphism $\psi\colon F(\rho) \longrightarrow F(\rho)$ such that $\varphi \oplus \psi \oplus \psi^{-1}$ is a commutator in $\mathrm{Aut}(F(\sigma \amalg \rho))$, say $\varphi \oplus \psi \oplus \psi^{-1} = \xi \eta \xi^{-1} \eta^{-1}$, $\xi, \eta \in \mathrm{Aut}(F(\sigma \amalg \rho \amalg \rho))$. The following identities in $\mathrm{Aut}(F(\sigma \amalg \rho \amalg \rho \amalg \sigma \amalg \rho \amalg \rho))$

$$(3.5) \qquad \begin{bmatrix} \xi\eta\xi^{-1}\eta^{-1} & 0 \\ 0 & 1 \end{bmatrix} = \begin{bmatrix} \xi & 0 \\ 0 & \xi^{-1} \end{bmatrix} \begin{bmatrix} \eta & 0 \\ 0 & \eta^{-1} \end{bmatrix} \begin{bmatrix} \xi^{-1}\eta^{-1} & 0 \\ 0 & \eta\xi \end{bmatrix}$$

$$(3.6) \qquad \begin{bmatrix} \xi & 0 \\ 0 & \xi^{-1} \end{bmatrix} = \begin{bmatrix} 1 & \xi \\ 0 & 1 \end{bmatrix} \begin{bmatrix} 1 & 0 \\ 1-\xi^{-1} & 1 \end{bmatrix} \begin{bmatrix} 1 & -1 \\ 0 & 1 \end{bmatrix} \begin{bmatrix} 1 & 0 \\ 1-\xi & 1 \end{bmatrix}$$

show that $\xi\eta\xi^{-1}\eta^{-1} \oplus 1$ is a product of 12 elementary automorphisms in $\mathrm{Aut}(F(\sigma \amalg \rho \amalg \rho \amalg \sigma \amalg \rho \amalg \rho))$. A formula like (3.6) shows that $\psi^{-1} \oplus \psi$ is a product of 4 elementary automorphisms in $\mathrm{Aut}(F(\rho \amalg \rho))$. Now note that

$$\varphi \oplus 1 = (\xi\eta\xi^{-1}\eta^{-1} \oplus 1)(1 \oplus \psi^{-1} \oplus \psi \oplus 1)$$

in $\mathrm{Aut}(F(\sigma \amalg \rho \amalg \rho \amalg \sigma \amalg \rho \amalg \rho))$.

Conversely if $\varphi \oplus 1$ is a product of elementary automorphisms, then $[F(\sigma), \varphi] = 0$ because of (3.3) and because $[F(\sigma), \varepsilon] = 0$ for every elementary automorphism. To prove the latter, notice the identity in $F(\sigma_1 \amalg \sigma_2 \amalg \sigma_2)$

$$\begin{bmatrix} 1 & \gamma & 0 \\ 0 & 1 & 0 \\ 0 & 0 & 1 \end{bmatrix} = \begin{bmatrix} 1 & 0 & \gamma \\ 0 & 1 & 0 \\ 0 & 0 & 1 \end{bmatrix} \begin{bmatrix} 1 & 0 & 0 \\ 0 & 1 & 0 \\ 0 & 1 & 1 \end{bmatrix} \begin{bmatrix} 1 & 0 & -\gamma \\ 0 & 1 & 0 \\ 0 & 0 & 1 \end{bmatrix} \begin{bmatrix} 1 & 0 & 0 \\ 0 & 1 & 0 \\ 0 & -1 & 1 \end{bmatrix}$$

which shows that stably any elementary automorphism is a commutator.

The Whitehead group of $R\mathcal{G}G$ will be defined as a quotient group of $K_1(R\mathcal{G}G)$.

Definition 3.7 $Wh(R\mathcal{J}G)$ *is the quotient group defined by requiring that the kernel of the natural map* $K_1(R\mathcal{J}G) \longrightarrow Wh(R\mathcal{J}G)$ *be the subgroup generated by all elements of the form* $[F(\sigma),u_{F(\sigma)}]$ *or* $[F(\sigma),F(\alpha,\nu)]$ *where* (S,σ) *is any bfg basis over* $\mathcal{J}G$, $u_{F(\sigma)}$ *is multiplication by a unit* $u\in R$, *and* $F(\alpha,\nu)\colon F(\sigma) \longrightarrow F(\sigma)$ *is the automorphism induced by some automorphism of bases* $(\alpha,\nu)\colon (S,\sigma) \longrightarrow (S,\sigma)$.

Since in $K_1(R\mathcal{J}G)$ one has the identities

$$-[F(\sigma),u_{F(\sigma)}] = [F(\sigma),u^{-1}_{F(\sigma)}] \text{ and } -[F(\sigma),F(\alpha,\nu)] = [F(\sigma),F(\alpha,\nu)^{-1}],$$

each element in the kernel of $K_1(R\mathcal{J}G) \longrightarrow Wh(R\mathcal{J}G)$ is represented by an automorphism of the form

$$(3.8) \qquad u^{(1)}_{F(\sigma_1)}\oplus\cdots\oplus u^{(\ell)}_{F(\sigma_\ell)}\oplus F(\alpha,\nu)\colon F(\sigma_1\amalg\cdots\amalg\sigma_{\ell+1}) \longrightarrow F(\sigma_1\amalg\cdots\amalg\sigma_{\ell+1})$$

for some $\ell\geq 0$, some set of units $u^{(1)},\ldots,u^{(\ell)} \in R$, and some basis automorphism (α,ν). It follows easily from 3.4 that one has the following criterion for the vanishing of an element $[F(\sigma),\varphi]$ in $Wh(R\mathcal{J}G)$.

Proposition 3.9 *Let* $\varphi\colon F(\sigma) \longrightarrow F(\sigma)$ *be an automorphism of a bfg free* $R\mathcal{J}G$ *module. Then* $[F(\sigma),\varphi] = 0$ *in* $Wh(R\mathcal{J}G)$ *if and only if there exist a bfg basis* (T,ρ), *an integer* $k \in [0,16]$, *and* k *elementary automorphisms* $\varepsilon_i\colon F(\sigma\amalg\rho) \longrightarrow F(\sigma\amalg\rho)$ *such that*

$$(\varphi\oplus 1_{F(\rho)})\varepsilon_1\varepsilon_2\cdots\varepsilon_k\colon F(\sigma\amalg\rho) \longrightarrow F(\sigma\amalg\rho)$$

has the form (3.8) *for some decomposition of* $\sigma\amalg\rho$ *as* $\sigma_1\amalg\sigma_2\amalg\cdots\amalg\sigma_{\ell+1}$.

Remark 3.10 If $\mathcal{J} = \mathcal{J}_0$ consists of exactly one object and one morphism then a functor $G\colon \mathcal{J}_0 \longrightarrow \mathcal{G}poid$ is simply a groupoid. In this case we shall write RG instead of $R\mathcal{J}_0G$. As a special case of the above definitions, we therefore have defined $K_1(RG)$ and $Wh(RG)$ for any groupoid G. If the groupoid G happens to be just a group π, then $K_1(R\pi)$ and $Wh(R\pi)$ coincide with the classically defined groups with these names. In fact the category of $R\mathcal{P}_0\pi$ modules is simply the category of $R[\pi]$ modules. Note also that an automorphism of the form $F(\alpha,\nu): F(\sigma) \longrightarrow F(\sigma)$ corresponds classically to the product of a permutation matrix and a diagonal matrix with group elements along the diagonal.

Returning to a general groupoid G, we note that it is not hard to show the following:

(3.11) $\quad K_1(RG) \cong \underset{(x)}{\oplus} K_1(R[G(x,x)])\quad$ and $\quad Wh(RG) \cong \underset{(x)}{\oplus} Wh(R[G(x,x)])$

where $G(x,x)$ is the group of automorphisms in G of $x \in |G|$ and x varies over a set of representatives of the components of G. We leave the proof to the reader noticing only that one gets a sum rather than a product because any bfg basis (S,σ) over $\mathcal{P}_0 G$ has S finite.

The following example describes an "amalgamation" construction which will be used in our proof of the combinatorial invariance of bc Whitehead torsion, cf. V.4.

Example 3.12 Let $\mathcal{H}$ be a family of sets from $|\mathcal{P}|$ and $G: \mathcal{P} \longrightarrow \mathcal{G}poid$ a functor. We form the category $\mathcal{B} = \underset{K \in \mathcal{H}}{\coprod} G(K)$ and endow $\mathcal{B}$ with the trivial endomorphism structure $C = I_\mathcal{B}$ and $\tau = 1$. There is a functor $A_\mathcal{H}: \mathcal{B} \longrightarrow \mathcal{P}G$ with $A_\mathcal{H}(x) = (x,K)$ for $x \in |G(K)|$ and $A_\mathcal{H}(\omega) = (\omega,1)$ for $\omega \in G(K)(x,y)$. There are also obvious identifications

(3.13) $\qquad\qquad R\mathcal{B}\text{-}mod = \underset{K}{\Pi} R[G(K)]\text{-}mod$

and

(3.14) $\qquad\qquad \mathcal{B}as(\mathcal{B}) = \underset{K}{\amalg} \mathcal{B}as(G(K)).$

Therefore, by 1.20, we have induced functors

(3.15) $$A_{\mathcal{K}!}: \Pi_K\ R[G(K)]\text{-}\mathit{mod} \longrightarrow R\mathcal{G}G\text{-}\mathit{mod},$$

and

(3.16) $$A_{\mathcal{K}!}: \amalg_K\ \mathcal{B}as(G(K)) \longrightarrow \mathcal{B}as(\mathcal{G}G).$$

We call these the *amalgamation functors* over $\mathcal{K}$. If $M = (M_K)$ is in $\Pi_K\ R[G(K)]\text{-}\mathit{mod}$ then it is easily seen that

$$A_{\mathcal{K}}(M) = \oplus_K\ \psi_{K!}(M_K)$$

where $\psi_K = A_{\mathcal{K}}|G(K)$ and the direct sum is in the sense of the discussion preceding 1.21. Remark 1.20 also shows that $A_{\mathcal{K}}$ restricts to a functor

$$\Pi_K \text{ free } R[G(K)]\text{-}\mathit{mod} \longrightarrow \text{free } R\mathcal{G}G\text{-}\mathit{mod}.$$

Actually, if for each $K\in\mathcal{K}$, we have a basis $\sigma_K: S_K \longrightarrow G(K)$, then the identifications of (3.13) and (3.14) make $F(\amalg_K\sigma_K) = (F(\sigma_K))$, and the natural isomorphism of 1.20 becomes an isomorphism

$$A_{\mathcal{K}!}((F(\sigma_K))) \cong F(\amalg_K\psi_{K!}\sigma_K).$$

Now suppose that $\mathcal{K}$ is a bounded and locally finite family and that each basis (S_K,σ_K) is bfg (equivalently, that each S_K is a finite set). Then it is easily seen that $\amalg_K\psi_{K!}\sigma_K$ is a bfg basis over $\mathcal{G}G$. Hence it follows that $A_{\mathcal{K}!}$ restricts to a functor

$$A_{\mathcal{K}!}: \Pi_K \text{ bfgfree}R[G(K)]\text{-}\mathit{mod} \longrightarrow \text{bfgfree}R\mathcal{G}G\text{-}\mathit{mod}.$$

This functor preserves direct sums, so it induces

$$A_{\mathcal{K}*}: K_1(\Pi_K \text{ bfgfree}R[G(K)]\text{-}\mathit{mod}) \longrightarrow K_1(R\mathcal{G}G).$$

We notice now that the projections

$$p_L: \Pi_K \text{ bfgfree}R[G(K)]\text{-}\mathit{mod} \longrightarrow \text{bfgfree}R[G(L)]\text{-}\mathit{mod}$$

$(L\in\mathcal{K})$, induce a homomorphism

$$p_*: K_1(\Pi_K \text{ bfgfree}R[G(K)]\text{-}\mathit{mod}) \longrightarrow \Pi_K\ K_1(R[G(K)])$$

with $p_* = \Pi_K p_{K*}$. We defer the proof of the following result to the end of the present example.

Lemma 3.17 *The homomorphism* p_* *is an isomorphism.*

It follows that $A_{\mathcal{K}*}$ can be viewed as a homomorphism

$$A_{\mathcal{K}*}\colon \Pi_K\, K_1(R[G(K)]) \longrightarrow K_1(R\mathcal{G}G). \tag{3.18}$$

It is easy to check that this homomorphism induces a homomorphism

$$\bar{A}_{\mathcal{K}*}\colon \Pi_K\, Wh(R[G(K)]) \longrightarrow Wh(R\mathcal{G}G). \tag{3.19}$$

Both of these homomorphisms will be referred to as *amalgamation homomorphisms*. They can be described nore intuitively as follows:

Let $(x_K) \in \Pi_{K\in\mathcal{K}} K_1(R[G(K)])$. Then for each $K\in\mathcal{K}$, $x_K = [F(\sigma_K),\varphi_K]$ for some basis $\sigma_K\colon S_K \longrightarrow G(K)$ and some automorphism $\varphi_K\colon F(\sigma_K) \longrightarrow F(\sigma_K)$. If $s\in S_K$, then $\sigma_K(s)$ is simply an object of the groupoid $G(K)$, so we can define a basis (S,σ) over $\mathcal{G}G$ by letting $S = \amalg_K S_K$ and $\sigma(s) = (\sigma_K(s),K)$ for $s\in S_K$. Since $\mathcal{K}$ is bounded and locally finite, (S,σ) is bfg. The R module $F(\sigma_K)(\sigma_K(s))$ is a direct summand in the R module $F(\sigma)(\sigma_K(s),K) = F(\sigma)(\sigma(s))$ (the two R modules actually coincide if K is such that no $K'\in\mathcal{K}$ has $K'\subseteq K$ and $K'\neq K$). Hence the elements $\varphi_K(1,s) \in F(\sigma_K)(\sigma_K(s))$ which determine φ_K also are elements of $F(\sigma)(\sigma(s))$. We define φ by requiring that $\varphi(1,s) = \varphi_K(1,s)$ for all $s \in S_K$ and all $K\in\mathcal{K}$. Since the same construction applied to the automorphisms φ_K^{-1}, yields a morphism which is an inverse of φ, φ is an automorphism, and $A_{\mathcal{K}*}((x_K))$ is just $[F(\sigma),\varphi]$.

Proof of 3.17: It is obvious that p_* is onto. To see that p_* is one-to-one, let $\sigma\colon S \longrightarrow \amalg_K G(K)$ be a bfg basis and suppose that

$$x = [F(\sigma),\varphi] \in K_1(\Pi_K \text{ bfgfreeR}[G(K)]\text{-mod})$$

has $p_*(x)=0$. Then for every $K\in\mathcal{K}$, $p_{K*}(x) = [F(\sigma_K),\varphi_K] = 0$ in $K_1(R[G(K)])$ where $S_K = \{s\in S \mid \sigma(s)\in|G(K)|\}$, $\sigma_K = \sigma|S_K$ and $\varphi_K = \varphi|F(\sigma_K)\colon F(\sigma_K) \longrightarrow F(\sigma_K)$. By 3.4, there exist bfg bases (U_K,ξ_K) over $G(K)$ and elementary automorphisms

$$\varepsilon_K^{(i)}\colon F(\sigma_K \amalg \xi_K) \longrightarrow F(\sigma_K \amalg \xi_K) \quad (1\leq i\leq 16)$$

so that

$$\varphi_K \oplus 1_{F(\xi_K)} = \Pi_{i=1}^{16}\varepsilon_K^{(i)} \quad (K\in\mathcal{K}).$$

Now the family $\varepsilon_K^{(i)}$ $(K\in\mathcal{K})$ defines a morphism $\varepsilon_K^{(i)}\colon F(\xi) \longrightarrow F(\xi)$, where $\xi = \amalg_K \xi_K$. Since it is easily seen that $\varepsilon^{(i)}$ is elementary and that $\varphi\oplus 1_{F(\xi)} = \Pi_{i=1}^{16}\varepsilon^{(i)}$, $[F(\sigma),\varphi] = 0$, and p_* is one-to-one.

Let $\mathcal{BLF}(\mathcal{P})$ be the set of all bounded, locally finite families $\mathcal{K}$. Write $\mathcal{K}\leq\mathcal{L}$ if for each $K\in\mathcal{K}$ there is some $L\in\mathcal{L}$ such that $K\subseteq L$. As usual, view $\mathcal{BLF}(\mathcal{P})$ as a category where there is precisely one morphism from $\mathcal{K}$ to $\mathcal{L}$ if $\mathcal{K}\leq\mathcal{L}$. (Hence, if $\mathcal{K}\leq\mathcal{L}$ and $\mathcal{L}\leq\mathcal{K}$ then $\mathcal{K}$ and $\mathcal{L}$ are isomorphic). If $\mathcal{K}\leq\mathcal{L}$ then there is a homomorphism

$$\mathcal{I}_{\mathcal{K},\mathcal{L}}\colon \prod_{K\in\mathcal{K}} Wh(R[G(K)]) \longrightarrow \prod_{L\in\mathcal{L}} Wh(R[G(L)])$$

with

$$\mathcal{I}_{\mathcal{K},\mathcal{L}}((x_K)) = ((y_L)) \text{ where } y_L = \sum_{i:K\subseteq L} i_*(x_K)$$

(with $K\in\mathcal{K}$, $L\in\mathcal{L}$). Note that $\mathcal{L}$ being locally finite guarantees that the sum in question is finite. This definition makes $\prod_{K\in\mathcal{K}} Wh(R[G(K)])$ functorial in $\mathcal{K} \in |\mathcal{BLF}(\mathcal{P})|$ and it is easily seen that $A_{\mathcal{K}*}$ and $\bar{A}_{\mathcal{K}*}$ are natural transformations into the corresponding constant functors. Hence one has

Proposition 3.20 *The homomorphisms* $A_{\mathcal{H}*}$ *and* $\bar{A}_{\mathcal{H}*}$, *respectively, induce homomorphisms*

$$A\colon \operatorname{colim}_{\mathcal{H}}\Big(\prod_{K\in\mathcal{H}} K_1(R[G(K)])\Big) \longrightarrow K_1(R\mathcal{P}G),$$

$$\bar{A}\colon \operatorname{colim}_{\mathcal{H}}\Big(\prod_{K\in\mathcal{H}} Wh(R[G(K)])\Big) \longrightarrow Wh(R\mathcal{P}G),$$

where the colimits are over the category $\mathcal{BFL}(\mathcal{P})$.

Remark 3.21 Suppose that K_i, $i\in I$, is a locally finite family of minimals of $|\mathcal{P}|$ such that for some $d_0\in\mathbb{Z}_+$, each minimal K of $|\mathcal{P}|$ is contained in $C^{d_0}K_i$ for some i. For any $d\geq 0$ the family $\mathcal{H}_d = \{C^dK_i \mid i\in I\}$ is bounded and locally finite. Also $\mathcal{H}_1 \leq \mathcal{H}_2 \leq\cdots\leq \mathcal{H}_d \leq\cdots$ and for each bounded and locally finite family $\mathcal{H}$, there is some $d\geq 0$ such that $\mathcal{H}\leq\mathcal{H}_d$. Therefore, in 3.20 one can replace the general colim by the direct limit

$$\varinjlim_{d} \prod_{K\in\mathcal{H}_d} Wh(R[G(K)]).$$

3^*. The algebraic K-theory of R𝒫G.

Let $R,\mathcal{P},G$ be as in section 3. The full subcategory bfgpR𝒫G-*mod* of bfg projective R𝒫G modules is exact in the sense of Quillen, [Q1], so one may form the classifying space of its Q-construction and define

$$(3^*.1)\qquad K_i'(R\mathcal{P}G) = \pi_{i+1}\,(BQ(\text{bfgpR}\mathcal{P}G\text{-}mod))\ ,\ i\geq 0.$$

Alternatively write $\mathcal{S} = S(\text{bfgpR}\mathcal{P}G\text{-}mod)$ for the subcategory consisting of all isomorphisms in bfgpR𝒫G-*mod* and form $\mathcal{S}^{-1}\mathcal{S}$ in the sense of Grayson, [Gr]. One then lets

$$(3^*.2)\qquad K_i'' = \pi_i(B(\mathcal{S}^{-1}\mathcal{S})),\ i\geq 0.$$

Presumably one can mimic the proof in [Gr] to show that the functors K_i' and K_i'' are isomorphic, but we have had no reason to check the details.

For geometric purposes the group $K_0(R\mathcal{F}G)$ defined as the Grothendieck group of bfgp$R\mathcal{F}G$-*mod* is of interest in a study of finiteness properties in $\mathcal{CW}^c/Z$. However, such a study is not included in the present volume.

4. Functoriality of $K_1(R\mathcal{F}G)$ and Wh$(R\mathcal{F}G)$.

We let R be a ring satisfying (2.1), $\mathcal{F}$ and $\mathcal{F}'$ be categories with endomorphism arising from spaces Z and Z' with boundedness control structures, and $G\colon \mathcal{F} \longrightarrow \mathcal{G}poid$ and $G'\colon \mathcal{F}' \longrightarrow \mathcal{G}poid$ be functors. We write $\mathcal{B} = \mathcal{F}G$, $\mathcal{B}' = \mathcal{F}'G'$. We want to know which functors $\phi\colon \mathcal{B} \longrightarrow \mathcal{B}'$ induce homomorphisms $\phi_*\colon K_1(R\mathcal{F}G) \longrightarrow K_1(R\mathcal{F}'G')$, and $\phi_*\colon \mathrm{Wh}(R\mathcal{F}G) \longrightarrow \mathrm{Wh}(R\mathcal{F}'G')$ and especially when these homomorphisms are isomorphisms.

Proposition 4.1 *If* ϕ *is endomorphism preserving and satisfies* (2.8) *and* (2.9), *then* $\phi_!$ *maps bfg free* $R\mathcal{F}G$ *modules to bfg free* $R\mathcal{F}'G'$ *modules. Consequently,* ϕ *induces homomorphisms*

$$\phi_*\colon K_1(R\mathcal{F}G) \longrightarrow K_1(R\mathcal{F}'G'), \text{ and } \phi_*\colon \mathrm{Wh}(R\mathcal{F}G) \longrightarrow \mathrm{Wh}(R\mathcal{F}'G').$$

These make K_1 *and* Wh *functorial wrt functors* ϕ *having the above properties.*

Proof: The first part is immediate in view of 1.11 and 2.7. Define ϕ_* by $\phi_*([F,\varphi]) = [\phi_! F, \phi_! \varphi]$. Since $\phi_!$ preserves direct sums, in order to check that this defines a homomorphism at the K_1-level, one only has to verify that $\phi_!$ "preserves" elementaries. We leave it to the reader to check that

$$\begin{array}{ccc} \phi_!(F\oplus G) & \cong & \phi_! F\oplus\phi_! G \\ \phi_!\begin{bmatrix}1 & \gamma\\ 0 & 1\end{bmatrix}\Big\downarrow & & \Big\downarrow\begin{bmatrix}1 & \phi_!\gamma\\ 0 & 1\end{bmatrix} \\ \phi_!(F\oplus G) & \cong & \phi_! F\oplus\phi_! G \end{array}$$

commutes.

To see that ϕ_* at the K_1-level induces a homomorphism ϕ_* at the Wh-level, first let $u \in R$ be a unit. An easy check shows that $\phi_!$ takes $u_F\colon F \to F$ into $u_{\phi_! F}\colon \phi_! F \longrightarrow \phi_! F$. Next, if $(\alpha,\nu)\colon (S,\sigma) \longrightarrow (S,\sigma)$ is an automorphism of the basis (S,σ), then 1.11 shows that

$$\begin{array}{ccc} F(\phi\sigma) & \cong & \phi_! F(\sigma) \\ F\phi_!(\alpha,\nu)\Big\downarrow & & \Big\downarrow\phi_! F(\alpha,\nu) \\ F(\phi\sigma) & \cong & \phi_! F(\sigma) \end{array}$$

commutes. Hence $\phi_*[F(\sigma),F(\alpha,\nu)] = [F(\phi\sigma),F\phi_!(\alpha,\nu)]$. Altogether, ϕ_* at the K_1-level respects the relations which define Wh, and induces ϕ_* at the Wh-level.

We leave it to the reader to check the functoriality.

If $f\colon (X,p) \longrightarrow (Y,q)$ is a bc map (with bound d, say) of bc spaces over Z, then one gets an endomorphism preserving functor satisfying (2.8) and (2.9)

(4.2) $$\phi = f_*\colon \mathcal{G}_1(X,p) \longrightarrow \mathcal{G}_1(Y,q)$$

given by

$$\phi((\omega,i)\colon (x,K) \longrightarrow (y,L)) = ((f_*(\omega),i)\colon (f(x),C^dK) \longrightarrow (f(y),C^dL))$$

The restriction of the induced functor $\phi_!$ to the category of bfg free $R\mathcal{G}_1(-,-)$ modules will be denoted

(4.3) $$f_!\colon \text{bfgfree}R\mathcal{G}_1(X,p)\text{-}\mathit{mod} \longrightarrow \text{bfgfree}R\mathcal{G}_1(Y,q)\text{-}\mathit{mod}.$$

The following result is basic for the geometric applications of $WhR\mathcal{G}G_1(X,p)$.

Theorem 4.4 *Assume that* $f\colon (X,p) \longrightarrow (Y,q)$ *has the following properties*

(i) (X,p) *and* (Y,q) *are coextensive; i.e. for some* $d'\geq 0$, *if* $q^{-1}(K) \neq \emptyset$, *then* $p^{-1}(C^{d'}K) \neq \emptyset$.

(ii) $f_*\colon \pi_i^c(X,p) \longrightarrow f^!\pi_i^c(Y,q)$ *is an isomorphism for* $i=0,1$.

Then the functor ϕ *of* (4.2) *induces an equivalence of categories* (4.3) *and isomorphisms*

$$f_*\colon K_1R\mathcal{G}G_1(X,p) \longrightarrow K_1R\mathcal{G}G_1(Y,q),$$

$$f_*\colon WhR\mathcal{G}G_1(X,p) \longrightarrow WhR\mathcal{G}G_1(Y,q).$$

Proof: It follows from II.5.3. and the definition of π_i^c in II.9 that $\phi_!\colon R\mathcal{G}G_1(X,p)\text{-}mod \longrightarrow R\mathcal{G}G_1(Y,q)\text{-}mod$ is an equivalence of categories. Therefore, it is a full and faithful functor. Hence the same is true about the functor $f_!$ of (4.3). Therefore, it suffices to prove that every bfg free $R\mathcal{G}G_1(Y,q)$ module M is isomorphic to $f_!(N)$ for some bfg free $R\mathcal{G}G_1(X,p)$ module N.

We may assume that $M = F(\sigma)$ for some bfg basis (S,σ) over $\mathcal{G}G_1(Y,q)$. In view of 1.11 it suffices to construct a bfg basis (S,σ') over $\mathcal{G}G_1(X,p)$ such that $(S,\phi\sigma')$ is isomorphic to (S,σ) as a basis over $\mathcal{G}G_1(Y,q)$.

Let $\sigma(s) = (y_s,K_s)$ for $s\in S$. Condition (i) guarantees that there exists an element $x_s \in p^{-1}(C^{d'}K_s)$. Recalling that d is a bound for f we conclude that $[y_s] \in \pi_0^c(Y,q)(f(x_s),C^{d+d'}(K_s)) = f^!\pi_0^c(Y,q)(x_s,C^{d'}K_s)$. Condition (ii) for i=0 now ensures that there is some $d''\geq 0$ such that

$$[y_s] \in \mathrm{Im}[f_*: \pi_0(p^{-1}(C^{d'+d''}K_s),x_s) \longrightarrow \pi_0(q^{-1}(C^{d+d'+d''}K_s),f(x_s))]$$

for each $s \in S$. We pick $y_s' \in p^{-1}(C^{d'+d''}K_s)$ with $[y_s] = f_*[y_s']$ and put $\sigma'(s) = (y_s', C^{d'+d''}K_s)$ for each $s \in S$. This defines a bfg basis (S,σ') over $\mathcal{P}G_1(X,p)$. Since $[y_s] = f_*[y_s']$ there is a morphism

$$\nu_s: \phi\sigma'(s) = f_*(y_s', C^{d'+d''}K_s) = (f(y_s'), C^{d+d'+d''}K_s) \longrightarrow$$

$$(y_s, C^{d+d'+d''}K_s) = C^{d+d'+d''}\sigma(s)$$

for each $s \in S$. By definition, this collection of morphisms defines a morphism $(1,\nu): (S,\sigma) \longrightarrow (S,\phi\sigma')$, which is an isomorphism by 2.11.

This finishes the proof that (4.3) is an equivalence of categories. The rest of the theorem is now obvious.

In some geometric applications, we will need more general criteria than those of 4.4 under which the homomorphisms of 4.1 are isomorphisms. We give the needed definitions.

Definition 4.5 *A function* $d: |\mathcal{B}| \longrightarrow \mathbb{Z}_+$ *is called* weakly bounded, *if every family* $\{(x_s,K_s) \mid s \in S\}$ *in* $|\mathcal{B}|$ *for which* $\{K_s \mid s \in S\}$ *is bounded in* $|\mathcal{P}|$, *has* $\{d(x_s,K_s) \mid s \in S\}$ *bounded in* $\mathbb{Z}_+$.

We remark that this definition also makes sense for $\mathcal{B} = \mathcal{P}$.

Definition 4.6 *An endomorphism preserving functor* $\phi: \mathcal{B} = \mathcal{P}G \longrightarrow \mathcal{P}'G' = \mathcal{B}'$ *is called* weakly eventually full [*respectively,* weakly eventually faithful] *if there is a weakly bounded function* $d: |\mathcal{B}| \longrightarrow \mathbb{Z}_+$ *such that* (i) [*respectively,* (ii)] *below holds*

(i) *For any* $\varphi' \in \mathcal{B}'(\phi b_1, \phi b_2)$, *there exists* $\varphi \in \mathcal{B}(b_1, C^{d(b_2)}b_2)$ *such that*

$$\begin{array}{ccc} \phi b_1 & \xrightarrow{\ \varphi'\ } & \phi b_2 \\ & \searrow^{\phi\varphi} & \downarrow{\scriptstyle \phi\tau^{d(b_2)}} \\ & & \phi C^{d(b_2)} b_2 \end{array}$$

commutes.

(ii) *If* $\varphi_1, \varphi_2 \in \mathcal{B}(b_1, b_2)$ *have* $\phi\varphi_1 = \phi\varphi_2$, *then*

$$\tau^{d(b_2)}\varphi_1 = \tau^{d(b_2)}\varphi_2 \colon b_1 \longrightarrow C^{d(b_2)} b_2.$$

The following theorem admits of an easy, though lengthy, proof which is given towards the end of this section.

Theorem 4.7 *Assume that* $\phi\colon \mathcal{P}G \longrightarrow \mathcal{P}'G'$ *is endomorphism preserving, weakly eventually full and faithful, and satisfies* (2.8) *and* (2.9). *Then* $\phi_!\colon$ bfgfreeR$\mathcal{P}$G-*mod* $\longrightarrow$ bfgfreeR$\mathcal{P}$'G'-*mod is full and faithful. If, in addition, every bfg basis* $\sigma'\colon S' \longrightarrow \mathcal{P}'G'$ *is isomorphic to* $\phi\sigma\colon S' \longrightarrow \mathcal{P}G$ *for some bfg basis* $\sigma\colon S \longrightarrow \mathcal{P}G$ *then the above functor* $\phi_!$ *is an equivalence of categories and induces isomorphisms*

$$\phi_*\colon K_1(R\mathcal{P}G) \longrightarrow K_1(R\mathcal{P}'G') \quad \text{and} \quad \phi_*\colon Wh(R\mathcal{P}G) \longrightarrow Wh(R\mathcal{P}'G').$$

There are two special cases of 4.7 which are important in applications. First let $\mathcal{P}$ and $\mathcal{P}'$ be as above, and let $G'\colon \mathcal{P}' \longrightarrow \mathcal{G}poid$. Then any functor $\psi\colon \mathcal{P} \longrightarrow \mathcal{P}'$ induces a functor $\hat{\psi}\colon \mathcal{P}(G'\psi) \longrightarrow \mathcal{P}'G'$ by setting

$$\hat{\psi}[(\omega, i)\colon (x, K) \longrightarrow (y, L))] = [(\omega, \psi(i))\colon (x, \psi(K)) \longrightarrow (y, \psi(L))].$$

Corollary 4.8 *In the above situation suppose that* ψ *has the following properties:*

(i) $\psi C \subseteq C^n\psi$ *for some* $n \in \mathbb{Z}_+$,

(ii) $C\psi \subseteq \psi C^m$ *for some* $m \in \mathbb{Z}_+$,

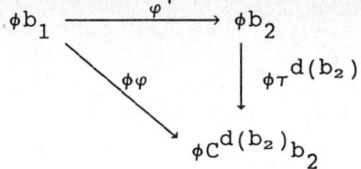

(iii) *For some weakly bounded function* d: $|\mathcal{F}| \longrightarrow \mathbb{Z}_+$, *if* $\psi L \subseteq \psi K$ *then* $L \subseteq C^{d(K)}K$,

(iv) $\{\psi(K_0)|K_0$ *minimal in* $\mathcal{F}\}$ *is bounded in* $|\mathcal{F}'|$,

(v) *There exists a weakly bounded function* $d'\colon |\mathcal{F}'| \longrightarrow \mathbb{Z}_+$ *such that for any* $K' \in |\mathcal{F}'|$, *there exists some* $K \in |\mathcal{F}|$ *with* $K' \subseteq \psi K \subseteq C^{d'(K')}K'$,

(vi) *For any family* $\{K_s|s\in S\}$ *in* $|\mathcal{F}|$, *if* $\{\psi(K_s)|s\in S\}$ *is bounded (in* $|\mathcal{F}'|$*), then so is* $\{K_s|s\in S\}$ *(in* $|\mathcal{F}|$*).*

Then $\hat{\psi}$ *satisfies the hypotheses of* 4.7. *Hence it induces an equivalence of categories*

$$\hat{\psi}_!\colon \text{bfgfreeR}\mathcal{F}(G'\psi)\text{-}mod \longrightarrow \text{bfgfreeR}\mathcal{F}'G'\text{-}mod$$

and isomorphisms

$$\hat{\psi}_*\colon K_1(R\mathcal{F}(G'\psi)) \longrightarrow K_1(R\mathcal{F}'G'), \text{ and } \hat{\psi}_*\colon Wh(R\mathcal{F}(G'\psi)) \longrightarrow Wh(R\mathcal{F}'G').$$

The proof is temporarily deferred.

In the second corollary we consider a natural transformation $\nu\colon G \longrightarrow H$ between functors $G,H\colon \mathcal{F} \longrightarrow \mathcal{G}poid$. Let $\mathcal{F}\nu\colon \mathcal{F}G \longrightarrow \mathcal{F}H$ be the functor of II.5 which maps the morphism $(\omega,i)\colon (x,K) \longrightarrow (y,L)$ to $(\nu_L(\omega),i)\colon (\nu_K(x),K) \longrightarrow (\nu_L(y),L)$.

We say that G and H are *weakly coextensive*, if there is a weakly bounded function $d\colon |\mathcal{F}| \longrightarrow \mathbb{Z}_+$ such that for every $K\in|\mathcal{F}|$, if H(K) is non empty, then so is $G(C^{d(K)}K)$, and if G(K) is non empty, then so is $H(C^{d(K)}K)$.

Definition 4.9 *Let* $G,H\colon \mathcal{F} \longrightarrow \mathcal{G}poid$ *be functors and let* $\nu\colon G \longrightarrow H$ *be a natural transformation. We say that* ν *is* weakly eventually epimorphic, *respectively,* weakly eventually monomorphic, *on* π_i $(i\in\mathbb{Z}_+)$, *if there exists a bounded function* d: $|\mathcal{P}| \longrightarrow \mathbb{Z}_+$ *such that* (i), *respectively* (ii), *holds for every* $(x,K) \in |\mathcal{P}G|$

(i) *In the diagram*

$$\begin{array}{ccc} \pi_i(\mathcal{P}G)(C^{d(K)}(x,K)) & \xrightarrow{\ \nu_*\ } & \nu^*\pi_i(\mathcal{P}H)(C^{d(K)}(x,K)) \\ & & \uparrow{\scriptstyle \tau_*^{d(K)}} \\ & & \nu^*\pi_i(\mathcal{P}H)(x,K) \end{array}$$

the image of $\tau_*^{d(K)}$ *is contained in the image of* ν_*,

(ii) *If in the diagram*

$$\begin{array}{ccc} \pi_i(\mathcal{P}G)(x,K) & \xrightarrow{\ \nu_*\ } & \nu^*\pi_i(\mathcal{P}H)(x,K) \\ \downarrow{\scriptstyle \tau_*^{d(K)}} & & \\ \pi_i(\mathcal{P}G)(C^{d(K)}(x,K)). & & \end{array}$$

$\alpha_1, \alpha_2 \in \pi_i(\mathcal{P}G)(x,K)$ *are elements with* $\nu_*(\alpha_1) = \nu_*(\alpha_2)$, *then* $\tau_*^{d(K)}(\alpha_1) = \tau_*^{d(K)}(\alpha_2)$.

In this definition we have written ν^* and ν_* in place of the more elaborate $(\mathcal{P}\nu)^*$ and $\pi_i(\mathcal{P}\nu)$.

Corollary 4.10 *Let* ν: G $\longrightarrow$ H *be a natural transformation between functors* $\mathcal{P} \longrightarrow \mathcal{G}\mathit{poid}$. *If* G *and* H *are weakly coextensive, and* ν *is weakly eventually epi- and monomorphic on* π_0 *and* π_1, *then* $\mathcal{P}\nu\colon \mathcal{P}G \longrightarrow \mathcal{P}H$ *satisfies the hypotheses of* 4.7. *Hence*

$$\mathcal{P}\nu_!\colon \text{bfgfree}R\mathcal{P}G\text{-}\mathit{mod} \longrightarrow \text{bfgfree}R\mathcal{P}H\text{-}\mathit{mod}$$

is an equivalence of categories and

$$\mathcal{P}\nu_*\colon K_1(R\mathcal{P}G) \longrightarrow K_1(R\mathcal{P}H) \text{ and } \mathcal{P}\nu_*\colon \mathrm{Wh}(R\mathcal{P}G) \longrightarrow \mathrm{Wh}(R\mathcal{P}H)$$

are isomorphisms.

The proof follows the same strategy as the proof of II.5.1, the main technical difference being that objects of the form C^dK with d constant are replaced by objects of the form $C^{d(K)}K$ where d is a weakly bounded function. We leave the details to the reader.

We remark that if G and H are coextensive and $\upsilon: G \longrightarrow H$ induces isomorphisms $\upsilon_*: \pi_i(\mathcal{P}G) \longrightarrow \upsilon^*\pi_i(\mathcal{P}H)$, then the hypotheses, and hence the conclusions, of 4.10 hold.

Example 4.11 Let $Z = \mathbb{R}^k$. Corresponding to the euclidean metric, ρ_1, and the box metric, ρ_2, are the two boundedness control structures (P_i,C), i=1,2, where $P_i = \{B_i(z,r) \mid z\in\mathbb{R}^k,\ r\in\mathbb{R}_+\}$, $CB_i(z,r) = B_i(z,r+1)$, and $B_i(z,r) = \{w\in\mathbb{R}^k \mid \rho_i(w,z)\leq r\}$, cf. II.1.3. There are also the boundedness control structures (P_i',C) with $P_i' = \{B_i(z,r) \mid z\in\mathbb{Z}^k,\ r\in\mathbb{Z}_+\}$ (i=1,2). Finally let (P_c,C) be the cubical control structure on $\mathbb{R}^k$, cf. II.1.5.

The corresponding categories with endomorphisms fit into a commutative diagram of functors

$$\begin{array}{ccccc} \mathcal{P}_1' & \xrightarrow{\psi'} & \mathcal{P}_2' & \xrightarrow{j_c} & \mathcal{P}_c \\ \downarrow{\scriptstyle j_1} & & \downarrow{\scriptstyle j_2} & & \\ \mathcal{P}_1 & \xrightarrow[\psi]{} & \mathcal{P}_2 & & \end{array}$$

where $j_1, j_2,$ and j_c are inclusions of subcategories while $\psi' = \psi|\mathcal{P}_1'$ and ψ maps $B_1(z,r)$ to $B_2(z,r)$.

Suppose now that $p: X \longrightarrow \mathbb{R}^k$, and consider the functors $G_i': \mathcal{P}_i \longrightarrow \mathcal{G}\mathit{poid}$ defined for i = 1,2,c to have $G_i'(K) = G_1(p^{-1}K)$ for $K\in|\mathcal{P}_i|$. Define the functors $G_i'': \mathcal{P}_i' \longrightarrow \mathcal{G}\mathit{poid}$ similarly for i = 1,2 and observe that $G_1'j_1 = G_1''$ and $G_2'' = G_2'j_2 = G_cj_c$. Furthermore, since $B_1(x,r) \subseteq B_2(x,r)$, there are natural transformations $\upsilon': G_1' \longrightarrow G_2'\psi$ and $\upsilon'': G_1'' \longrightarrow G_2''\psi'$. By combining this information we get the following commutative diagram of functors

$$\begin{array}{ccccccc}
\mathscr{P}'_1 G''_1 & \xrightarrow{\mathscr{P}_1 \nu''} & \mathscr{P}'_1(G''_2\psi') & \xrightarrow{\hat{\psi}'} & \mathscr{P}'_2 G''_2 & \xrightarrow{\hat{j}_c} & \mathscr{P}_c G'_c \\
\downarrow{\scriptstyle \hat{j}_1} & & \downarrow{\scriptstyle \hat{j}_1} & & \downarrow{\scriptstyle \hat{j}_2} & & \\
\mathscr{P}_1 G'_1 & \xrightarrow{\mathscr{P}_1 \nu'} & \mathscr{P}_1(G'_2\psi) & \xrightarrow{\hat{\psi}} & \mathscr{P}_2 G'_2 & &
\end{array}$$

Proposition 4.12 *All the functors in the above diagram induce equivalences on the categories of bfg free* $R\mathscr{P}G$ *modules. Hence all the induced homomorphisms at the* K_1*- and the* Wh*-levels are isomorphisms.*

Proof: We only have to show that each of the functors j_1, j_2, j_c, ψ, and ψ' satisfies the conditions of 4.8 and that each of the natural transformations ν', and ν'' satisfy the hypotheses of 4.10. We treat only j_2, ψ, and ν', and leave the other, very similar, cases to the reader.

For j_2, the proofs of (i) (with $n=0$), (ii) (with $m=0$), (iii) (with $d(K)=0$ for all K), (iv), and (vi) are obvious. To verify (v), let $[r]$ be the integer part of $r\in\mathbb{R}$ as usual, and for $x = (x_1,x_2,\ldots,x_k) \in \mathbb{R}^k$, let $[x] = ([x_1],[x_2],\ldots,[x_k]) \in \mathbb{Z}^k$. The inclusions

$$B_2(x,r) \subseteq B_2([x],[r]+2) \subseteq B_2(x,r+3)$$

show that (v) holds with $d'(K') = 3$ for all K'.

For ψ, conditions (i) (with $n=0$), (ii) (with $m=0$), (iv), and (vi) are equally obvious. Since $B_1(x,r) \subseteq B_2(x,r) \subseteq B_1(x,r\sqrt{k})$, condition (iii) holds with $d(B_1(x,r)) = [r(\sqrt{k}-1)]+1$, and (v) holds with $d'(B_2(x,r)) = [r(\sqrt{k}-1)]+1$.

Finally, we show that ν' satisfies all the hypotheses of 4.10. The inclusions $B_1(x,r) \subseteq B_2(x,r) \subseteq B_1(x,r\sqrt{k})$ suffice to prove that G'_1 and $G'_2\psi$ are weakly coextensive. Let $d(B_1(z,r)) = [r(\sqrt{k}-1)]+1$ as above. The typical object (x,K) of $\mathscr{P}_1 G'_1$ has the form $(x,B_1(z,r))$ with $x\in p^{-1}B_1(z,r)$, and for that object the diagram of 4.9.(i) takes the form

$$\begin{array}{ccc}
\pi_i(p^{-1}B_1(z,r[\sqrt{k}]+1),x) & \xrightarrow{\ \nu_*\ } & \pi_i(p^{-1}B_2(z,r[\sqrt{k}]+1),x) \\
 & & \Big\uparrow \tau_*^{d(B_1(z,r))} \\
 & & \pi_i(p^{-1}B_2(z,r),x)
\end{array}$$

where both maps are induced by inclusions. Since $B_2(z,r) \subseteq B_1(z,r[\sqrt{k}]+1)$, the conclusion in 4.9.(i) is seen to hold for every $i\in\mathbb{Z}_+$. In particular, ν is weakly eventually epimorphic on π_0 and π_1. We leave to the reader the similar proof that ν is weakly eventually monomorphic on π_0 and π_1.

We return now to the proofs that were deferred.

Proof of 4.7: To prove that the functor $\phi_!$ of (4.7) is full and faithful it suffices to check

(4.13) For every pair (S,σ) and (T,ρ) of bfg bases over $\mathscr{G}G$ the map $\phi_!\colon R\mathscr{G}G\text{-}mod(F(\sigma),F(\rho)) \longrightarrow R\mathscr{G}'G'\text{-}mod(F\phi(\sigma),F\phi(\rho))$ is bijective.

Here we have identified $F\phi(\sigma)$ with $\phi_!F(\sigma)$ by means of the isomorphism in 1.11. This identification permits us to give an easy description of $\phi_!$. In fact, from 1.3 we know that any $\varphi\colon F(\sigma) \longrightarrow F(\rho)$ is given by a set of elements

$$(4.14) \qquad \varphi_d(\sigma(s))(1,s) = \sum_{I_s} r_{is}(\beta_{is},t_{is}) \in F(\rho)(C^d\sigma(s)), \quad s\in S,$$

where $d\in\mathbb{Z}_+$, I_s is a finite indexing set, $r_{is}\in R$, $t_{is}\in T$, and $\beta_{is} \in \mathscr{G}G(\rho(t_{is}),C^d\sigma(s))$. For this φ the morphism $\phi_!\varphi\colon F(\phi\sigma) \longrightarrow F(\phi\rho)$ is given by the elements

$$(4.15) \qquad (\phi_!\varphi)_{nd}(\phi\sigma(s))(1,s) = \sum_{I_s} r_{is}(\hat{\beta}_{is},t_{is}) \in F(\phi\rho)(C^{nd}\phi\sigma(s)),$$

where $\hat{\beta}_{is}$ is the composition

$$\phi\rho(t_{is}) \xrightarrow[\phi(\beta_{is})]{} \phi C^d\sigma(s) \xrightarrow[\alpha^{(d)}]{} C^{nd}\phi\sigma(s)$$

with $\alpha^{(d)}$ the dth "iteration" of the natural transformation $\alpha\colon \phi C \longrightarrow C^n\phi$ entering into the definition of ϕ being endomorphism preserving (one has $\alpha^{(d)} = (C^{(d-1)n}\phi\alpha)\alpha^{(d-1)}C)$).

To check that the map in (4.13) is monic, we assume that φ, given by (4.14), has $\phi_!\varphi=0$. This means that there is some $d' \geq d$ such that the elements of (4.15) vanish under the maps

$$\tau_*^{n(d'-d)}\colon F(\phi\rho)(C^{nd}\sigma(s)) \longrightarrow F(\phi\rho)(C^{nd'}\sigma(s)) .$$

In (4.14) we can replace d by d' and β_{is} by $\tau^{d'-d}\beta_{is}$. Since the diagram

$$\begin{array}{ccccc} \phi\rho(t_{is}) & \xrightarrow{\phi(\beta_{is})} & \phi C^d\sigma(s) & \xrightarrow{\alpha^{(d)}} & C^{dn}\phi(s) \\ & & \downarrow{\scriptstyle \tau^{d'-d}} & & \downarrow{\scriptstyle \tau^{n(d'-d)}} \\ & & \phi C^{d'}\sigma(s) & \xrightarrow[\alpha^{(d')}]{} & C^{d'n}\phi(s) \end{array}$$

is easily seen to commute, in (4.15) this will replace nd by nd' and $\hat{\beta}_{is}$ by $\tau^{n(d'-d)}\hat{\beta}_{is}$. Altogether we may assume that *the elements of* (4.15) *vanish*. The precise meaning of this vanishing is the following. For each $t\in T$, each $s\in S$, and each $\beta' \in \mathcal{F}'G'(\rho(t),C^{nd}\sigma(s))$ we let $I_{s,t,\beta'} = \{i\in I_s \mid \hat{\beta}_{is}=\beta'\}$. Then

$$(4.16) \qquad \sum \{r_{is} \mid i\in I_{s,t,\beta'}\} = 0 \quad \text{for all} \quad s,t,\beta'.$$

Now recall the natural transformation $\beta\colon C\phi \longrightarrow \phi C^m$ which is also part of the definition of ϕ being endomorphism preserving. It has iterates $\beta^{(d)}\colon C^d\phi \longrightarrow \phi C^{md}$ given by $\beta^{(d)} = (\beta C^{m(d-1)})(C\beta^{(d-1)})$. If $i \in I_{s,t,\beta'}$, then there is a commutative diagram

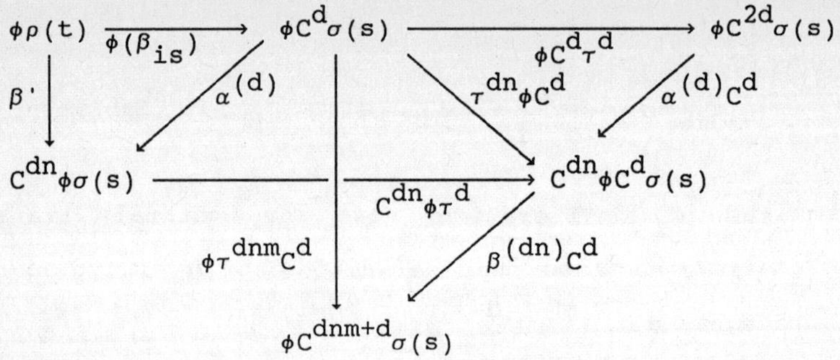

In fact $(\alpha^{(d)}C^d)(\phi C^d\tau^d) = (\alpha^{(d)}C^d)(\phi\tau^d C^d) = (\alpha^{(d)}(\phi\tau^d))C^d = \tau^{dn}\phi C^d$ by the characteristic property of $\alpha^{(d)}$. Similarly the characteristic property of $\beta^{(d)}$ shows that $(\beta^{(dn)}C^d)(\tau^{dn}\phi C^d) = \phi\tau^{dnm}C^d$. Finally, $\alpha^{(d)}\phi(\beta_{is}) = \hat{\beta}_{is} = \beta'$ because $i \in I_{s,t,\beta'}$ and $(\alpha^{(d)}C^d)(\phi C^d\tau^d) = (C^{dn}\phi\tau^d)\alpha^{(d)}$ by naturality of $\alpha^{(d)}$.

We use the commutative diagram to conclude that

(4.17) For fixed s,t,β', the map $\phi(\tau^{dnm}C^d\circ\beta_{is})\colon \phi\rho(t) \longrightarrow \phi C^{dnm+d}\sigma(s)$ is independent of $i\in I_{s,t,\beta'}$.

Since $\{\sigma(s) \mid s\in S\}$ is bounded, so is $d(\sigma(s))$. Let $d(\sigma(s)) \leq N$ for all s. By 4.6.(ii), it then follows that

(4.18) For fixed s,t,β', the map $\tau^{N+dnm}C^d\circ\beta_{is}\colon \rho(t) \longrightarrow C^{N+dnm+d}\sigma(s)$ is independent of $i\in I_{s,t,\beta'}$.

Finally, in the description of φ (see 4.14), we can replace each $\varphi_d(\sigma(s))(1,s) \in F(\rho)(C^d\sigma(s))$ by its image under τ_*^{N+dnm} in $F(\rho)(C^{d+N+dnm}\sigma(s))$. But from (4.18) and (4.16) it is seen that each of these images vanishes. Hence $\varphi=0$ and we have shown that $\phi_!$ is a monomorphism.

Next we show that $\phi_!$ is epic. The idea is similar to the one above. Any $\psi'\colon F(\phi\sigma) \longrightarrow F(\phi\rho)$ is given by a set of elements

(4.19) $\quad \psi_e'(\phi\sigma(s))(1,s) = \sum_{J_s} r_{js}'(\gamma_{js}', t_{js}) \quad \in \quad F(\phi\rho)(C^e\sigma(s)) \; , \quad s\in S \; ,$

where J_s is a finite index set, $e \in \mathbb{Z}_+$, $r_{js}' \in R$, $t_{js} \in T$, and $\gamma_{js}' \in \mathcal{G}'G'(\phi\rho(t_{js}), C^e\phi(\sigma(s))$. For any $f \in \mathbb{Z}_+$, $\phi C^{me}\tau^f = \phi\tau^f C^{me}$, so

$$\begin{array}{ccccc}
\phi\rho(t_{js}) & \xrightarrow{\gamma_{js}'} & C^e\phi\sigma(s) & \xrightarrow{C^e\phi\tau^f} & C^e\phi C^f\sigma(s) \\
 & & \Big\downarrow \beta^{(e)} & & \Big\downarrow \beta^{(e)}C^f \\
 & & \phi C^{me}\sigma(s) & \xrightarrow[\phi\tau^f C^{me}]{} & \phi C^{me+b}\sigma(s)
\end{array}$$

commutes and by 4.5.(ii), we can choose f (independent of j,s) so big that

(4.20) $\quad (\phi\tau^f C^{me})\beta^{(e)}\gamma_{js}' = \phi(\beta_{js})$ for some $\beta_{js} \in \mathcal{G}G(\rho(t_{js}), C^{me+f}\sigma(s))$

Now the elements

(4.21) $\psi_{me+f}(\sigma(s))(1,s) = \sum_{J_s} r_{js}'(\beta_{js}, t_{js}) \quad \in \quad F(\rho)(C^{me+f}\sigma(s)) \; , \quad s\in S$

define a morphism $\psi\colon F(\sigma) \longrightarrow F(\rho)$. We claim that $\phi_!(\psi) = \psi'$ (so that $\phi_!$ is epic). To prove this, it suffices to check that the diagram

$$\begin{array}{ccc}
\phi\rho(t_{js}) & \xrightarrow{\phi(\beta_{js})} & \phi C^{me+f}\sigma(s) \\
\Big\downarrow \gamma_{js}' & & \Big\downarrow \alpha^{(me+f)} \\
C^e\phi\sigma(s) & & C^{(me+f)n}\phi\sigma(s) \\
 & \searrow^{\tau^{M-e}C^e\phi} \quad \swarrow_{\tau^{M-(me+f)n}C^{(me+f)n}\phi} & \\
 & C^M\phi\sigma(s) &
\end{array}$$

commutes for some sufficiently large M (independent of j and s).

The definition of $\phi(\beta_{js})$ in (4.20) is a composition starting with γ_{js}' so it suffices to check that the composition of the unbroken arrows in the diagram

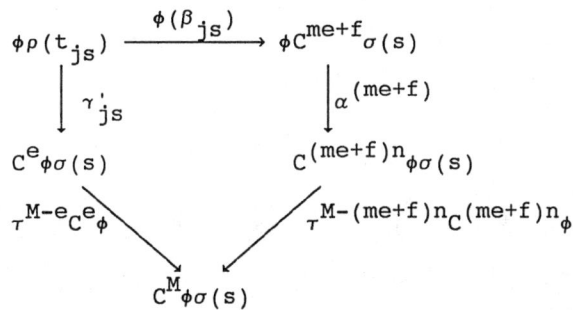

$$
\begin{array}{ccccc}
C^e\phi & \xrightarrow{\beta^{(e)}} & \phi C^{me} & \xrightarrow{\phi\tau^f C^{me}} & \phi C^{me+f} \\
 & & \Big\downarrow{\scriptstyle \alpha^{(me)}} & & \Big\downarrow{\scriptstyle \alpha^{(me+f)}} \\
 & & C^{men}\phi & \xrightarrow{\tau^{fn}C^{men}\phi} & C^{(me+f)n}\phi \\
 & & & \searrow{\scriptstyle \tau^{M-men}C^{men}\phi} & \Big\downarrow{\scriptstyle \tau^{M-(me+f)n}C^{(me+f)n}\phi} \\
 & & & & C^M\phi
\end{array}
$$

equals $\tau^{M-e}C^e\phi$ for a suitable M. The dotted arrows make the diagram commute (use the definition of iterates of α). In fact we claim that $M = men+e$ will do, i.e. we claim that

$$
\begin{array}{ccc}
C^e\phi & \xrightarrow{\tau^{men}C^e\phi} & C^{men+e}\phi \\
\Big\downarrow{\scriptstyle \beta^{(e)}} & & \Big\uparrow{\scriptstyle \tau^e C^{men}\phi} \\
\phi C^{me} & \xrightarrow[\alpha^{(me)}]{} & C^{men}\phi
\end{array}
$$

commutes. For e=1 this is seen as follows: The diagram

$$
\begin{array}{ccc}
C\phi & \xrightarrow{\tau C\phi} & C^2\phi \\
\Big\downarrow{\scriptstyle \beta} & & \Big\downarrow{\scriptstyle C\beta} \\
\phi C^m & \xrightarrow{\tau\phi C^m} & C\phi C^m \\
\Big\downarrow{\scriptstyle \alpha^{(m)}} & & \Big\downarrow{\scriptstyle C\alpha^{(m)}} \\
C^{mn}\phi & \xrightarrow{\tau C^{mn}\phi} & C^{mn+1}\phi
\end{array}
$$

commutes by naturality of τ. Also $C\beta\circ\tau C\phi = C\beta\circ C\tau\phi = C(\beta\circ\tau\phi) = C\phi\tau^m$ by the characteristic property of β. Hence $C\alpha^{(m)}\circ C\beta\circ\tau C\phi = C(\alpha^{(m)}\circ\phi\tau^m) = C(\tau^{mn}\phi) = \tau^{mn}C\phi$ as desired (using the characteristic property of $\alpha^{(m)}$). The case for e>1 is left to the reader.

This finishes the proof of (4.13). Thus we have established that $\phi_!$ is full and faithful.

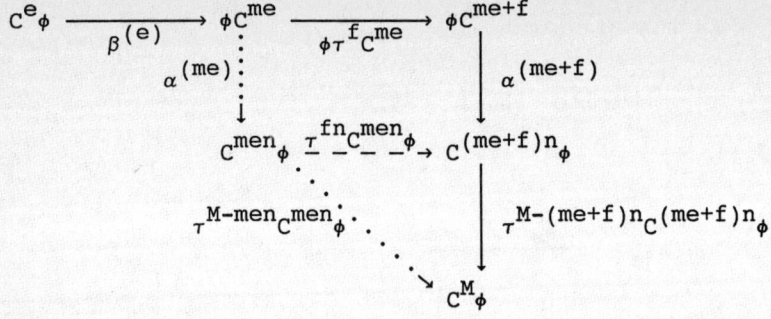

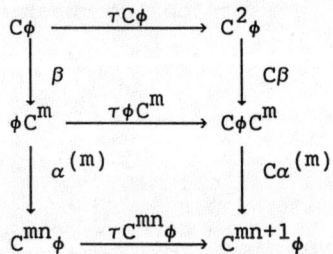

If the extra hypothesis of 4.7 holds, then every bfg free $R\mathcal{G}'G'$ module is isomorphic to one of the form $\phi_! F(\sigma)$ where (S,σ) is bfg. Therefore, $\phi_!$ is an equivalence of categories. The rest of Theorem 4.7 follows easily.

Proof of 4.8: Properties (i) and (ii) simply state that ψ is endomorphism preserving. It follows easily that $\hat{\psi}$ is endomorphism preserving.

Properties (iv) and (i), together with the functoriality of ψ, guarantee that (2.8) holds for $\hat{\psi}$. Also, (v) and (iii) imply that (2.9) holds for $\hat{\psi}$.

From (iii), one sees that ψ is weakly eventually full and faithful (the latter being trivial). Since $(G'\psi)(K) = G'(\psi K)$, the same is true of $\hat{\psi}$.

Finally, let (S,σ') be a bfg basis with $\sigma'(s) = (x_s', K_s')$. Using (v), pick $K_s \in |\mathcal{G}|$ with $K_s' \subseteq \psi K_s \subseteq C^{d'(K_s')} K_s'$ and define $\sigma\colon S \longrightarrow |\mathcal{G}(G'\psi)|$ by $\sigma(s) = (\bar{x}_s', K_s)$ where $\bar{x}_s' \in G'\psi(K_s)$ is the image of x_s' under $G'(K_s') \longrightarrow G'(\psi K_s)$. By (vi), the basis (S,σ) is bounded. Also, for any $K \in |\mathcal{G}|$ $\{s \mid K_s \subseteq K\} \subseteq \{s \mid K_s' \subseteq \psi K\}$, so (S,σ) is locally finite. Hence (S,σ) is bfg. Therefore, by (2.11), the obvious map of bases $(S,\sigma') \longrightarrow (S,\psi\sigma)$ is an isomorphism.

We have verified that all the hypotheses of 4.7 hold for $\hat{\psi}$. Hence 4.8 is a corollary of 4.7.

5. Involutions on $K_1(R\mathcal{P}G)$ and $Wh(R\mathcal{P}G)$.

Let $R,\mathcal{P}$ and $G: \mathcal{P} \longrightarrow \mathcal{G}poid$ be as in section 2. In this section we define the notion of a Stiefel-Whitney class on $\mathcal{P}G$ and show that every such class gives rise to involutions, denoted $(\ )^*$, on $K_1(R\mathcal{P}G)$ and $Wh(R\mathcal{P}G)$. Such involutions are used in Chapter VI in the proof of the Realization Theorem and the statement of the Duality Theorem.

Classically, the geometrically relevant involution on $Wh(\pi)$, π a group, involves a Stiefel-Whitney class, $w_1: \pi \longrightarrow \mathbb{Z}_2$, and the antiautomorphism $g \longrightarrow g^{-1}$ on π. In our setup the antiautomorphism becomes the map $\gamma \longrightarrow \gamma^{-1}$ from $G(K)(x,y)$ to $G(K)(y,x)$ and w_1 generalizes as follows. (We view $\mathbb{Z}_2 = \{\pm 1\}$ as a groupoid with one object and two morphisms as usual).

Definition 5.1 *A* Stiefel-Whitney class *on* $\mathcal{P}G$ *is a functor* $w_1: \mathcal{P}G \longrightarrow \mathbb{Z}_2$ *such that*

$$w_1(\tau_{(x,K)}: (x,K) \longrightarrow C(x,K)) = 1$$

for each $(x,K) \in |\mathcal{P}G|$.

We remark that Stiefel-Whitney classes occur naturally in the study of bc manifolds and refer the reader to VI.7 for such a discussion. We also note that since Stiefel-Whitney classes on $\mathcal{P}G$ are functors, there is the notion of two Stiefel-Whitney classes being naturally equivalent.

Theorem 5.2 *Every Stiefel-Whitney class on* $\mathcal{P}G$ *gives rise to involutions* $*$ *on* $K_1(R\mathcal{P}G)$ *and* $Wh(R\mathcal{P}G)$. *Naturally equivalent Stiefel-Whitney classes give rise to identical involutions.*

The involutions on $K_1(R\mathcal{P}G)$ and $Wh(R\mathcal{P}G)$ will be induced by an operation $*$ which assigns to any morphism $\varphi: F(\sigma) \longrightarrow F(\rho)$ between bfg based $R\mathcal{F}G$ modules, an adjoint morphism $\varphi^*: F(\rho) \longrightarrow F(\sigma)$.

For concreteness, we shall only treat the case where $\mathcal{F}$ *is associated with a boundedness control structure* (P,C) *on a space* Z *and* $G = G_1(X,p)$ *for some bc space* (X,p) *over* Z. *The results of this section, however, are valid for any functor* $G: \mathcal{F} \longrightarrow \mathcal{G}poid$. The definition of the adjoint morphism φ^* uses a new description of the $R\mathcal{F}G$ module morphisms from $F(\sigma)$ to $F(\rho)$ for which a little bit of notation is needed.

For any two bfg bases (S,σ), (T,ρ) over $\mathcal{F}G$ with $\sigma(s) = (x_s,K_s)$ and $\rho(t) = (y_t,L_t)$, we let $M(\sigma,\rho)$ be the set

$$\{(s,t,\gamma,n) \mid s\in S,\ t\in T,\ \gamma \text{ is a path in } X \text{ from } y_t \text{ to } x_s,\ n\in R\}$$

A pair (I,a), where I is an indexing set and $a: I \longrightarrow M(\sigma,\rho)$ is a function, will be called d-*admissible* if $a(i) = (s_i,t_i,\gamma_i,n_i)$ satisfies

(5.3) $\quad L_{t_i} \subseteq C^d K_{s_i}$ and γ_i is a path in $p^{-1}(C^d K_{s_i})$; and

(5.4) $\quad$ For all $s\in S$, the set $I_s = \{i\in I \mid s_i = s \text{ and } n_i \neq 0\}$ is finite.

The pair (I,a) is called *admissible*, if it is d-admissible for some $d\geq 0$. We denote by $A(\sigma,\rho)$ (respectively $A_d(\sigma,\rho)$) the class of all admissible (respectively d-admissible) pairs. Then $A_d(\sigma,\rho) \subseteq A_{d+1}(\sigma,\rho)$ and $A(\sigma,\rho) = \cup_d A_d(\sigma,\rho)$.

By 1.3, any $(I,a) \in A_d(\sigma,\rho)$ determines a morphism $\varphi^{(a)}: F(\sigma) \longrightarrow F(\rho)$ by setting

(5.5) $$\varphi^{(a)}(1,s) = \Sigma_i n_i(\hat{\gamma}_i,t_i) \in F(\rho)(C^d\sigma(s)) \ , \ s\in S$$

where the sum extends over the finite set I_s of (5.4), and $\hat{\gamma}_i = (\text{cls}\ \gamma_i,\ \text{incl}): \rho(t_i) \longrightarrow C^d\sigma(s)$. (As usual, if the set I_s is empty the sum is interpreted to be 0.)

It follows easily from 1.3 that every $\varphi: F(\sigma) \longrightarrow F(\rho)$ is of the form $\varphi^{(a)}$ for some admissible $a: I \longrightarrow M(\sigma,\rho)$. Hence, in order to get a complete description of $R\mathcal{F}G\text{-}\mathcal{mod}(F(\sigma),F(\rho))$ in terms of admissible pairs, it suffices to decide when a d-admissible pair (I,a) and an e-admissible pair (J,b) have $\varphi^{(a)} = \varphi^{(b)}$.

For any d-admissible pair (I,a) and any integer $f \geq d$, there is an equivalence relation $\sim_f$ on I, defined as follows:

$$(5.6) \qquad i \sim_f j \Leftrightarrow s_i = s_j \ (= s, \text{ say}),\ t_i = t_j \text{ and } \mathrm{cls}\gamma_i = \mathrm{cls}\gamma_j \text{ in } p^{-1}C^f K_s.$$

The equivalence classes in I under the relation $\sim_f$ are the subsets of the form

$$I_f(s,t,\beta) = \{i \in I \mid s_i = s,\ t_i = t,\ \mathrm{cls}\gamma_i = \mathrm{cls}\beta \text{ in } p^{-1}C^f K_s\}$$

where $s \in S$, $t \in T$ and β is any path from y_t to x_s in $p^{-1}C^f K_s$. (Of course, for most choices of (s,t,β), the equivalence class $I_f(s,t,\beta)$ will be empty.)

We note that by (5.4), the sum $\Sigma\{n_i \mid i \in I_f(s,t,\beta)\}$ is well defined for each equivalence class. If (I,a) is given as above, while (J,b) has $b(j) = (s_j', t_j', \gamma_j', m_j)$, the desired criterion for the equality of $\varphi^{(a)}$ and $\varphi^{(b)}$ can be expressed in terms of such sums as follows:

Proposition 5.7 *Let* $a\colon I \longrightarrow M(\sigma,\rho)$ *and* $b\colon J \longrightarrow M(\sigma,\rho)$ *be* d- *and* e-*admissible, respectively. Then* $\varphi^{(a)} = \varphi^{(b)}\colon F(\sigma) \longrightarrow F(\rho)$ *if and only if there exists an integer* $f \geq d,e$ *such that*

$$(5.8) \qquad \Sigma\{n_i \mid i \in I_f(s,t,\beta)\} = \Sigma\{m_j \mid j \in J_f(s,t,\beta)\}$$

holds for all $s \in S$, *all* $t \in T$, *and all paths* β *from* y_t *to* x_s *in* $p^{-1}C^f K_s$.

Proof: By 1.3, $\varphi^{(a)} = \varphi^{(b)}$ if and only if there exists some $f \geq d,e$ such that for every $s \in S$, $\varphi^{(a)}(1,s)$ and $\varphi^{(b)}(1,s)$ meet under the homomorphisms

$$F(\rho)(C^d\sigma(s)) \xrightarrow{\tau_*^{f-d}} F(\rho)(C^f\sigma(s)) \xleftarrow{\tau_*^{f-e}} F(\rho)(C^e\sigma(s)).$$

A straightforward computation shows that for every $s \in S$, the element $\varphi^{(a)}(1,s)$ defined in (5.5) satisfies

$$\tau_*^{f-d}(\varphi^{(a)}(1,s)) = \Sigma\, n(s,t,\beta)(\hat{\beta},t) \in F(\rho)(C^f(\sigma(s))$$

where $s \in S$ is fixed, t ranges over T, β ranges over a set of representatives for all path classes from x_s to y_t in $p^{-1}C^fK_s$, $\hat{\beta} = (\mathrm{cls}\beta,\ \mathrm{incl})\colon (y_t,L_t) \longrightarrow (x_s,C^fK_s)$, and $n(s,t,\beta)$ is the sum on the left hand side of (5.8). Since there is a similar formula for $\tau_*^{f-e}(\varphi^{(b)}(1,s))$, the conclusion of 5.7 follows.

We need a description of compositions in our present terms. To give it, let (U,ψ) be another bfg basis over $\mathcal{P}G$ and let $b\colon J \longrightarrow M(\rho,\psi)$ with $b(j) = (t_j',u_j,\delta_j,m_j)$ define an e-admissible pair. We define $c\colon K \longrightarrow M(\sigma,\psi)$ by the following formulas:

$$(5.9)\qquad K = \{(i,j) \in I{\times}J \mid t_j = t_j'\};\ \text{and}$$
$$c(i,j) = (s_i,u_j,\delta_j\cdot\gamma_i,m_jn_i)\ ,\quad (i,j) \in K.$$

A straightforward argument then proves the following proposition:

Proposition 5.10 *If* (I,a) *and* (J,b) *are* d- *and* e-*admissible, respectively, then* (K,c) *is* (d+e)-*admissible and*

$$\varphi^{(c)} = \varphi^{(b)}\varphi^{(a)}\colon F(\sigma) \longrightarrow F(\psi).$$

With the above description of morphisms $\varphi\colon F(\sigma) \longrightarrow F(\rho)$ at our disposal, we can finally turn to the definition of the adjoint morphism. Let $w_1\colon \mathcal{P}G \longrightarrow \mathbb{Z}_2$ be a Stiefel-Whitney class and $(I,a) \in A_d(\sigma,\rho)$ be as above. We define a function $a^*\colon I^* \longrightarrow M(\rho,\sigma)$ by setting

$$(5.11)\qquad I^* = I,\ \text{and}\ a^*(i) = (t_i,s_i,\gamma_i^{-1},w_1(\hat{\gamma}_i)n_i)\ ,\ i \in I^*.$$

Here $\hat{\gamma}_i$ is given as in (5.7). We note that even though $\hat{\gamma}_i$ depends on the integer d, $w_1(\hat{\gamma}_i)$ does not. The essential properties of this construction are collected in the following proposition:

Proposition 5.12 *If* (I,a) *is admissible, then so is* (I^*,a^*), *and if* $\varphi^{(a)} = \varphi^{(b)}$, *then* $\varphi^{(a^*)} = \varphi^{(b^*)}$. *The assignment* $\varphi^{(a)} \longrightarrow \varphi^{(a^*)}$ *defines a function*

$$*: R\mathcal{G}G\text{-}mod(F(\sigma),F(\rho)) \longrightarrow R\mathcal{G}G\text{-}mod(F(\rho),F(\sigma)).$$

which has the following properties:

$$(1_{F(\sigma)})^* = 1_{F(\sigma)}, \quad (\xi\varphi)^* = \varphi^*\xi^*, \quad \text{and } \varphi^{**} = \varphi$$

for any $\varphi: F(\sigma) \longrightarrow F(\rho)$ *and any* $\xi: F(\rho) \longrightarrow F(\psi)$.

Proof: Let (I,a) satisfy (5.3) and (5.4) and let $\{\text{rad}K_s \mid s\in S\}$ be bounded by b. Since $L_{t_i} \subseteq C^d K_{s_i}$, an easy application of II.1.1.(iv) shows that $K_{s_i} \subseteq C^{b+\theta(d+b)} L_{t_i}$. Since γ_i is a path in $p^{-1}C^d K_{s_i}$, γ_i^{-1} is a path in $p^{-1}C^{d+b+\theta(d+b)} L_{t_i}$. Thus (5.3) holds for (I^*,a^*) with $d^* = d+d'+\theta(d+d')$.

To see that (5.4) holds for (I^*,a^*), we fix some $t\in T$. If $i\in I$ has $t_i=t$ then $K_{s_i} \subseteq C^{d_*} L_t$. Since (S,σ) is locally finite, this leaves only finitely many choices for the value of s_i. If also $w_1(\hat{\gamma}_i)n_i \neq 0$, then $n_i \neq 0$ and, by (5.4), each of the finitely many possible values for s_i leaves only finitely many choices for i. Thus (5.4) holds for (I^*,a^*).

Next let $\varphi^{(a)} = \varphi^{(b)}$, where $b: J \longrightarrow M(\sigma,\rho)$ is e-admissible and has $b(j) = (s'_j,t'_j,\gamma'_j,m_j)$. Let $f\geq d,e$ be chosen such that (5.8) holds. By the above, (I^*,a^*) is d^*-admissible, and (J^*,b^*) is e^*-admissible. We let $f^* = f + \max(d^*,e^*)$. By 5.7, in order to show that $\varphi^{(a^*)} = \varphi^{(b^*)}$, it suffices to prove

(5.13) $\Sigma\{w_1(\hat{\gamma}_i)n_i \mid i\in I^*_{f^*}(t,s,\delta)\} = \Sigma\{w_1(\hat{\gamma}'_j)m_j \mid j\in J^*_{f^*}(t,s,\delta)\}$

for every $t\in T$, every $s\in S$ and every path δ from x_s to y_t in $p^{-1}C^{f^*}L_t$. The proof of (5.13) for a fixed choise of (t,s,δ) relies on the following observation which follows directly from the definitions:

(5.14) If $i\in I^*_{f^*}(t,s,\delta)$ and $i\sim_f i'$, then $i'\in I^*_{f^*}(t,s,\delta)$ and $w_1(\hat{\gamma}_i) = w_1(\hat{\gamma}_{i'})$.

It follows that $I^*_{f^*}(t,s,\delta)$ decomposes as a disjoint union

$$I^*_{f^*}(t,s,\delta) = \coprod_\beta I_e(s,t,\beta)$$

where β ranges over a set, D, of representatives of all path classes in $p^{-1}C^fK_s$ which map to the class of δ^{-1} under the inclusion $p^{-1}C^fK_s \subseteq p^{-1}C^{f*}L_t$. Moreover, $w_1(\hat{\gamma}_i) = w_1(\hat{\beta})$ is constant for $i \in I_f(s,t,\beta)$. Hence

$$\sum \{w_1(\hat{\gamma}_i)n_i \mid i \in I^*_{f^*}(t,s,\delta)\} = \sum_\beta w_1(\hat{\beta})[\Sigma\{n_i \mid i \in I_f(s,t,\beta)\}]$$

where β ranges over D as before. Similarly,

$$\sum \{w_1(\hat{\gamma}'_j)m_j \mid j \in J^*_{f^*}(t,s,\delta)\} = \sum_\beta w_1(\hat{\beta})[\Sigma\{m_j \mid j \in J_f(s,t,\beta)\}],$$

and (5.8) now shows that (5.13) holds.

We leave to the reader the easy proof of the functoriality results for the assignment $\varphi \longrightarrow \varphi^*$.

Lemma 5.15 *The assignment* $[F(\sigma),\varphi] \longrightarrow [F(\sigma),\varphi]^* = [F(\sigma),\varphi^*]$ *defines an involution* $*$ *on* $K_1(R\mathcal{G}G)$ *and on* $Wh(R\mathcal{G}G)$.

Proof: Since $[F,\varphi] \longrightarrow [F,\varphi^*]$ preserves the relations by 5.12 and the easy observation that $\varphi = \varphi_1\oplus\varphi_2\colon F(\sigma_1\amalg\sigma_2) \longrightarrow F(\sigma_1\amalg\sigma_2)$ has $\varphi^* = \varphi_1^*\oplus\varphi_2^*\colon F(\sigma_1\amalg\sigma_2) \longrightarrow F(\sigma_1\amalg\sigma_2)$, it induces a homomorphism $*$ on

$K_1(R\mathcal{P}G)$. Since $\varphi^{**} = \varphi$, this homomorphism is an involution. To see that there is an induced involution on $Wh(R\mathcal{P}G)$, one notes that $u^*_{F(\sigma)} = u_{F(\sigma)}: F(\sigma) \longrightarrow F(\sigma)$ for every unit $u \in R$, and that $F(\alpha,\nu)^* = F(\alpha^{-1},\nu^*): F(\sigma) \longrightarrow F(\sigma)$ for every automorphism of bases $(\alpha,\nu): (S,\sigma) \longrightarrow (S,\sigma)$. Here, if ν is represented by the morphisms $\nu_n(s): \sigma\alpha(s) \longrightarrow C^n\sigma(s)$, $s \in S$, then ν^* is represented by the morphisms,

$$\sigma(\alpha^{-1}(s)) \xrightarrow[\tau_n]{} C^n\sigma(\alpha^{-1}(s)) \xrightarrow[C^m\nu_n'(\alpha^{-1}(s))]{} C^{d+n}\sigma(s)$$

where $\nu_n'(\alpha^{-1}(s))$ is given as in 2.10.

This finishes the construction of the involutions on $K_1(R\mathcal{P}G)$ and $Wh(R\mathcal{P}G)$. We still need to show that naturally equivalent Stiefel-Whitney classes give identical involutions.

Let $\nu: w_1 \longrightarrow w_1'$ be a natural equivalence of Stiefel-Whitney classes. Then for every $\hat{\gamma}: (x,K) \longrightarrow (y,L)$ in $\mathcal{P}G$, the diagram

$$\begin{array}{ccc} w_1(x,K) & \xrightarrow{w_1(\hat{\gamma})} & w_1(y,L) \\ \downarrow{\scriptstyle \nu_{(x,K)}} & & \downarrow{\scriptstyle \nu_{(y,L)}} \\ w_1'(x,K) & \xrightarrow[w_1'(\hat{\gamma})]{} & w_1'(y,L) \end{array}$$

commutes. Let (S,σ) be a bfg basis over $\mathcal{P}G$ and consider an automorphism $\varphi: F(\sigma) \longrightarrow F(\sigma)$. By 5.7, $\varphi = \varphi^{(a)}$ for some $(I,a) \in A(\sigma,\sigma)$. Let $a(i) = (s_i', s_i, \gamma_i, n_i)$. Then the "adjoints", ψ and ψ', obtained from φ by using w_1 and w_1', respectively, are given by

$$\psi(1,s) = \Sigma\, w_1(\hat{\gamma}_i) n_i(\hat{\gamma}_i^{-1}, s_i) \text{ and } \psi'(1,s) = \Sigma\, w_1'(\hat{\gamma}_i) n_i(\hat{\gamma}_i^{-1}, s_i)$$

where both sums run over $I_s = \{i \in I \mid s_i' = s \text{ and } n_i \neq 0\}$ and $\hat{\gamma}_i^{-1} = (\text{cls } \gamma_i^{-1}, \text{incl}): \sigma(s_i) \longrightarrow C^d\sigma(s)$ for some suitable d. Then

$$w_1'(\hat{\gamma}_i)\nu_{\sigma(s_i)} = \nu_{C^d\sigma(s)} w_1(\hat{\gamma}_i)$$

in $\mathbb{Z}_2$ and if we define Φ: $F(\sigma) \longrightarrow F(\sigma)$ by setting

$$\Phi(1,s) = \nu_{\sigma(s)}(1,s) \ , \quad s \in S,$$

an easy calculation shows that $\psi\Phi = \Phi\psi'$. Therefore, $[F(\sigma),\psi] = [F(\sigma),\psi']$, and the involutions induced by w_1 and w_1' coincide.

CHAPTER V
THE ISOMORPHISM BETWEEN THE GEOMETRIC AND ALGEBRAIC WHITEHEAD GROUPS

Let Z be a boundedness control space and $(X,p) \in |\mathcal{CW}^c_\ell/Z|$. The main results of this chapter are Theorem 3.1, which states that there is a natural isomorphism $\tau\colon Wh^c(X,p) \longrightarrow Wh\mathbb{Z}\mathcal{G}G_1(X,p)$, and Theorem 4.1, which states that $\tau([Y,X,q])$ is a combinatorial invariant of $(Y,X,q) \in DR^c(X,p)$. The proofs of these theorems follow the proofs of the corresponding results in the classical compact case quite closely. Thus, in section 1, we follow Milnor's beautiful account [Mi2] and study the torsion of *based* $R\mathcal{G}G$ module chain complexes with based homology. In particular, the Algebraic Subdivision Theorem carries over to the bc setting without change. In section 2 we show how the cellular chains $C_*^{F,cell}(\tilde{Y},\bar{X})$ on the universal cover of a bc CW pair (Y,X,q) become based over $\mathbb{Z}\mathcal{G}G_1(Y,q)$ and establish a similar result for $\pi^c_r(Y^{(r)},Y^{(r-1)},q)$ where $Y^{(r)} = X \cup (r\text{-skeleton of } Y)$. Section 3 is devoted to the proof of the above mentioned Theorem 3.1 and section 4, to the proof of Theorem 4.1. Finally in the brief section 5 we compare the approach taken to defining the geometric Whitehead group in this book with that of our earlier paper [AM2]. In particular we show that our present, simple definition of *elementary expansions* leads to the same geometric Whitehead group as our earlier, more complicated definition.

1. Whitehead torsion for based chain complexes.

In this section we define Whitehead torsion for based $R\mathcal{G}G$ module chain complexes and establish its basic properties. Since the approach taken here follows that of Milnor, [Mi2, Sections 1-5], quite closely, we will assume the reader is familiar with that material.

Let M be a bfg free $R\mathcal{F}G$ module. A *basing* of M is a pair $b = ((S,\sigma),\varphi)$ consisting of a basis (S,σ) over $\mathcal{F}G$ and an isomorphism $\varphi\colon F(\sigma) \longrightarrow M$. If $b_i = ((S_i,\sigma_i),\varphi_i)$ $(i=1,2)$ are two basings of M, then by IV.2.8 there exists an isomorphism $(\alpha,\nu)\colon (S_1,\sigma_1) \longrightarrow (S_2,\sigma_2)$ and we let

$$[b_2/b_1] = [F(\sigma_2), F(\alpha,\nu)\varphi_1^{-1}\varphi_2] \in Wh(R\mathcal{F}G)$$

It is clear that this element is independent of the choice of (α,ν). If $[b_2/b_1] = 0$, then we call b_1 *equivalent* to b_2 and we write $b_1 \sim b_2$. The following, easily established, identities show that $\sim$ is an equivalence relation:

$$[b_3/b_1] = [b_3/b_2] + [b_2/b_1] \text{ and } [b/b] = 0,$$

By a *based* $R\mathcal{F}G$ module, we shall understand a bfg free $R\mathcal{F}G$ module M which is equipped with an equivalence class of basings. Any $b = ((S,\sigma),\varphi)$ in this equivalence class is called a *preferred basing*; while (S,σ) and φ are called a *preferred basis* and a *preferred isomorphism*, respectively.

Let $0 \longrightarrow M_0 \xrightarrow{j} M_1 \xrightarrow{k} M_2 \longrightarrow 0$ be a short exact sequence of bfg free $R\mathcal{F}G$ modules and let $b_i = ((S_i,\sigma_i),\varphi_i)$ be a basing of M_i for i=0,2. We then define a basing $b_0b_2 = ((S_1,\sigma_1),\varphi_1)$ of M_1 by setting $S_1 = S_0 \amalg S_2$, $\sigma_1 = \sigma_0 \amalg \sigma_2$, and $\varphi_1 = j\varphi_0 \oplus s\varphi_2$ where $s\colon M_2 \longrightarrow M_1$ splits the exact sequence. Although b_0b_2 depends on a choice of splitting, its equivalence class depends only on those of b_0 and b_2.

More generally, given bfg free modules $M_0 \subseteq M_1 \subseteq \cdots \subseteq M_k$ and basings b_i of the quotients M_i/M_{i-1} $(i=1,\ldots,k)$, we obtain a basing $b_1b_2\ldots b_k$ of M_k/M_0 which is well defined up to equivalence. As in [Mi2, section 2], we see that this construction is associative and commutative up to equivalence.

Let $C_n \xrightarrow{\partial} C_{n-1} \xrightarrow{\partial} \cdots \xrightarrow{\partial} C_0$ be a chain complex, C_*, of based R$\mathcal{G}$G modules and suppose also that each $H_i(C_*)$ is based. Let c_i be a preferred basing for C_i and h_i be a preferrred basing for $H_i(C_*)$. We then define the Whitehead torsion $\tau(C_*)$ as follows:

Case 1: Assume that the boundary module $B_i = \text{Im}(\partial\colon C_{i+1} \longrightarrow C_i)$ is bfg free for each i. In this case choose a basing b_i for B_i and set

$$\tau(C_*) = \Sigma_i (-1)^i [b_i h_i b_{i-1}/c_i]$$

where the basing $b_i h_i b_{i-1}$ comes from the exact sequences

$$0 \longrightarrow Z_i \longrightarrow C_i \longrightarrow B_{i-1} \longrightarrow 0 \quad \text{and} \quad 0 \longrightarrow B_i \longrightarrow Z_i \longrightarrow H_i(C_*) \longrightarrow 0.$$

That $\tau(C_*)$ does not depend on the choice of b_i follows as in [Mi2, Section 2].

Case 2: In general B_i is not bfg free but the above exact sequences show that B_i is stably bfg free, i.e. there exist bfg free R$\mathcal{G}$G modules $F(\rho_i)$ such that $B_i \oplus F(\rho_i)$ is bfg free. Let $\bar{C}_*$ be the chain complex

$$\bar{C}_n \xrightarrow{\bar{\partial}} \bar{C}_{n-1} \xrightarrow{\bar{\partial}} \cdots \xrightarrow{\bar{\partial}} \bar{C}_0$$

with

$$\bar{\partial} = \begin{bmatrix} \partial & 0 & 0 \\ 0 & 0 & 1 \\ 0 & 0 & 0 \end{bmatrix} : \bar{C}_i = C_i \oplus F(\rho_i) \oplus F(\rho_{i-1}) \longrightarrow C_{i-1} \oplus F(\rho_{i-1}) \oplus F(\rho_{i-2}) = \bar{C}_{i-1} .$$

Then $\bar{B}_i$ is bfg free and we define $\tau(C_*)$ by setting $\tau(C_*) = \tau(\bar{C}_*)$. Again, the arguments of [Mi2, Sections 3-4] show that $\tau(C_*)$ is well defined.

Example 1.1 Suppose that the based acyclic chain complex, C_*, is concentrated in two dimensions, i and i-1. Then C_* looks like $\partial_i\colon C_i \longrightarrow C_{i-1}$ and ∂_i is an isomorphism. It follows that $B_i = 0$ and $B_{i-1} = C_{i-1}$. Thus we may take $b_{i-1} = c_{i-1}$ and we get

$$\tau(C_*) = (-1)^i[\partial_i^{-1}(c_{i-1})/c_i]$$

where $\partial_i^{-1}(c_{i-1}) = ((S_{i-1},\sigma_{i-1}),\partial_i^{-1}\varphi_{i-1})$ for $c_{i-1} = ((S_{i-1},\sigma_{i-1}),\varphi_{i-1})$.

Example 1.2 Let $f\colon M_1 \longrightarrow M_0$ be an isomorphism between based $R\mathcal{G}G$ modules. We can view f as a based acyclic chain complex, C_*, concentrated in degrees i and i-1. We define $\tau(f)$ to be $(-1)^{i-1}\tau(C_*)$. If $\mu_k\colon F(\sigma_k) \longrightarrow M_k$ (k=1,0) is a preferred isomorphism and $(\alpha,\nu)\colon (S_1,\sigma_1) \longrightarrow (S_0,\sigma_0)$ is an isomorphism of bases, then it follows from 1.1 that $\tau(f)$ is represented by the automorphism

$$F(\sigma_0) \xrightarrow{F(\alpha,\nu)^{-1}} F(\sigma_1) \xrightarrow{\mu_1} M_1 \xrightarrow{f} M_0 \xrightarrow{\mu_0^{-1}} F(\sigma_0).$$

We say that the isomorphism $f\colon M_1 \longrightarrow M_0$ of based $R\mathcal{G}G$ modules is *simple* if $\tau(f) = 0$.

The following theorem can be proved exactly as Milnor proves Theorem 3.2 in [Mi2].

Theorem 1.3 *Let* $0 \longrightarrow C'_* \longrightarrow C_* \longrightarrow C''_* \longrightarrow 0$ *be a short exact sequence of based chain complexes of finite length. Assume that there are preferred basings* c'_i, c_i, *and* c''_i *with* $c_i = c'_i c''_i$. *Also assume that all the homology modules* $H_iC'_*$, H_iC_*, $H_iC''_*$ *are based and view the homology long exact sequence*

$$\cdots \longrightarrow H_iC'_* \longrightarrow H_iC_* \longrightarrow H_iC''_* \longrightarrow H_{i-1}C'_* \longrightarrow \cdots$$

as a based, acyclic chain complex $\mathcal{H}_*$ *with* $\mathcal{H}_0 = H_0C''_*$. *Then*

$$\tau(C_*) = \tau(C'_*) + \tau(C''_*) + \tau(\mathcal{H}_*).$$

Let C_* be a based chain complex of $\mathbb{Z}\mathcal{G}G$ modules with $\varphi_i\colon F(\sigma_i) \longrightarrow C_i$ a preferred isomorphism and with $\sigma_i\colon S_i \longrightarrow \mathcal{G}G$. A filtration by subcomplexes

$$(1.4) \qquad 0 = C_*^{(-1)} \subseteq C_*^{(0)} \subseteq \cdots \subseteq C_*^{(m)} = C_*$$

is said to be *adapted* to the bases $\varphi_*: F(\sigma_*) \longrightarrow C_*$, if there exist filtrations

$$\emptyset = S_i^{(-1)} \subseteq S_i^{(0)} \subseteq S_i^{(1)} \subseteq \cdots \subseteq S_i^{(m)} = S_i$$

$(i=0,1,\ldots,n)$ such that the restrictions

$$\varphi_i^{(\lambda)} = \varphi_i | F(\sigma_i^{(\lambda)}): F(\sigma_i^{(\lambda)}) \longrightarrow C_i^{(\lambda)}$$

are isomorphisms for all i and all λ. Here, of course, we have put $\sigma_i^{(\lambda)} = \sigma_i | S_i^{(\lambda)}$.

We note that when a filtration is adapted to the given bases then the quotient complexes $C_*^{(\lambda)}/C_*^{(\lambda-1)}$ become based by means of the isomorphisms

$$\varphi_{i\lambda} = (F(\sigma_{i\lambda}) \xrightarrow{\varphi_i|} C_i^{(\lambda)} \xrightarrow{\pi} C_i^{(\lambda)}/C_i^{(\lambda-1)})$$

where π is the projection, $\sigma_{i\lambda} = \sigma_i | S_{i\lambda}$, and $S_{i\lambda} = S_i^{(\lambda)} - S_i^{(\lambda-1)}$.

The filtration (1.4) will be called *simply adapted* to the bases if it is adapted to the bases for C_* and has the following additional properties:

(1.5) $\quad H_i(C_*^{(\lambda)}/C_*^{(\lambda-1)}) = 0$ for $i \neq \lambda$,

(1.6) $\quad H_\lambda(C_*^{(\lambda)}/C_*^{(\lambda-1)})$ is based, with preferred basing $((T_\lambda,\rho_\lambda),\chi_\lambda)$ $(0 \leq \lambda \leq m)$,

(1.7) $\quad$ The torsion of $C_*^{(\lambda)}/C_*^{(\lambda-1)}$ vanishes for $0 \leq \lambda \leq m$.

If the filtration (1.4) satisfies condition (1.5), then one defines a chain complex $\bar{C}_*$ having $\bar{C}_\lambda = H_\lambda(C^{(\lambda)}/C^{(\lambda-1)})$ and boundary operator $\partial: \bar{C}_\lambda \longrightarrow \bar{C}_{\lambda-1}$ obtained from the exact homology sequence of the triple $(C_*^{(\lambda)}, C_*^{(\lambda-1)}, C_*^{(\lambda-2)})$. The proof of Lemma 5.1 of [Mi2] carries over verbatim to prove that there is a canonical isomorphism

(1.8) $$H_*(C_*) \cong H_*(\bar{C}_*).$$

If $H_*(C_*)$ is based, then this isomorphism will be used to base $H_*(\bar{C}_*)$. The proof of Theorem 5.2 of [Mi2] can be used verbatim to prove the following theorem:

Theorem 1.9 *(Algebraic Subdivision Theorem)* *Let* C_* *be a based* $\mathbb{Z}\mathcal{P}G$ *module chain complex with based homology. If* (1.4) *is a simply adapted filtration of* C_*, *then* $\tau(C_*) = \tau(\bar{C}_*)$.

This theorem will be used in section 4 to show that $[Y,X,q] \in Wh^c(X,p)$ is a combinatorial invariant and in chapter VI to show that the torsion of a bc PL h-cobordism can be computed from the handle structure. In both cases the real work lies in verifying condition (1.7). The verification relies on a proposition whose context we now set.

Let $\mathcal{K}$ be a bounded, locally finite family of sets in $|\mathcal{P}|$. For each $K \in \mathcal{K}$, let C_{*K} be a based $R[G(K)]$ module chain complex with based homology. Recall from IV.3.15, IV.3.16 and IV.3.19 that there are amalgamation functors

$$A_{\mathcal{K}!}: \amalg_K \mathcal{B}as(G[K]) \longrightarrow \mathcal{B}as(\mathcal{P}G)$$

and

$$A_{\mathcal{K}!}: \Pi_K\ R[G(K)]\text{-}mod \longrightarrow R\mathcal{P}G\text{-}mod$$

and an induced amalgamation homomorphism

$$\bar{A}_{\mathcal{K}*}: \Pi_K\ Wh(R[G(K)]) \longrightarrow Wh(R\mathcal{P}G).$$

Proposition 1.10 *Amalgamations of basings define basings for* $C_* = A_{\mathcal{K}!}((C_{*K}))$ *and for* $H_*(C_*)$. *With these basings*

$$\tau(C_*) = \bar{A}_{\mathcal{K}*}((\tau(C_{*K}))).$$

Proof: For each $K \in \mathcal{H}$, let $c_{iK} = ((S_{iK}, \sigma_{iK}), \varphi_{iK})$ be a preferred basing for C_{iK}. Then

$$A_{\mathcal{H}!}((\varphi_{iK})): A_{\mathcal{H}!}((F(\sigma_{iK}))) \longrightarrow A_{\mathcal{H}!}((C_{iK})) = C_i$$

is an isomorphism. By IV.1.20 and IV.3.12, there is a natural isomorphism

$$\hat{\mu}: A_{\mathcal{H}!}((F(\sigma_{iK}))) \longrightarrow F(A_{\mathcal{H}!}(\amalg_K \sigma_{iK})).$$

The composite $A_{\mathcal{H}!}((\varphi_{iK}))\hat{\mu}^{-1}$ is then a preferred isomorphism which we use to base C_i.

In light of the argument above, to base $H_*(C_*)$ by amalgamation, it suffices to prove that $H_i(C_*) = A_{\mathcal{H}!}((H_i(C_{*\mathcal{H}})))$. For fixed $K \in \mathcal{H}$, since C_{iK} and $H_i(C_{*K})$ are free, a standard induction argument shows that the exact sequences of $R[G(K)]$ modules

$$0 \longrightarrow Z_{iK} \longrightarrow C_{iK} \longrightarrow B_{i-1K} \longrightarrow 0 \text{ and } \quad 0 \longrightarrow B_{iK} \longrightarrow Z_{iK} \longrightarrow H_{iK} \longrightarrow 0$$

(where $H_{iK} = H_i(C_{*K})$) are split exact. When we form the product over all $K \in \mathcal{H}$ and apply the amalgamation functor $A_{\mathcal{H}!}$ the split exact property is preserved. Hence we have the split exact sequences of $R\mathcal{G}G$ modules

$$0 \longrightarrow A_{\mathcal{H}!}((Z_{iK})) \longrightarrow A_{\mathcal{H}!}((C_{iK})) \longrightarrow A_{\mathcal{H}!}((B_{i-1K})) \longrightarrow 0$$

$$0 \longrightarrow A_{\mathcal{H}!}((B_{iK})) \longrightarrow A_{\mathcal{H}!}((Z_{iK})) \longrightarrow A_{\mathcal{H}!}((H_{iK})) \longrightarrow 0$$

and an elementary, although tedious, argument using these sequences now shows that $H_i(C_*) = A_{\mathcal{H}!}((H_{iK}))$ as claimed. Hence $H_i(C_*)$ may be based by amalgamation.

The proof that with these basings $\tau(C_*) = A_{\mathcal{H}!}((\tau(C_{*K})))$ is a straightforward exercise in the definitions.

2. Based $\mathbb{Z}\mathcal{F}G$ modules in bc algebraic topology.

In this section we show that based $\mathbb{Z}\mathcal{F}G$ modules occur naturally in the algebraic topology of bc spaces, namely as cellular chains on universal covers, and as homotopy modules of relative r-skeleta.

Let (Y,X,q) be a pair of finite, bc CW complexes, $p=q|X$, $Y^{(r)} = X \cup (\text{r-skeleton of } Y)$ and $q^{(r)}=q|Y^{(r)}$ $(r=-1,0,\ldots,n=\dim Y)$. Then

$$(Y^{(r)},q^{(r)}) = (Y^{(r-1)},q^{(r-1)}) \cup_{f_r} (S_r \times I^r, q_r)$$

for some finite, bc r-cell $(S_r \times I^r, q_r)$ over Z. Let

$$F_r\colon (S_r \times I^r, q_r) \longrightarrow (Y,q)$$

be a characteristic map for the r-cell. We assume F_r chosen so that $qF_r = q_r$. We recall that $I^r=[0,1]^r$ so that $0\in\partial I^r$.

Let $(A,q|A)$ be a bc subcomplex of (Y,q) and suppose that $F_r(s,0) \in A$ for every $s\in S_r$. Let $\sigma_r\colon S_r \longrightarrow \mathcal{F}G_1(A,q|A)$ be a bfg basis with $\sigma_r(s) = (a_s,K_s)$. We say that (S_r,σ_r) is *geometrically prescribed* if it comes with a set of path classes $\Omega_r = \{\omega_s \mid s\in S_r\}$ and $((S_r,\sigma_r),\Omega_r)$ satisfies

(2.1) For some $d\geq 0$, $F_r(s\times I^r) \subseteq q^{-1}C^dK_s$ for each $s\in S_r$; and

(2.2) For each $s\in S_r$, ω_s is a path class from a_s to $F_r(s,0)$ in $q^{-1}C^dK_s \cap A$.

Lemma 2.3 (i) *For each $r\geq 0$ there exists a geometrically prescribed basis $((S_r,\sigma_r),\Omega_r)$ over $\mathcal{F}G_1(Y,q)$. Any two such bases are isomorphic in* $\mathcal{B}as(\mathcal{F}G_1(Y,q))$.

(ii) *For each $r\geq 1$ there exists a geometrically prescribed basis $((S_r,\sigma_r),\Omega_r)$ over $\mathcal{F}G_1(Y^{(r-1)},q^{(r-1)})$. Any two such bases are isomorphic in* $\mathcal{B}as(\mathcal{F}G_1(Y^{(r-1)},q^{(r-1)}))$.

Proof: Since (Y,q) is a finite bc CW complex there exists a bounded family $\{K_s \mid s\in S_r\}$ of sets in $|\mathcal{F}|$ for which (2.1) holds with $d=0$. If we let $y_s = F_r(s,0)$, $\sigma_r(s) = (y_s,K_s)$, and ω_s be the constant path class at y_s, then the existence parts of (i) and (ii) follow. For the uniqueness proof IV.2.11 is useful. We leave the details to the reader.

Remark 2.4 There is a converse to the uniqueness parts of 2.3. Namely, if $((S_r,\sigma_r),\Omega_r)$ is geometrically prescribed and $(1,\upsilon)\colon (S_r,\sigma_r) \longrightarrow (S_r,\sigma_r')$ is an isomorphism of bases, then $((S_r,\sigma_r'),\Omega_r')$ is also geometrically prescribed, where for $s\in S_r$, $\omega_s' = \upsilon(s)\omega_s$.

We next turn to the universal cover of (Y,q). We recall from II.9.3 that this is the fragmented space $\tilde{Y}\colon \mathcal{F}G_1(Y,q) \longrightarrow \mathcal{CW}$ with

$$\tilde{Y}(y,K) = P(Y_K,y)/\!\sim \;, \quad y \in q^{-1}K,\ K \in |\mathcal{F}| \;,$$

where Y_K is the smallest subcomplex of Y containing $q^{-1}K$, $P(Y_K,y)$ is the space of paths in Y_K, starting at y, and $\sim$ denotes the relation of homotopy relative to endpoints. Also recall that the endpoint map $p_{(y,K)}\colon P(Y_K,y)/\!\sim \longrightarrow Y_K$ presents $P(Y_K,y)/\!\sim$ as "the" universal cover of the component $Y_{y,K}$ of Y_K containing y.

The subspace (X,p) of (Y,q) determines a sub-fragmented-space $\bar{X}$ of $\tilde{Y}$ by requiring that

$$\begin{array}{ccc} \bar{X}(y,K) & \subseteq & \tilde{Y}(y,K) \\ \big\downarrow & & \big\downarrow{\scriptstyle p_{(y,K)}} \\ X\cap Y_K & \subseteq & Y_K \end{array}$$

be a pullback diagram for any $K \in |\mathcal{F}|$.

We denote by $C_* = C_*^{F,\mathrm{cell}}(\tilde{Y},\bar{X})$ the cellular chains of the pair $(\tilde{Y},\bar{X})$. It is the chain complex of $\mathbb{Z}\mathcal{F}G_1(Y,q)$ modules having

$$C_r(y,K) = H_r(\tilde{Y}(y,K)^{(r)} \cup \bar{X}(y,K),\ \tilde{Y}(y,K)^{(r-1)} \cup \bar{X}(y,K))$$

for any $(y,K) \in |\mathscr{F}G_1(Y,q)|$, cf. II.3.7.

The significance of geometrically prescribed bases is described in the following proposition whose proof is deferred:

Proposition 2.5 (i) *Let $r \geq 0$. Any geometrically prescribed basis $((S_r,\sigma_r),\Omega_r)$ over $\mathscr{F}G_1(Y,q)$ determines an isomorphism*

$$\varphi_r\colon F(\sigma_r) \longrightarrow C_r^{F,\mathrm{cell}}(\tilde{Y},\bar{X}).$$

Moreover, $((S_r,\sigma_r),\varphi_r)$ is a well defined basing for $C_r^{F,\mathrm{cell}}(\tilde{Y},\bar{X})$.

(ii) *Let $r \geq 2$, and if $r=2$ assume that the attaching map for the bc 2-cell is bc null homotopic. Then any geometrically prescribed basis $((S_r,\sigma_r),\Omega_r)$ over $\mathscr{F}G_1(Y^{(r-1)},q^{(r-1)})$ determines an isomorphism*

$$\varphi_r\colon F(\sigma_r) \longrightarrow \pi_r^c(Y^{(r)},Y^{(r-1)},q^{(r)}).$$

Moreover, $((S_r,\sigma_r),\varphi_r)$ is a well defined basing of $\pi_r^c(Y^{(r)},Y^{(r-1)},q^{(r)})$.

We describe the construction of the maps φ_r of 2.4 (i) and (ii). In the first case the characteristic map F_r induces maps

$$F_r(s,-)\colon (I^r,\partial I^r) \longrightarrow (Y_{c^dK_s}, X \cap Y_{c^dK_s})$$

where $\sigma_r(s) = (y_s,K_s)$ as in the discussion preceding (2.1). The choice of path class ω_s in (2.2) specifies a lift $\tilde{F}_r(s,-)$ such that

$$\begin{array}{ccc} & \overset{\tilde{F}_r(s,-)}{\nearrow} & \tilde{Y}(y_s,c^dK_s) = \tilde{Y}(c^d\sigma(s)) \\ & & \downarrow \\ I^r & \xrightarrow{\;F_r(s,-)\;} & Y_{c^dK_s} \end{array}$$

commutes and such that $\tilde{F}_r(s,-)(0) = \omega_s$. Obviously $\tilde{F}_r(s,-)$ maps $(I^r,\partial I^r)$ into

$$(\tilde{Y}(C^d\sigma(s))^{(r)} \cup \bar{X}(C^d\sigma(s)),\ (\tilde{Y}(C^d\sigma(s))^{(r-1)} \cup \bar{X}(C^d\sigma(s))).$$

Therefore, if we let $z_r \in H_r(I^r,\partial I^r)$ be the standard generator, we get the elements

$$z_{rs} = \tilde{F}_r(s,-)_*(z) \in C_r(C^d\sigma(s))\ ,\quad s\in S_r,$$

$(r=0,1,\ldots,n)$. The morphism φ_r is now defined by setting $\varphi_r(1,s) = z_{rs}$ and invoking IV.1.3.

In the case of homotopy, we have $\sigma_r(s) = (y_s,K_s)$ with $y_s\in Y^{(r-1)}$ and ω_s is a path class in $Y^{(r-1)}$ for each $s\in S_r$. Since $F_r(s,-)$ maps $(I^r,\partial I^r,0)$ into $(q^{-1}C^dK_s\cap Y^{(r)},\ q^{-1}C^dK_s\cap Y^{(r-1)},\ F_r(s,0))$, one has the elements

$$[F_r(s,\ )] \in \pi_r(q^{-1}C^dK_s\cap Y^{(r)},\ q^{-1}C^dK_s\cap Y^{(r-1)},\ F_r(s,0)),$$

and their images

$$\begin{aligned}\omega_{s\#}[F_r(s,\ )] &\in \pi_r(q^{-1}C^dK_s\cap Y^{(r)},\ q^{-1}C^dK_s\cap Y^{(r-1)},y_s)\\ &= \pi_r^c(Y^{(r)},Y^{(r-1)},q^{(r)})(C^d\sigma(s)).\end{aligned}$$

In this case $\varphi_r\colon F(\sigma_r) \longrightarrow \pi_r^c(Y^{(r)},Y^{(r-1)},q^{(r)})$ is determined by setting $\varphi_r(1,s) = \omega_{s\#}[F_r(s,-)]$ and invoking IV.1.3.

Remark 2.6 Suppose that the geometrically prescribed basis in 2.5 (i) can be chosen to have $y_s \in X$ for every $s \in S_r$. Then σ_r can be regarded as a function $\sigma_r'\colon S_r \longrightarrow \mathcal{P}G_1(X,p)$ and the construction described above can be used to define a morphism $\varphi_r'\colon F(\sigma_r') \longrightarrow i^!C_r^{F,\mathrm{cell}}(\tilde{Y},\bar{X})$ where $i\colon \mathcal{P}G_1(X,p) \longrightarrow \mathcal{P}G_1(Y,q)$ is induced by the inclusion. In this case it is not hard to show that φ_r' is just the composite

$$F(\sigma_r') \xrightarrow[\eta]{} i^! i_! F(\sigma_r') \xrightarrow[i^!(\hat{\mu})]{} i^! F(\sigma_r) \xrightarrow[i^!(\varphi_r)]{} i^! C_r^{F,cell}(\tilde{Y},\bar{X})$$

where $\eta\colon I \longrightarrow i^! i_!$ is the unit of the adjunction and $\hat{\mu}$ is the natural equivalence of IV.1.11. In particular, if

$$i_!\colon \mathbb{Z}\mathcal{G}G_1(X,p)\text{-}mod \longrightarrow \mathbb{Z}\mathcal{G}G_1(Y,q)\text{-}mod$$

is an equivalence of categories, then φ_r' is an isomorphism and $i^! C_r^{F,cell}(\tilde{Y},\bar{X})$ becomes a based $\mathbb{Z}\mathcal{G}G_1(X,p)$ module.

A similar remark applies to homotopy. Namely, the above construction can be used to define a morphism $\varphi_r'\colon F(\sigma_r') \longrightarrow i^! \pi_r^C(Y^{(r)},Y^{(r-1)},q^{(r)})$ and φ_r' is an isomorphism (making $i^! \pi_r^C(Y^{(r)},Y^{(r-1)},q^{(r)})$ into a based $\mathbb{Z}\mathcal{G}G_1(X,p)$ module) when

$$i_!\colon \mathbb{Z}\mathcal{G}G_1(X,p)\text{-}mod \longrightarrow \mathbb{Z}\mathcal{G}G_1(Y^{(r-1)},q^{(r-1)})\text{-}mod$$

is an equivalence of categories.

The situation described in 2.6 arises when (Y,X,q) has cells only in two adjacent dimensions, say $n-1$ and n, and either $n\geq 4$ or $n=3$ and the attaching map for the bc 2-cell is bc null homotopic. In this case, we write (U,t) for the relative $(n-1)$-skeleton and have the inclusions

$$(X,p) \xrightarrow{i} (U,t) \xrightarrow{j} (Y,q).$$

We observe that at the level of $\mathbb{Z}\mathcal{G}G_1(-,-)\text{-}mod$ both $i_!$ and $(ji)_!$ are equivalences of categories by II.5.3. Hence $i^! j^! C_r^{F,cell}(\tilde{Y},\bar{X})$ and $i^! \pi_n^C(Y,U,q)$ are based $\mathbb{Z}\mathcal{G}G_1(X,p)$ modules (as is, of course, also $\pi_{n-1}^C(U,X,t)$).

The following proposition, whose proof is deferred, is important by allowing us to replace cellular chain modules by homotopy modules:

Proposition 2.7 *In the above situation there are simple isomorphisms* h_r (r=n-1,n) *such that the following diagram commutes:*

$$\begin{array}{ccc} i^!\pi_n^c(Y,U,q) & \xrightarrow{\partial} & \pi_{n-1}^c(U,X,t) \\ \downarrow h_n & & \downarrow h_{n-1} \\ i^!j^!C_n^{F,cell}(\tilde{Y},\bar{X}) & \xrightarrow{\partial} & i^!j^!C_{n-1}^{F,cell}(\tilde{Y},\bar{X}) \end{array}$$

The rest of this section is occupied by the proofs of 2.5 and 2.7.

Proof of 2.5.(i): We first consider the case of the cellular chains and study the behaviour of the morphism φ_r under variation of the choices involved. A different choice is always indicated by attaching a prime to the quantity in question. In each case one only has to find a simple isomorphism $\xi\colon F(\sigma) \longrightarrow F(\sigma')$ with $\varphi_r = \varphi_r'\xi$.

First replace y_s by y_s' and pick path classes ω_s and ω_s' as in (2.2). The path classes $\omega_s'\omega_s^{-1}$ from y_s' to y_s in $q^{-1}C^dK_s$ ($s\in S_r$) specify an isomorphism of bases $(1,\nu)\colon (S_r,\sigma_r) \longrightarrow (S_r,\sigma_r')$ and one easily verifies that $\varphi_r = \varphi_r' F(1,\nu)$.

The same argument, with $y_s' = y_s$ but allowing $\omega_s' \neq \omega_s$, takes care of variations in the choice of ω_s.

If $\{K_s' \mid s\in S_r\}$ is a different, bounded family satisfying (2.1) for some $d'\geq 0$ then II.1.1 (iv) shows that there exists a bounded family $\{K_s'' \mid s\in S_r\}$ satisfying (2.1) for some $d''\geq 0$ and such that $K_s' \cup K_s \subseteq K_s''$ for each $s\in S_r$. We may now compare each of the families $\{K_s\}$ and $\{K_s'\}$ to $\{K_s''\}$, i.e. we may assume that $K_s \subseteq K_s'$ for each $s\in S$. In that case we can keep $y_s' = y_s$ for each s. There is then an obvious morphism of bases $(1,\nu)\colon (S,\sigma) \longrightarrow (S,\sigma')$ where each ν_s is an identity morphism. By IV.2.11, $(1,\nu)$ is an isomorphism, and a simple calculation shows that $\varphi_r'F(1,\nu) = \varphi_r$ as required.

Finally, consider variations in the choice of $F_r(s,-)$. The only possibility is to compose $F_r(s,-)$ with a relative homeomorphism $h\colon (I^r,\partial I^r) \longrightarrow (I^r,\partial I^r)$. All this can do, is to change the sign on $\varphi_r(\sigma(s))(1,s)$ for some $s\in S_r$, i.e. to compose φ_r with an automorphism of the form $-1\oplus 1\colon F(\sigma_r')\oplus F(\sigma_r'') \longrightarrow F(\sigma_r')\oplus F(\sigma_r'')$ for some decomposition of σ_r.

We now only have to show that φ_r is an isomorphism, and when doing so, we can assume that d=0 in (2.1) (else replace each K_s by C^dK_s). In that case φ_r is represented by a natural transformation

$$\varphi_{r0}\colon F(\sigma_r) \longrightarrow C_r = C_r^{F,\mathrm{cell}}(\tilde{Y},\bar{X}) \ .$$

which we shall show to be eventually monomorphic and eventually epimorphic, cf. I.4.6.

Let $(y,K) \in |\mathscr{P}G_1(Y,q)|$. To describe $F(\sigma_r)(y,K)$ and $C_r(y,K)$ we need the following sets:

(2.8) $\quad S_r(K) = \{s\in S_r \mid K_s\subseteq K\}$,

(2.9) $\quad S_r'(K) = \{s\in S_r \mid F_r(s\times I^r)\subseteq Y_K\}$,

(2.10) $\quad \Omega(y_s,y,K) = \{\omega|\omega \text{ path class in } q^{-1}K \text{ from } y \text{ to } y_s\}$,
for each $s\in S_r(K)$.

We note that if $s\in S_r(K)$ and $i_s\colon K_s\subseteq K$ is the inclusion, then the set of morphisms from (y_s,K_s) to (y,K) in $\mathscr{P}G_1(Y,q)$ is $\{(i_s,\omega) \mid \omega\in\Omega(y_s,y,K)\}$. Therefore, $F(\sigma_r)(y,K)$ is the free $\mathbb{Z}$ module with basis

(2.11) $$\{((i_s,\omega),s)|\ s\in S_r(K),\ \omega\in\Omega(y_s,y,K)\}.$$

On the other hand it is a classical fact that $C_r(y,K)$ is the free $\mathbb{Z}$ module with basis

(2.12) $$\{\omega_*(i_{rs}) \mid s\in S_r'(K),\ \omega\in\Omega(y_s,y,K)\}.$$

Moreover, if $\omega_*(i_{rs}) = \omega_*'(i_{rs'})$, then $\omega=\omega'$ and $s=s'$. Also $S_r(K) \subseteq S_r'(K)$, because (2.1) holds with d=0. Finally, $\varphi_{r0}(y,K)$ maps $((i_s,\omega),s)$

to $\omega_*(i_{rs})$. Hence, $\varphi_{r0}(y,K)$ is a monomorphism (actually onto a direct summand).

To see that φ_{r0} is eventually onto, one only has to prove that there exists an integer $d'\geq 0$ such that $S_r'(K) \subseteq S_r(C^dK)$ for each $K\in|\mathcal{F}|$. In fact, if this holds, then it follows from the above discussion that

$$\mathrm{Im}[\tau_*^{d'}: C_r(y,K) \longrightarrow C_r(C^d(y,K)] \subseteq$$
$$\mathrm{Im}[\varphi_{r0}(C^{d'}(y,K)): F(\sigma_r)(y,K) \longrightarrow C_r(C^{d'}(y,K))).$$

To construct d', we proceed as follows: Since $\{K_s \mid s\in S_r\}$ is bounded, and since the smallest subcomplex fragmentation is equivalent to the inverse image fragmentation (cf. II.2.5), there exists an integer $d''\geq 0$ such that $\mathrm{rad}\ K_s \leq d''$ for $s\in S_r$ and $Y_K \subseteq q^{-1}(C^{d''}K)$ for $K\in|\mathcal{F}|$. Now let $s\in S_r'(K)$. Then $F_r(s\times I^r) \subseteq q^{-1}(K_s) \cap q^{-1}(C^{d''}K)$. By II.1.1 (iv), it follows that $K_s \subseteq C^{d'}K$ where $d' = 2d''+\theta(d'')$. Hence, $s\in S_r(C^{d'}K)$ as required.

Proof of 2.5.(ii): To simplify notation, we write Y for $Y^{(r)}$, X for $Y^{(r-1)}$, q for $q^{(r)}$, S for S_r, σ for σ_r, and φ for φ_r. We think of $\pi_r^c(Y,X,q)$ as defined in terms of the smallest subcomplex fragmentation, Y_K, $K\in|\mathcal{F}|$, of (Y,q). Thus, for $(x,K) \in |\mathcal{F}G_1(X,p)|$

$$\pi_r^c(Y,X,q)(x,K) = \pi_r(Y_K, X\cap Y_K, x) .$$

Then the morphism $\varphi: F(\sigma) \longrightarrow \pi_r^c(Y,X)$ is given by the collection of elements

$$\kappa_s = \omega_{s\#}[F_r(s,-)] \in \pi_r(Y_{C^dK_s}, X\cap Y_{C^dK_s}, y_s) = \pi_r^c(Y,X,q)(C^d\sigma(s))$$

$(s\in S)$, where $d\geq 0$ satisfies (2.1).

As in the proof of 2.5.(i), one must first show that different choices of $((S,\sigma),\Omega)$ and charateristic homeomorphism F only replace φ by a composition $\varphi' = \varphi\xi$ where $\xi: F(\sigma) \longrightarrow F(\sigma')$ is a simple isomorphism. We leave this part to the reader.

In the proof that φ is an isomorphism, we may now assume that d=0, $y_s = F_r(s,0)$, and ω_s is trivial for each $s\in S$.

As in the proof of 2.5.(i) we start by defining for each $(x,K) \in |\emptyset G_1(X,p)|$ some sets:

(2.13) $S(K) = \{s\in S \mid K_s \subseteq K\}$,

(2.14) $S'(K) = \{s\in S \mid qF(s\times I^n) \subseteq K\}$,

(2.15) $\Omega(y_s,x,K) = \{\omega \mid \omega$ path class in $p^{-1}K$ from x to $y_s\}$, for each $s\in S(K)$.

Now $F(\sigma)(x,K)$ is the free $\mathbb{Z}$ module with basis

(2.16) $((i_s,\omega),s) \mid s\in S(K),\ \omega\in\Omega(y_s,x,K)\}$.

where as above $i_s\colon K_s \subseteq K$ is the inclusion.

If $r\geq 3$, then [Wh, Theorem V.1.1] shows that $\pi_r^c(Y,X,q)(x,K) = \pi_r(Y_K, X\cap Y_K, x)$ is the free $\mathbb{Z}$ module with basis

(2.17) $\{\omega_*\kappa_s \mid s\in S'(K),\ \omega\in\Omega(y_s,x,K)\}$.

Moreover, if $\omega_*\kappa_s = \omega'_*\kappa_{s'}$, then $\omega=\omega'$ and $s=s'$. Finally, the natural transformation $\varphi_0\colon F(\sigma) \longrightarrow \pi_n^c(Y,X,q)$, which represents φ, maps $((i_s,\omega),s)$ to $\omega_*\kappa_s$. Therefore $\varphi_0(x,K)$ maps $F(\sigma)(x,K)$ monomorphically onto a direct summand of $\pi_n^c(Y,X,q)(x,K)$ and it follows that φ is eventually monomorphic.

The proof that φ is eventually epimorphic is precisely like the corresponding part of the proof of 2.5.(i). We leave the details to the reader.

If r=2, the proof is completely analogous, using [Wh, Remark, p.215] rather than the above reference.

Remark 2.18 We have phrased Proposition 2.5.(i) in terms of the smallest subcomplex fragmentation of (Y,q). In applications it will be convenient to replace Y_K, $K\in|\mathcal{F}|$ by a more general fragmentation. Suppose that $W: \mathcal{F} \longrightarrow \mathcal{CW}$ is a fragmentation of (Y,q) which has the following properties:

(2.19) $Y_K \subseteq W_K$ for each $K\in|\mathcal{F}|$.

(2.20) For some $d\geq 0$, $W_K \subseteq Y_{C^d K}$ for each $K\in|\mathcal{F}|$.

In that case there is an obvious isomorphism of fragmented spaces over $\mathcal{F}G_1(Y,q)$, $\tilde{Y} \cong \tilde{W}$, induced by liftings of the inclusions $Y_K \subseteq W_K$ (the inverse is induced by liftings of the inclusions $W_K \subseteq Y_{C^d K}$). This isomorphism induces a canonical isomorphism

$$\alpha_r: C_r^{F,cell}(\tilde{Y},\bar{X}) \longrightarrow C_r^{F,cell}(\tilde{W},\bar{X}).$$

Moreover, the proof of 2.5.(i) adapts very simply to prove that $C_*^{F,cell}(\tilde{W},\bar{X})$ is based by means of isomorphisms

$$\varphi_r': F(\sigma_r) \longrightarrow C_r^{F,cell}(\tilde{W},\bar{X}) \ , \quad r=0,1,\ldots,n \ .$$

Since (S_r,σ_r) is constructed solely from the cell structure of (Y,q) without mentioning of the fragmentation, the preferred bases for $C_*^{F,cell}(\tilde{Y},\bar{X})$ and $C_*^{F,cell}(\tilde{W},\bar{X})$ can be taken to coincide. Moreover the isomorphisms

$$\varphi_r: F(\sigma_r) \longrightarrow C_r^{F,cell}(\tilde{Y},\bar{X}),$$

respectively,

$$\varphi_r': F(\sigma_r) \longrightarrow C_r^{F,cell}(\tilde{W},\bar{X})$$

can be chosen to have $\varphi_r' = \varphi_r\alpha_r$; i.e. the isomorphism α_r is simple.

Proof of 2.7: Let $(x,K) \in |\mathcal{F}G_1(X,p)|$. The desired commutative diagram in $\mathbb{Z}\mathcal{F}G_1(X,p)$-*mod* is represented by a commutative diagram in $\mathbb{Z}\text{-}mod^{\mathcal{F}G_1(X,p)}$, the instance of which at the object (x,K) is as follows:

$$\begin{array}{ccc}
\pi_n(Y_K, U\cap Y_K, x) & \xrightarrow{\partial} & \pi_{n-1}(U\cap Y_K, X\cap Y_K, x) \\
\downarrow \cong & & \downarrow \cong \\
\pi_n(\tilde{Y}(x,K), \bar{U}(x,K), \varepsilon_x) & \xrightarrow{\partial} & \pi_{n-1}(\bar{U}(x,K), \bar{X}(x,K), \varepsilon_x) \\
\downarrow h & & \downarrow h \\
H_n(\tilde{Y}(x,K), \bar{X}(x,K)) & \xrightarrow{\partial} & H_{n-1}(\tilde{Y}(x,K), \bar{X}(x,K))
\end{array}$$

Here the two unlabelled isomorphisms are the inverses of the isomorphisms induced by the covering maps $\tilde{Y}(x,K) \longrightarrow Y_K$, $\bar{U}(x,K) \longrightarrow U\cap Y_K$, and $\bar{X}(x,K) \longrightarrow X\cap Y_K$, the canonical basepoint in $\bar{X}(x,K)$ is called ε_x, and h denotes the Hurewicz map.

We note that the resulting morphisms

$$h_n\colon i^!\pi_n^c(Y,U,q) \longrightarrow i^!j^!C_n^{F,\mathrm{cell}}(\tilde{Y},\bar{X}),$$

and

$$h_{n-1}\colon \pi_{n-1}^c(U,X,t) \longrightarrow i^!j^!C_{n-1}^{F,\mathrm{cell}}(\tilde{Y},\bar{X}),$$

are simple isomorphisms because they make the diagrams

$$\begin{array}{ccc}
 & & i^!\pi_n^c(Y,U,q) \\
 & \overset{\varphi_n'}{\nearrow} & \\
F(\sigma_n') & & \downarrow h_n \\
 & \underset{\varphi_n'}{\searrow} & \\
 & & i^!j^!C_n^{F,\mathrm{cell}}(\tilde{Y},\bar{X})
\end{array}
\quad \text{and} \quad
\begin{array}{ccc}
 & & \pi_{n-1}^c(U,X,t) \\
 & \overset{\varphi_{n-1}'}{\nearrow} & \\
F(\sigma_{n-1}') & & \downarrow h_{n-1} \\
 & \underset{\varphi_{n-1}'}{\searrow} & \\
 & & i^!j^!C_{n-1}^{F,\mathrm{cell}}(\tilde{Y},\bar{X})
\end{array}$$

commute. In these diagrams, the notations σ_i' and φ_i' $(i=n-1,n)$ are those of 2.6. This completes the proof of 2.7.

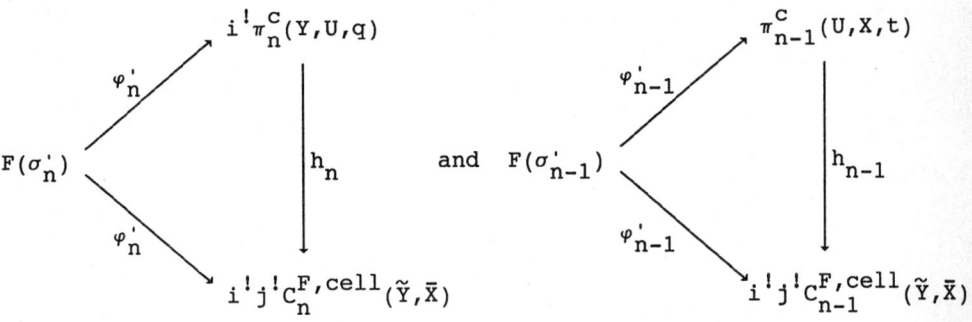

3. The isomorphism τ: $Wh^c(X,p) \longrightarrow Wh\mathbb{Z}\mathcal{F}G_1(X,p)$.

This section is totally occupied by the proof of the following theorem:

Theorem 3.1 *Let* $(Y,X,q) \in DR^c(X,p)$, $(X,p) \in |\mathcal{CW}^c_\ell/Z|$, *and* $i: (X,p) \longrightarrow (Y,q)$ *be the inclusion. Then* $i^! C_*^{F,cell}(\tilde{Y},\bar{X})$ *is a based, acyclic* $\mathbb{Z}\mathcal{F}G_1(X,p)$ *module chain complex and the formula*

$$\tau([Y,X,q]) = \tau(i^! C_*^{F,cell}(\tilde{Y},\bar{X}))$$

defines an isomorphism

$$\tau: Wh^c(X,p) \longrightarrow Wh\mathbb{Z}\mathcal{F}G_1(X,p)$$

which is natural in $(X,p) \in |\mathcal{CW}^c_\ell/Z|$.

Proof that $i^! C_*^{F,cell}(\tilde{Y},\bar{X})$ is based and acyclic: Remark 2.6 shows that $i^! C_*^{F,cell}(\tilde{Y},\bar{X})$ is based. The given deformation retraction from (Y,q) to (X,p) induces a deformation retraction from the (smallest subcomplex) fragmented space $\{Y_K \mid K \in |\mathcal{F}|\}$ to its fragmented subspace $\{X \cap Y_K \mid K \in |\mathcal{F}|\}$. This deformation retraction in turn lifts to a deformation retraction from $\tilde{Y}$ to $\bar{X}$, and it follows that $i_*: H_*^F(\bar{X}) \longrightarrow H_*^F(\tilde{Y})$ is an isomorphism. Then the homology long exact sequence of $(\tilde{Y},\bar{X})$ shows that $H_*(\tilde{Y},\bar{X})$ vanishes, i.e. $C_*^{F,cell}(\tilde{Y},\bar{X})$ is acyclic. Since $i^!$ is an exact functor by I.5.2, it follows that $i^! C_*^{F,cell}(\tilde{Y},\bar{X})$ is acyclic.

Proof that τ is well defined: Let $(Y,X,q) \in DR^c(X,p)$ and let $j: (Y,q) \nearrow (W,u)$ be an elementary expansion. We drop the superscript "cell" from our notation and prove

$$\tau(i^! j^! C_*^F(\tilde{W},\bar{X})) = \tau(i^! C_*^F(\tilde{Y},\bar{X})). \tag{3.2}$$

There is the fragmented space $\bar{Y}: \mathcal{F}G_1(W,u) \longrightarrow \mathcal{CW}$ formed from $\tilde{W}$ by pull back; i.e. for each $(w,K) \in |\mathcal{F}G_1(W,u)|$ there is a pull back diagram

$$\begin{array}{ccc} \bar{Y}(w,K) & \subseteq & \tilde{W}(w,K) \\ \big\downarrow & & \big\downarrow \\ Y \cap W_K & \subseteq & W_K \end{array}$$

Clearly there is a short exact sequence of $\mathbb{Z}\mathcal{F}G_1(X,p)$ module chain complexes

$$i^! j^! C_*^F(\bar{Y},\bar{X}) \longrightarrow i^! j^! C_*^F(\tilde{W},\bar{X}) \longrightarrow i^! j^! C_*^F(\tilde{W},\bar{Y}).$$

Each of these chain complexes is acyclic and based and the hypothesis of 1.3 holds. Therefore

$$\tau(i^! j^! C_*^F(\tilde{W},\bar{X})) = \tau(i^! j^! C_*^F(\bar{Y},\bar{X})) + \tau(i^! j^! C_*^F(\tilde{W},\bar{Y}))$$

and we can finish our proof by showing

$$(3.3) \qquad \tau(i^! j^! C_*^F(\bar{Y},\bar{X})) = \tau(i^! C_*^F(\tilde{Y},\bar{X}))$$

and

$$(3.4) \qquad \tau(i^! j^! C_*^F(\tilde{W},\bar{Y})) = 0.$$

We first prove (3.3). The inclusions $Y_K \subseteq Y \cap W_K$, $K \in |\mathcal{F}|$, define an isomorphism between different fragmentations of (Y,q). This isomorphism lifts to an isomorphism α: $\tilde{Y} \longrightarrow j^! \bar{Y}$ of fragmented spaces over $\mathcal{F}G_1(Y,q)$. Actually, α gives an isomorphism of fragmented pairs, α: $(\tilde{Y},\bar{X}) \longrightarrow j^!(\bar{Y},\bar{X})$. We can identify $C_*^F(j^!(\bar{Y},\bar{X}))$ with $j^! C_*^F(\bar{Y},\bar{X})$, and when we do so the induced isomorphism

$$\alpha_*: C_*^F(\tilde{Y},\bar{X}) \longrightarrow j^! C_*^F(\bar{Y},\bar{X})$$

is simple in each degree. Hence so is $i^! \alpha_*$: $i^! C_*^F(\tilde{Y},\bar{X}) \longrightarrow i^! j^! C_*^F(\bar{Y},\bar{X})$, and (3.3) follows.

We next prove (3.4). Because of the obvious naturality we only have to show that $\tau(C_*^F(\tilde{W},\bar{Y})) = 0$. Let $(W,u) \longrightarrow (Y,q)$ have its two bc cells in dimensions r and r-1, where $r \geq 0$. We can choose the preferred bases in the two dimensions equal to one another, say $(S_r,\sigma_r) = (S_{r-1},\sigma_{r-1}) = (S,\sigma)$ where $\sigma\colon S \longrightarrow |\mathscr{P}G_1(Y,q)|$. Now $\tau(C_*^F(\tilde{W},\bar{Y}))$ is represented, up to a sign, by the automorphism

$$F(\sigma) \xrightarrow[\varphi_r]{} C_r^F(\tilde{W},\bar{Y}) \xrightarrow[\partial]{} C_{r-1}^F(\tilde{W},\bar{Y}) \xrightarrow[\varphi_{r-1}^{-1}]{} F(\sigma)$$

and it is an easy consequence of the remark preceding III.2.4 that this automorphism equals $1_{F(\sigma)}$.

Proof that τ is a homomorphism: For j=1,2, let $(Y_j,X,q_j) \in DR^c(X,p)$ with inclusion $i_j\colon (X,p) \leq (Y_j,q_j)$. If

$$\begin{array}{ccc} (X,p) & \xrightarrow{i_1} & (Y_1,q_1) \\ {\scriptstyle i_2}\downarrow & & \downarrow{\scriptstyle k_2} \\ (Y_2,q_2) & \xrightarrow[k_1]{} & (Y,q) \end{array}$$

is the corresponding push out, then $Y_K = Y_{1K} \cup Y_{2K}$ with $X \cap Y_K = (X \cap Y_{1K}) \cap (X \cap Y_{2K})$ for each $K \in |\mathscr{P}|$. If we let $i = k_2 i_1 = k_1 i_2$, then it easily follows that

$$i^! C_*^F(\tilde{Y},\bar{X}) \cong i_1^! C_*^F(\tilde{Y}_1,\bar{X}) \oplus i_2^! C_*^F(\tilde{Y}_2,\bar{X})$$

by an isomorphism which is simple in each degree. Therefore by 1.3,

$$\tau(i^! C_*^F(\tilde{Y},\bar{X})) = \tau(i_1^! C_*^F(\tilde{Y}_1,\bar{X})) + \tau(i_2^! C_*^F(\tilde{Y}_2,\bar{X}))$$

as claimed.

Proof that τ is monic: Let $(Y,X,q) \in DR^c(X,p)$ have torsion zero. By III.4.4, we can assume that (Y,X,q) is in simplified form with cells in dimensions r-1 and r with $r \geq 3$. By 2.7, $\tau([Y,X,q])$ is represented by the isomorphism of based modules

$$i^! \pi_r^c(Y,U,q) \xrightarrow{\partial} \pi_{r-1}^c(U,X,t)$$

where (U,t) is the relative r-1 skeleton of (Y,X,q).

We conclude from IV.2.6 that the two preferred bases, (S_r,σ_r) and (S_{r-1},σ_{r-1}), are isomorphic. Hence we can replace (S_r,σ_r) by (S_{r-1},σ_{r-1}). For brevity call this common preferred basis (S,σ), and let (as before)

$$\varphi_r\colon F(\sigma) \longrightarrow i^! \pi_r^c(Y,U,q) \text{ and } \varphi_{r-1}\colon F(\sigma) \longrightarrow \pi_{r-1}^c(U,X,t)$$

be preferred isomorphisms. The net effect of these choices is that $\tau([Y,X,q])$ is represented, up to sign, by the automorphism

$$\delta = \varphi_{r-1}^{-1}\partial\varphi_r\colon F(\sigma) \longrightarrow F(\sigma).$$

By IV.3.9, the vanishing hypothesis means that after a possible stabilization δ is a product of (at most 16) elementary automorphisms of $F(\sigma)$ and automorphisms of the form IV.3.8. Hence, we only need to prove the following 3 lemmas, whose proofs are deferred:

Lemma 3.5 *If δ has the form* IV.3.8 *for some decomposition of σ, then* $[Y,X,q] = 0$ *in* $Wh^c(X,p)$.

Lemma 3.6 *For any bfg basis (T,ρ) over $\mathcal{G}_1(X,p)$ one can find an elementary expansion* $(Y,q) \nearrow_e (W,s)$ *with cells in dimensions* r-1 *and* r *such that $\tau([W,X,s])$ is represented by $\delta \oplus 1_{F(\rho)}\colon F(\sigma \amalg \rho) \longrightarrow F(\sigma \amalg \rho)$.*

Lemma 3.7 *For any elementary automorphism $\varepsilon\colon F(\sigma) \longrightarrow F(\sigma)$ there exists a formal deformation* $(Y,q) \curvearrowright (Y',q')$ rel(X,p) *such that* (Y',X,q') *is in simplified form with cells of dimension* r-1 *and* r, *and $\tau([Y',X,q'])$ is represented by $\delta\varepsilon\colon F(\sigma) \longrightarrow F(\sigma)$.*

Proof that τ is epic: Let $\delta\colon F(\sigma) \longrightarrow F(\sigma)$ represent an arbitrary element $x \in \mathrm{Wh}\mathbb{Z}\mathcal{P}G_1(X,p)$. We want to construct a pair $(Y,X,q) \in DR^c(X,p)$ with $\tau([Y,X,q]) = x$. Let $\sigma\colon S \longrightarrow \mathcal{P}G_1(X,p)$ have $\sigma(s) = (x_s,K_s)$ for $s\in S$. We can assume that each x_s is a 0-cell of X. We first form (U,t) by wedging a 2-sphere S_s^2 onto X at the point x_s, for each $s\in S$, and by letting t be constant on each S_s^2. Clearly $(U,t) \in |\mathcal{C}\mathcal{W}_{\ell}^{c}/Z|$. Also, by 2.5, (S,σ) is a preferred basis for $\pi_2^c(U,X,t)$ with a preferred isomorphism $\varphi_2\colon F(\sigma) \longrightarrow \pi_2^c(U,X,t)$.

We next want to construct (Y,q) from (U,t) by attaching a finite, bc 3-cell $(S\times I^3,q_3)$. We start by defining the attaching map. Since the bc 2-cell of (U,X,t) is trivially attached, the composition

$$F(\sigma) \xrightarrow[\varphi_2]{} \pi_2^c(U,X,t) \xrightarrow[\partial]{} \pi_1^c(X,p)$$

vanishes, so there exists a commutative diagram

$$\begin{array}{ccc} F(\sigma) & \xrightarrow{\varphi_2} & \pi_2^c(U,X,t) \\ & \searrow^{\hat{\varphi}_2} & \uparrow \ell_* \\ & & i^!\pi_2^c(U,t) \end{array}$$

where ℓ_* is part of the homotopy long exact sequence.

The composition $\hat{\varphi}_2\delta\colon F(\sigma) \longrightarrow i^!\pi_2^c(U,t)$ is represented, in the sense of IV.1.3, by a family of elements

$$m(s) \in \pi_2(t^{-1}c^dK_s,x_s)\ , \quad s\in S$$

for some integer $d\geq 0$. We choose maps $f(s,-)\colon \partial I^3 \longrightarrow t^{-1}c^dK_s$ such that $[f(s,-)] = m(s)$, for each $s\in S$.

We construct Y by attaching to U a 3-cell, e_s^3, along $f(s,-)$ for each $s\in S$. We must show that $t\colon U \longrightarrow Z$ extends to a map $q\colon Y \longrightarrow Z$ such that $(Y,q) \in |\mathcal{C}\mathcal{W}_f^c/Z|$. We write $[S_s^2]$, $s\in S$, for the elements of $\pi_2(t^{-1}c^dK_s,x_s)$ represented by the 2-spheres S_s^2. The elements $m(s)$ can be chosen as integral linear combinations of elements of the form $\omega_{\#}[S^2_{s'}]$ where ω is a path class in $t^{-1}C^dK_s$ from x_s to $x_{s'}$. Therefore each $m(s)$ vanishes under $t_*: \pi_2(t^{-1}C^dK_s, x_s) \longrightarrow \pi_2(C^dK_s, t(x_s))$; this means that $t: t^{-1}C^dK_s \longrightarrow C^dK_s$ extends over the 3-cell e^3_s. A fixed choice of extension for each $s \in S$ now defines the desired map $q: Y \longrightarrow Z$. We leave it to the reader to check that $(Y,q) \in |\mathcal{C}\mathcal{W}^c_\ell/Z|$

By construction (S,σ) is a preferred basis for $i^!\pi^c_3(Y,U,q)$ and there is a preferred isomorphism φ_3 which makes the following diagram commute

$$\begin{array}{ccc} F(\sigma) & \xrightarrow{\ \delta\ } & F(\sigma) \\ \downarrow{\scriptstyle \varphi_3} & & \downarrow{\scriptstyle \varphi_2} \\ i^!\pi^c_3(Y,U,q) & \xrightarrow[\ \partial\ ]{} & \pi^c_2(U,X,t) \end{array}$$

Hence, in order to finish our proof, we only need to establish the following lemma, whose proof is deferred:

Lemma 3.8 *The pair* (Y,X,q) *constructed above is in* $DR^c(X,p)$.

Proof that τ is natural: Let $f: (X,p) \longrightarrow (X',p')$ be a cellular map in $\mathcal{C}\mathcal{W}^c_\ell/Z$, and let $(Y,X,q) \in DR^c(X,p)$ be in simplified form. We must show that $\tau(f_*[Y,X,q]) = f_*\tau([Y,X,q])$. As above, let (U,t) be the $(r-1)$-skeleton of (Y,X,q) and let the diagram

$$\begin{array}{ccccc} (X,p) & \xrightarrow{\ i\ } & (U,t) & \xrightarrow{\ j\ } & (Y,q) \\ \downarrow{\scriptstyle f} & & \downarrow{\scriptstyle g} & & \downarrow{\scriptstyle h} \\ (X',p') & \xrightarrow[\ k\ ]{} & (U',t') & \xrightarrow[\ \ell\ ]{} & (Y',q') \end{array}$$

consist of two successive pushouts. It follows easily that (Y',X',q') is in simplified form with $(r-1)$-skeleton equal to (U',t').

Let preferred bases and isomorphisms for $i^!\pi^c_3(Y,U,q)$ and $\pi^c_2(U,X,t)$ be given as before, so that $\delta = \varphi^{-1}_{r-1}\partial\varphi_r: F(\sigma) \longrightarrow F(\sigma)$ represents

$\tau([Y,X,q])$ up to a sign. There is a commutative diagram

$$\begin{array}{ccccccc}
F(f_!\sigma) & \xrightarrow[\cong]{\hat{\mu}^{-1}} & f_!F(\sigma) & \xrightarrow{f_!(\varphi_r)} & f_!i^!\pi_r^c(Y,U,q) & \xrightarrow{\hat{h}_*} & k^!\pi_r^c(Y',X',q') \\
 & & \Big\downarrow{\scriptstyle f_!\delta} & & \Big\downarrow{\scriptstyle f_!\delta} & & \Big\downarrow{\scriptstyle \partial} \\
F(f_!\sigma) & \xrightarrow[\hat{\mu}^{-1}]{\cong} & f_!F(\sigma) & \xrightarrow[f_!(\varphi_{r-1})]{} & f_!\pi_{r-1}^c(U,X,t) & \xrightarrow[\hat{g}_*]{} & \pi_{r-1}^c(U',X',t')
\end{array}$$

whose ingredients we proceed to explain. First, $\hat{\mu}$ is the isomorphism $f_!F(\sigma) \longrightarrow F(f_!\sigma)$ from IV.1.11. Next, h_* and g_* are morphisms

$$h_*\colon \pi_r^c(Y,U,q) \longrightarrow g^!\pi_r^c(Y',U',q'),$$

$$g_*\colon \pi_{r-1}^c(U,X,t) \longrightarrow f^!\pi_{r-1}^c(U',X',t'),$$

induced by g and h, respectively. Finally, $\hat{h}_*$ is the adjoint of

$$i^!h_*\colon i^!\pi_r^c(Y,U,q) \longrightarrow i^!g^!\pi_r^c(Y',U',q') = f^!k^!\pi_r^c(Y',U',q')$$

and $\hat{g}_*$ is the adjoint of g_*. The commutativity of the diagram can be shown by direct calculations.

As remarked above, δ represents $\tau([Y,X,q])$ up to a sign, so $f_*(\tau([Y,X,q]))$ is represented by $f_!\delta$, and also by $\hat{\mu}(f_!\delta)\hat{\mu}^{-1}$, up to the same sign. Therefore, we only have to show that the rows in the diagram above are preferred isomorphisms for $k^!\pi_r^c(Y',U',q')$ and $\pi_{r-1}^c(U',X',t')$, respectively.

We treat the lower row only. First recall that $\hat{\mu}$ is the adjoint of a morphism $\mu\colon F(\sigma) \longrightarrow f^!F(f_!\sigma)$, defined in IV.1.13. Thus $\hat{\mu}$ is the composition in the upper row of the diagram

$$\begin{array}{ccccc}
f_!F(\sigma) & \xrightarrow{f_!(\mu)} & f_!f^!F(f_!\sigma) & \xrightarrow{\varepsilon} & F(f_!\sigma) \\
\Big\downarrow{\scriptstyle f_!(\varphi_{r-1})} & & \Big\downarrow{\scriptstyle f_!f^!(\varphi'_{r-1})} & & \Big\downarrow{\scriptstyle \varphi'_{r-1}} \\
f_!\pi_{r-1}^c(U,X,t) & \xrightarrow[f_!(g_*)]{} & f_!f^!\pi_{r-1}^c(U',X',t') & \xrightarrow[\varepsilon]{} & \pi_{r-1}^c(U',X',t')
\end{array}$$

in which φ'_{r-1} is a preferred isomorphism for $\pi^{c}_{r-1}(U',X',t')$. The right hand square commutes because the counit ε is natural. The reader can easily check that the two morphisms

$$f^{!}(\varphi'_{r-1})\mu,\ g_{*}\varphi_{r-1}\colon F(\sigma) \longrightarrow f^{!}\pi^{c}_{r-1}(U',X',t')$$

are represented, in the sense of IV.1.3, by the same set of elements

$$m'(s) \in \pi^{c}_{r-1}(U',X',t')(C^{d'}\sigma(s))\ ,\quad s\in S.$$

Hence $f^{!}(\varphi'_{r-1})\mu = g_{*}\varphi_{r-1}$, and the left hand square commutes because $f_{!}$ is a functor. Since the lower row in the diagram is $\hat{g}_{*}$, we conclude that $\varphi'_{r-1} = \hat{g}_{*}f_{!}(\varphi_{r-1})\hat{\mu}^{-1}$ as claimed.

The rest of this section is occupied by proofs of Lemmas 3.5-3.8.

Proof of 3.5: By assumption there are a decomposition of a preferred basis, say $(S,\sigma) = (S',\sigma')\amalg(S'',\sigma'')$, preferred isomorphisms φ_r and φ_{r-1}, and an automorphism (α'',ν'') of the basis (S'',σ'') such that

$$\delta = \varphi^{-1}_{r-1}\partial\varphi_r = -1_{F(\sigma')}\oplus F(\alpha'',\nu'')\colon F(\sigma'\amalg\sigma'') \longrightarrow F(\sigma'\amalg\sigma'').$$

By changing the orientation of the characteristic map $F_r(s,-)$ for each $s\in S'$, we can change φ_r such that $-1_{F(\sigma')}$ gets replaced by $1_{F(\sigma')}$. Since $1_{F(\sigma')}$ is induced by the identity automorphism of the basis (S',σ'), we may actually assume that $\delta = F(\alpha,\nu)\ :\ F(\sigma) \longrightarrow F(\sigma)$ for some automorphism (α,ν) of (S,σ). Now $\varphi_r F(\alpha,\nu)^{-1}$ is also a preferred isomorphism and when we use that choice we get $\delta = 1\colon F(\sigma) \longrightarrow F(\sigma)$.

In terms of the attaching map $f_r\colon (S\times\partial I^r,q_r|) \longrightarrow (U,t)$ and the characteristic map $F_{r-1}\colon (S\times I^{r-1},q_{r-1}) \longrightarrow (U,t)$ this means that there is some $d\geq 0$ such that

$$f_r(s,-)\colon (\partial I^r,0) \longrightarrow (t^{-1}C^{d}K_s,x_s),$$

$$F_{r-1}(s,-)\colon (I^{r-1},\partial I^{r-1},0) \longrightarrow (t^{-1}C^dK_s,p^{-1}C^dK_s,x_s),$$

and

$$j_*\colon \pi_{r-1}(t^{-1}C^dK_s,x_s) \longrightarrow \pi_{r-1}(t^{-1}C^dK_s,p^{-1}C^dK_s,x_s)$$

maps $[f_r(s,-)]$ to $[F_{r-1}(s,-)]$. Hence $f_r(s,-)$ is homotopic to a map $f'_r(s,-)\colon (\partial I^r,0) \longrightarrow (t^{-1}C^dK_s,x_s)$ which satisfies

(3.9) $\quad f'_r(s,-)|I^{r-1} = F_{r-1}(s,-)$ and $f'_r(s\times J^{r-1}) \subseteq p^{-1}C^dK_s$.

(recall that $J^{r-1} = \partial I^{r-1}\times I \cup I^{r-1}\times\{0\}$). We choose homotopies $h(s,-)\colon \partial I^r\times I \longrightarrow t^{-1}C^dK_s$ from $f_r(s,-)$ to $f'_r(s,-)$. As s varies over S, $h(s,-)$ form a bc homotopy $h\colon (S\times\partial I^r\times I,(q_r|)\pi) \longrightarrow (U,t)$ from f_r to f'_r. It then follows from III.4.5 that $(Y,q) \curvearrowright (Y',q') = (U,t)\cup_{f'_r}(S\times I^r,q_r)$. Since, by (3.9), $(X,p) \nearrow_e (Y',q')$, this finishes the proof of 3.5.

Proof of 3.6: Let $\rho(t) = (x_t,L_t)$, $t\in T$. We form (W,s) by wedging onto Y one (r-1)-sphere, S_t^{r-1}, at the point x_t and filling S_t^{r-1} in by an r-disk, D_t^r, for each $t\in T$. We extend $q\colon Y \longrightarrow Z$ to a map $s\colon W \longrightarrow Z$ by letting $s|D_r^t$ be constant for each t. We leave all further details to the reader.

Proof of 3.7: Write (S_{r-1},σ_{r-1}) as (S,σ) and let

$$\varepsilon = \begin{bmatrix} 1 & \gamma \\ 0 & 1 \end{bmatrix}\colon F(\sigma_1)\oplus F(\sigma_2) \longrightarrow F(\sigma_1)\oplus F(\sigma_2)$$

where $S = S_1 \amalg S_2$, $\sigma_i = \sigma|S_i$ $(i=1,2)$, and $\gamma\colon F(\sigma_2) \longrightarrow F(\sigma_1)$. We let

$$(V,v) = (U,t)\cup_{f'_r}(S_1\times I^r,q_r|)$$

where $f'_r = f_r|S_1\times\partial I^r$. Then

$$(Y,q) = (V,v)\cup_g(S_2\times I^r,q_r|)$$

where g is the composition

$$S_2\times\partial I^r \subseteq S\times\partial I^r \xrightarrow{f_r} (U,t) \subseteq (V,v).$$

We shall construct (Y',q') as

$$(Y',q') = (V,v)\cup_{g'}(S_2\times I^r,q_r|)$$

where $g'\simeq g$ so that III.4.5 guarantees that $(Y,q) \curvearrowright (Y',q')$.

To construct g', we note that there is an integer $d\geq 0$ such that the composition

$$f(\sigma_2) \xrightarrow{\varphi_r|} i^!\pi_r^c(Y,U,q) \xrightarrow{\partial} \pi_{r-1}^c(U,t)$$

is represented by the collection of elements

$$[f_r''(s,-)] \in \pi_{r-1}(t^{-1}c^dK_s,x_s)\ ,\quad s\in S_2.$$

Moreover, d can be chosen so large that

$$F(\sigma_2) \xrightarrow{\gamma} F(\sigma_1) \xrightarrow{\varphi_r|} i^!\pi_r^c(Y,U,q) \xrightarrow{\partial} \pi_{r-1}^c(U,t)$$

is represented by a collection of elements

$$[h(s,-)] \in \pi_{r-1}(t^{-1}c^dK_s,x_s)\ ,\quad s\in S_2\ .$$

We choose $g''(s,-)\colon (\partial I^r,0) \longrightarrow (t^{-1}c^dK_s,x_s)$ to be a representative of the sum $[f_r''(s,-)] + [h(s,-)] \in \pi_{r-1}(t^{-1}c^dK_s,x_s)$ for $s\in S_2$. It is easily seen that $g''\colon (S_2\times\partial I^r,q_r|) \longrightarrow (U,t)$ is a bc map. We let g' be the composition

$$(S_2\times\partial I^r,q_r|) \xrightarrow{g''} (U,t) \overset{j}{\subseteq} (V,v).$$

To show that $g'\simeq g$, we note that $\varphi_r|F(\sigma_1)$ maps into $i^!\pi_r^c(V,U,v) \subseteq i^!\pi_r^c(Y,U,q)$ so that $j_*[h(s,-)] = 0$ in $\pi_{r-1}(v^{-1}c^dK_s,x_s)$. This means that $g'\simeq g\colon (S_2\times\partial I^r,q_r|) \longrightarrow (V,v)$.

From the construction of g"(s,-) we also see that

$$\partial\varphi_r'|F(\sigma_2) = \partial\varphi_r|F(\sigma_2) + \partial\varphi_r\gamma \text{ , and } \partial\varphi_r'|F(\sigma_1) = \partial\varphi_r|F(\sigma_1)$$

which means that $\varphi_{r-1}^{-1}\partial\varphi_r' = \varphi_{r-1}^{-1}\partial\varphi_r\varepsilon$ as desired.

Proof of 3.8: By II.10.2, it suffices to show that $\pi_i^c(Y,X,q) = 0$ for i=0,1,2,3. We can fragment (Y,X,q) by letting Y_K be the smallest subcomplex of Y containing $q^{-1}K$ and $X_K = X \cap Y_K$. Then Y_K is obtained from X_K by wedging on a number of 2-spheres and attaching to the resulting space a number of 3 disks. Hence $\pi_i(Y_K,X_K,x) = 0$ for all $x \in X_K$, and for i=0,1. Therefore, $\pi_i^c(Y,X,q) = 0$ for i=0,1.

A similar argument shows that $\pi_2^c(Y,U,q) = 0$ (recall that (U,t) is the 2-skeleton of (Y,q)). Now the long exact sequence for the triple (Y,U,X,q), together with the fact that $\partial\colon i^!\pi_3^c(Y,U,q) \longrightarrow \pi_2^c(U,X,t)$ is an isomorphism, shows that $\pi_2^c(Y,X,q) = 0$.

To see that $\pi_3^c(Y,X,q) = 0$, consider the sequence

$$0 \longrightarrow H_3^F(\tilde{Y},\bar{X}) \longrightarrow C_3^{F,\text{cell}}(\tilde{Y},\bar{X}) \xrightarrow{\partial} C_2^{F,\text{cell}}(\tilde{Y},\bar{X}) \longrightarrow H_2^F(\tilde{Y},\bar{X}) \longrightarrow 0.$$

which is exact by II.3.6. Since $j^!i^!\colon \mathcal{F}G_1(Y,q)\text{-}mod \longrightarrow \mathcal{F}G_1(X,p)\text{-}mod$ is an equivalence of categories and $j^!i^!(\partial)$ is an isomorphism by 2.7 and construction, ∂ is also an isomorphism. Hence $H_3^F(\tilde{Y},\bar{X}) = H_2^F(\tilde{Y},\bar{X}) = 0$. The continuation of the exact sequence shows that also $H_i^F(\tilde{Y},\bar{X}) = 0$ for i=0,1. Therefore, II.8.2 shows that $\pi_3^F(\tilde{Y},\bar{X}) = 0$. This means that there is some $d \geq 0$ such that for all $\bar{x} \in \bar{X}(y,K)$ and all $(y,K) \in |\mathcal{F}G_1(Y,q)|$

$$0 = \operatorname{Im}[\pi_3(\tilde{Y}(y,K),\bar{X}(y,K),\bar{x}) \longrightarrow \pi_3(\tilde{Y}(y,C^dK),\bar{X}(y,C^dK),\bar{\bar{x}})]$$

(where $\bar{\bar{x}}$ is the image of $\bar{x}$ in $\bar{X}(y,C^dK)$). The projection $\tilde{Y}(y,K) \longrightarrow Y_K$ induces an isomorphism

$$\pi_3(\tilde{Y}(y,K),\bar{X}(y,K),\bar{x}) \longrightarrow \pi_3(Y_K,X\cap Y_K,x)$$

and a similar result holds for $\tilde{Y}(y,C^dK) \longrightarrow Y_{C^dK}$. Hence it follows that

$$0 = \mathrm{Im}[\pi_3(Y_K, X\cap Y_K, x) \longrightarrow \pi_3(Y_{C^d K}, X\cap Y_{C^d K}, x)]$$

for any $(x,K) \in |\mathscr{G}G_1(X,p)|$ which shows that $\pi_3^c(Y,X,q) = 0$ by I.4.5.

This finishes the proof of 3.8.

The argument for the vanishing of $\pi_0^c(Y,X,q)$ and $\pi_1^c(Y,x,q)$, which was given in the above proof of 3.8, generalizes immediately to prove the following lemma which is needed in VI.4.1.

Lemma 3.10 *For any bc CW pair* (Y,X,q), $\pi_i^c(Y^{(r)},Y^{(r-1)},q^{(r)}) = 0$ *for* $i<r$.

4. Combinatorial invariance of Whitehead torsion.

Let $(Y,X,q) \in DR^c(X,p)$ and let sY be a CW complex obtained from Y by subdivision. For any subcomplex W of Y one has an induced subdivision, sW, of W. Clearly $(sY,sX,q) \in DR^c(sX,p)$, and the identity map $i: (X,p) \longrightarrow (sX,p)$ is bc. The combinatorial invariance theorem can be expressed as follows:

Theorem 4.1 *In the above situation* $i_*([Y,X,q]) = [sY,sY,q]$ *in* $Wh^c(sX,p)$.

Corollary 4.2 *The identity map* $i: (X,p) \longrightarrow (sX,p)$ *is simple.*

Proof of 4.2: Let $\pi: X\times I \longrightarrow X$ be the projection and $\rho: M(i) \longrightarrow s(X)$ be the standard retraction. Then $(M(i),X,q\rho)$ is obtained from $(X\times I,X,\rho\pi)$ by subdivision. Since $[X\times I,X,p\pi] = 0$, the theorem implies that $[M(i),X,q\rho] = 0$, i.e. that i is simple (cf. III.5.1).

Proof of 4.1: We first observe that $i_*([Y,X,q]) = [s_1Y,sX,q]$, where the subdivision s_1Y of Y consists of the cells of sX together with those of $Y-X$. We now only have to show that $[s_1Y,sX,q] = [sY,sX,q]$. Here sY is a subdivision of s_1Y which leaves the subcomplex sX intact. Hence, in the proof of 4.1 we may assume that $sX = X$.

By 3.1, it suffices to show

(4.3) $\quad\quad\quad \tau([Y,X,q]) = \tau([sY,X,q]) \quad$ in $\quad \mathrm{Wh}\mathbb{Z}\mathscr{P}G_1(X,p)$.

The proof of (4.3) runs as follows: We construct a based, acyclic chain complex C_* of $\mathbb{Z}\mathscr{P}G_1(X,p)$ modules for which

(4.4) $\quad\quad\quad\quad\quad \tau(C_*) = \tau([sY,X,q])$.

Furthermore, C_* will be equipped with a filtration

(4.5) $\quad\quad 0 = C_*^{(-1)} \subseteq C_*^{(0)} \subseteq \cdots \subseteq C_*^{(\lambda)} \subseteq \cdots \subseteq C_*^{(n)} = C_*$

which is simply adapted to the bases for C_* so that the Algebraic Subdivision Theorem, (1.9), implies

(4.6) $\quad\quad\quad\quad\quad \tau(C_*) = \tau(\bar{C}_*)$.

Finally, (4.5) will have $\bar{C}_* = C_*^{F,\mathrm{cell}}(\tilde{Y},X)$ so that

(4.7) $\quad\quad\quad\quad\quad \tau(\bar{C}_*) = \tau([Y,X,q])$.

Clearly (4.5)-(4.7) imply (4.3).

To construct C_*, we consider the smallest subcomplex fragmentations $\{Y_K \mid K \in |\mathscr{P}|\}$, respectively, $\{(sY)_K \mid K \in |\mathscr{P}|\}$, of (Y,q), respectively, of (sY,q). There is also a different fragmentation, $V: \mathscr{P} \to \mathscr{CW}$, of (sY,q) with $V_K = s(Y_K)$ (the subdivision of Y_K induced by sY). Clearly $(sY)_K \subseteq V_K$ for each $K \in |\mathscr{P}|$. By II.2.5, there is some integer $d \geq 0$ such that $V_K = s(Y_K) \subseteq q^{-1}c^d K$ for each $K \in |\mathscr{P}|$. Since $q^{-1}c^d K \subseteq (sY)_{c^d K}$, we see that

$(sY)_K$ and V_K, $K\in|\mathscr{F}|$, are equivalent fragmentations of (sY,q).

Let $U_K = X\cap V_K$ for $K\in|\mathscr{F}|$. Then (V,U) is a fragmented pair over $\mathscr{F}$. Let $(\tilde{V},\bar{U})$ be the universal cover of the fragmented pair (V,U). The inclusions

$$((sY)_K, X\cap(sY)_K) \subseteq (V_K,U_K) \subseteq ((sY)_{c^{d}K}, X\cap(sY)_{c^{d}K})$$

$(K\in|\mathscr{F}|)$ lift to morphisms which show that

$$(4.8) \qquad (s\tilde{Y},\bar{X}) \cong (\tilde{V},\bar{U})$$

as fragmented pairs over $\mathscr{F}G_1(Y,q)$.

We define C_* to be $i^! C_*^{F,\mathrm{cell}}(\tilde{V},\bar{U})$. The cells of (sY,X) give preferred bases, (S_r,σ_r), and preferred isomorphisms $\varphi_r\colon F(\sigma_r) \longrightarrow C_r$ as in 2.5.(i). Moreover, the isomorphism of (4.8) induces an isomorphism $C_*^{F,\mathrm{cell}}(s\tilde{Y},\bar{X}) \longrightarrow C_*^{F,\mathrm{cell}}(\tilde{V},\bar{U})$ which is simple in each degree, cf. 2.18. Therefore, (4.4) holds.

To define the filtration (4.5), we start by a filtration of $\tilde{Y}$

$$\bar{X} = \tilde{Y}^{(-1)} \subseteq \tilde{Y}^{(0)} \subseteq \cdots \subseteq \tilde{Y}^{(\lambda)} \subseteq \cdots \subseteq \tilde{Y}^{(n)} = \tilde{Y}$$

where $\tilde{Y}^{(\lambda)}(y,K)$ is the relative λ-skeleton of $(\tilde{Y},\bar{X})(y,K)$. The subdivision $s(Y_K)$ of Y_K lifts to a subdivision, $\tilde{s}(\tilde{Y}(y,K))$ of $\tilde{Y}(y,K)$, which, in turn, induces subdivisions $\tilde{s}(\tilde{Y}(y,K)^{(\lambda)})$, $(\lambda=-1,0,\ldots,n)$. We put $\tilde{V}^{(\lambda)}(y,K) = \tilde{s}(\tilde{Y}(y,K)^{(\lambda)})$. This defines a filtration

$$\bar{U} = \tilde{V}^{(-1)} \subseteq \tilde{V}^{(0)} \subseteq \cdots \subseteq \tilde{V}^{(\lambda)} \subseteq \cdots \subseteq \tilde{V}^{(n)} = \tilde{V}.$$

We note that $\bar{U} = \bar{X}$.

Since $\tilde{V}^{(\lambda)}(y,K)$ is a subcomplex of $\tilde{V}(y,K)$ for each $(y,K) \in |\mathscr{F}G_1(Y,q)|$, we can define the filtration (4.5) by putting

$$C_*^{(\lambda)} = i^! C_*^{F,\mathrm{cell}}(\tilde{V}^{(\lambda)},\bar{X}) \subseteq i^! C_*^{F,\mathrm{cell}}(\tilde{V},\bar{X}) = C_*.$$

Since the preferred basis for C_r is given by the cell structure and we are dealing with subcomplexes, it is clear that the filtration (4.5) is adapted to the preferred bases in the sense of section 1. We must prove that the filtration is *simply* adapted to the bases, i.e. that conditions (1.5)-(1.7) hold.

For any $(y,K) \in |\mathscr{P}G_1(Y,q)|$,

$$(C_*^{(\lambda)}/C_*^{(\lambda-1)})(y,K) = C_*^{F,cell}(\tilde{V}(y,K)^{(\lambda)},\tilde{V}(y,K)^{(\lambda-1)}).$$

Therefore,

$$H_i(C_*^{(\lambda)}/C_*^{(\lambda-1)})(y,K) = H_i(\tilde{V}(y,K)^{(\lambda)},\tilde{V}(y,K)^{(\lambda-1)}).$$

This shows that (1.5) holds. It also shows that (1.6) holds since

$$H_\lambda(C_*^{(\lambda)}/C_*^{(\lambda-1)}) = i^! C_\lambda^{F,cell}(\tilde{Y},\bar{X}).$$

Moreover, the boundary operator ∂ in $C_*^{F,cell}(\tilde{Y},\bar{X})$ is defined to come from the triple $(\tilde{Y}^{(\lambda)},\tilde{Y}^{(\lambda-1)},\tilde{Y}^{(\lambda-2)})$, so the chain complex $\bar{C}_*$ of 1.9 equals $i^! C_*^{F,cell}(\tilde{Y},\bar{X})$ and (4.7) holds.

The rest of this section is occupied by the proof of the following lemma, which shows that the filtration (4.5) satisfies (1.7), and completes the proof of 4.1:

Lemma 4.9 *For any* $\lambda = 0,1,\ldots,n$, *the chain complex* $D_* = C_*^{(\lambda)}/C_*^{(\lambda-1)}$ *has* $\tau(D_*) = 0$.

Proof: The proof uses 1.10. We fix λ and let E be the set of λ-cells of Y. Using II.2.5 and the fact that (Y,q) is a finite bc CW complex, it is easy to construct a bounded family $\{X_e \mid e \in E\}$ of subcomplexes of $Y^{(\lambda-1)}$ and a bounded, locally finite family $\mathscr{K} = \{K_e \mid e \in E\}$ such that for each $e \in E$, $Y_e = X_e \cup e$ is a subcomplex of $Y^{(\lambda)}$ and $q(Y_e) \subseteq K_e$.

We define a fragmented pair $(\bar{Y}_e,\bar{X}_e)$ on $G_1(q^{-1}K_e)$ by requiring that the following diagram

$$\begin{array}{ccccc} \bar{X}_e(z) & \subseteq & \bar{Y}_e(z) & \subseteq & \tilde{Y}(z,K_e) \\ \downarrow & & \downarrow & & \downarrow p_{(z,K_e)} \\ X_e & \subseteq & Y_e & \subseteq & Y_{K_e} \end{array}$$

be a pullback for each $z \in |G_1(q^{-1}K_e)|$ and letting the required maps

$$\zeta_*: (\bar{Y}_e(z),\bar{X}_e(z)) \longrightarrow (\bar{Y}_e(z'),\bar{Y}_e(z'))$$

corresponding to morphisms $\zeta \in G_1(q^{-1}K_e)(z,z')$ be restrictions in the obvious way.

We let $D_{*e} = C_*^{F,cell}(\bar{sY}_e,\bar{sX}_e)$ where $(\bar{sY}_e,\bar{sX}_e)$ is the subdivision of $(\bar{Y}_e,\bar{X}_e)$ covering the subdivision (sY_e,sX_e) of (Y_e,X_e). Then $H_i(D_{*e}) = 0$ for $i \neq \lambda$, and the results of section 2 show that $H_\lambda(D_{*e})$ is based by a lifting of the cell e, while D_{ie} is based by lifts of the i-cells of $s(Y_e)-s(X_e) = s(e)$. Thus $\tau(D_{*e}) \in Wh(\mathbb{Z}[G_1(q^{-1}K_e)])$ is defined. Since a straightforward study of the definitions involved shows that $D_* = A_{\mathcal{K}!}((D_{*e}))$, by 1.10 it how suffices to prove that $\tau(D_{*e}) = 0$.

Let $C_*^{cell}(\tilde{Y}_e,\bar{X}_e)$ be the (usual) cellular chains on the (usual) universal cover $(\tilde{Y}_e,\bar{X}_e)$. This is a based chain complex of $\mathbb{Z}\pi_1(Y_e,y_e)$ modules with based homology and $\tau(C_*^{cell}(\tilde{Y}_e,\bar{X}_e)) = 0$ by [Mi2, Lemma 7.3]. On the other hand there is a functor $\psi_e: \pi_1(Y_e,y_e) \longrightarrow G_1(q^{-1}K_e)$ which takes the unique object of $\pi_1(Y_e,y_e)$ to $y_e \in |G_1(q^{-1}K_e)|$ and the loop class γ in Y_e to its class in $G_1(q^{-1}K_e)$. This induces a functor

$$\psi_{e!}: \mathbb{Z}\pi_1(Y_e,y_e)\text{-}mod \longrightarrow \mathbb{Z}[G_1(q^{-1}K_e)]\text{-}mod,$$

and it is easily seen that $\tau(D_{*e}) = \psi_{e!}\tau(C_*^{cell}(\tilde{Y}_e,\bar{X}_e))$. Thus $\tau(D_{*e}) = 0$ and the proof of 4.9 is complete.

5. Bumpy homotopy equivalences.

Let $(Y,X,q) \in DR^c(X,p)$ when (X,p) is a finite, bc CW complex over Z. Following Siebenmann, [Si], we call (Y,X,q) *bumpy* if there exists a bounded locally finite family $\mathcal{K}$ of sets from $|\mathcal{F}|$ and for each $K \in \mathcal{K}$ a subcomplex Y_K of Y (not to be confused with the smallest subcomplex fragmentation) such that the following conditions hold

(5.1) $q(Y_K) \subseteq K$ for each $K \in \mathcal{K}$;

(5.2) $Y_K \cap Y_L \subseteq X$, if $K \neq L$ in $\mathcal{K}$;

(5.3) $Y = X \cup (\cup_K Y_K)$; and

(5.4) For each $K \in \mathcal{K}$, the inclusion $X \cap Y_K \subseteq Y_K$ is a homotopy equivalence.

If these conditions hold then for each $K \in \mathcal{K}$, we can form the universal cover of the pair (Y_K, X_K) as a pair of fragmented CW complexes

$$(\tilde{Y}_K, \bar{X}_K): G_1(q^{-1}K) \longrightarrow \mathcal{FCW} .$$

Since $\tilde{Y}_K$ deformation retracts to $\bar{X}_K$ the cellular chain complex $C_{*K} = C_*^{F,cell}(\tilde{Y}_K, \bar{X}_K)$ is a based, acyclic $\mathbb{Z}[G_1(q^{-1}K)]$ module chain complex with torsion invariant $\tau(C_{*K}) \in Wh\mathbb{Z}[G_1(q^{-1}K)]$. A check of the definitions shows that the amalgamation functor of IV.3.15

$$A_{\mathcal{K}!}: \Pi_K\, \mathbb{Z}[G_1(q^{-1}K)]\text{-}mod \longrightarrow \mathbb{Z}\mathcal{F}G_1(Y,q)\text{-}mod$$

takes $(C_{*K})_K$ to $C_*^{F,cell}(\tilde{Y},\bar{X})$ and by 1.10 we get the following result:

Proposition 5.1 *If* $(Y,X,q) \in DR^c(X,p)$ *is bumpy as described above, then* $\tau([Y,X,q]) = A_{\mathcal{K}*}((\tau(C_{*K})))$.

In our earlier exposition, [AM2], we defined an *elementary expansion* to be a bumpy pair as above, where each inclusion $X \cap Y_K \subseteq Y_K$ is an *elementary expansion* in the classical sense. The above proposition shows that such an expansion has $\tau([Y,X,q]) = A_{\mathcal{K}*}((0)_K)$. Hence, by 3.1,

$[Y,X,q] = 0$ in $Wh^c(X,p)$, and we conclude from these algebraic considerations that the more complicated elementary expansions of [AM2] lead to the same geometric Whitehead group as the present, simpler ones.

It seems conceivable that Siebenmann's complicated expansions in [Si] could be simplified in an analogous way.

CHAPTER VI
BOUNDEDLY CONTROLLED MANIFOLDS AND THE s-COBORDISM THEOREM

This chapter is devoted to the development of some basic techniques for the study of manifolds boundedly controlled over Z. The main objective is to establish an s-cobordism theorem for such manifolds. Throughout this discussion we will assume that the boundedness control structure (P,C) on Z (cf. II.1.1) is *tame* in the sense of the following definition:

Definition *The boundedness control structure* (P,C) *on* Z *is* tame *if there exists an integer* $d \geq 0$ *such that*

(i) $\{\text{Int } C^d(K_0) \mid K_0 \in |\mathcal{P}| \text{ is a minimal element}\}$ *is an open cover of* Z;

(ii) *For all* $K \in |\mathcal{P}|$, $\text{Cl}(K) \subseteq \text{Int } C^d K$; *and*

(iii) Cl(K) *is compact.*

All the examples of boundedness control structures given in Chapter II are tame.

Let $\mathcal{PL}^c/Z$ be the category whose objects are pairs (M,p) where M is a PL manifold, possibly with boundary, and $p: M \longrightarrow Z$ is proper, and whose morphisms are the boundedly controlled PL maps $f: (M_1,p_1) \longrightarrow (M_2,p_2)$. One should think of the objects in $\mathcal{PL}^c/Z$ as the boundedly controlled analogues of ordinary compact PL manifolds.

Let $(M^n,p) \in |\mathcal{PL}^c/Z|$ be a bc n-manifold and suppose $\partial M = \partial_- M \amalg \partial_+ M$. If the inclusions $(\partial_\pm M, p_\pm) \longrightarrow (M,p)$ $(p_\pm = p|\partial_\pm M)$ are homotopy equivalences in $\mathcal{PL}^c/Z$, we call $(M,\partial_- M,\partial_+ M,p)$ a *bc h-cobordism*. In section 6 we show how to assign torsion invariants $\tau(M,\partial_{\pm}M,p) \in Wh\mathbb{Z}\mathcal{G}_1(\partial_{\pm}M,p_{\pm})$ to a bc PL h-cobordism. The main results of this chapter then include the following theorems.

Theorem (*Boundedly Controlled s-Cobordism Theorem*). *Let* $(M^n,\partial_-M,\partial_+M,p)$ *be a bc PL h-cobordism. Let* $n\geq 6$. *Then there is a bc PL homeomorphism* $h\colon (\partial_-M\times I,\partial_-M\times 0,\partial_-M\times 1,p_-\pi) \longrightarrow (M,\partial_-M,\partial_+M,p)$ *with* $h|\partial_-M\times 0$ *the identity if and only if* $\tau(M,\partial_-M,p) = 0$.

In this theorem $\pi\colon \partial_-M\times I \longrightarrow \partial_-M$ is projection on the first factor. This notation will be used consistently throughout this chapter.

Theorem (*Realization Theorem*) *Let* $n\geq 6$ *and* (N^{n-1},q) *be a bc PL manifold. Then for any element* $\tau_0 \in Wh\mathbb{Z}\mathcal{G}_1(N,q)$ *there exists a bc PL h-cobordism* $(M^n,\partial_-M,\partial_+M,p)$ *such that* $(\partial_-M,p_-) = (N,q)$ *and* $\tau(M,\partial_-M,p) = \tau_0$.

By combining these theorems in the usual way, one sees that bc PL h-cobordisms are completely classified by their torsion invariants.

In section 7 we show how to assign an essentially unique Stiefel-Whitney class $w_1\colon \mathcal{G}_1(M,p) \longrightarrow \mathbb{Z}_2$ to a manifold (M,p) boundedly controlled over Z. Let $*\colon Wh\mathbb{Z}\mathcal{G}_1(M,p) \longrightarrow Wh\mathbb{Z}\mathcal{G}_1(M,p)$ be the associated involution constructed in IV.5.

Theorem (*Duality Theorem*) *Let* $(M^n,\partial_-M,\partial_+M,p)$ *be a bc PL h-cobordism. Then*

$$i_{+*}\,\tau(M,\partial_+M,p_+) = (-1)^{n-1}(i_{-*}\,\tau(M,\partial_-M,p_-))^*$$

where the homomorphisms $i_{\pm *}\colon Wh\mathbb{Z}\mathcal{G}_1(\partial_{\pm}M,p_{\pm}) \longrightarrow Wh\mathbb{Z}\mathcal{G}_1(M,p)$ *are induced by the inclusions.*

The proofs of (relative versions of) these theorems are given in sections 6 and 7.

Although the proofs of these theorems and of the results used to prove them carry over to the DIFF category almost without change, the same is not quite true for the TOP category. The problem in that case is how to assign to a bc TOP manifold (M,p) a *well-defined* boundedly controlled simple homotopy type (i.e. a bc homotopy equivalence f: (M,p) ⟶ (X,q) where $(X,q) \in |\mathcal{CW}^c_{\ell}/Z|$. Although there are several obvious approaches to doing this, the authors have not checked the details of any of those approaches.

Most of this chapter is devoted to establishing the basic results needed to prove the s-Cobordism and Realization Theorems. In section 1, we show that if $(M,p) \in |\mathcal{PL}^c/Z|$ and $\partial M \neq \emptyset$, then ∂M has a bc collar in M. In section 2 we examine the process of attaching boundedly controlled handles or boundedly controlled disks to $(M,p) \in |\mathcal{PL}^c/Z|$ and in section 3, we show such a manifold has a handle decomposition. Section 4 developes some basic techniques for modifying handle decompositions of bc PL manifolds. The geometric connectivity of a manifold is investi gated in section 5. Finally, the last two sections, 6 and 7, contain the proofs of the Boundedly Controlled s-Cobordism, the Realization Theorem, and the Duality Theorem.

1. Boundedly controlled collars.

Let $(M,p) \in |\mathcal{PL}^c/Z|$ and let $\partial M = \partial_0 M \cup \partial_1 M$ where $\emptyset \neq \partial_0 M \subseteq \partial M$ and $W = \partial_0 M \cap \partial_1 M$ is the (common) boundary of $\partial_i M$ (i=0,1). The main results of this section are Propositions 1.1 and 1.2 which show that

$(\partial_0 M, W, p|\partial_0 M)$ has a bc collar in $(M, \partial_1 M, p)$ and that p may be appropriately normalized on such a collar.

A *bc collar* of $(\partial_0 M, W, p|\partial_0 M)$ in $(M, \partial_1 M, p)$ is a bc homeomorphism $c: (\partial_0 M\times I, W\times I, (p|\partial_0 M)\pi) \longrightarrow (N, N_1, p|N)$ such that $c(x,0) = x$ where N (respectively, N_1) is a neighborhood of $\partial_0 M$ (respectively, W) in M (respectively, $\partial_1 M$). Note that when $\partial_0 M = \partial M$ and $W = \partial_1 M = \emptyset$, the definition specializes to define a bc collar of $(\partial M, p|\partial M)$ in (M,p).

Proposition 1.1. *Let* $(M,p) \in |\mathcal{PL}^c/Z|$ *and* $\partial M = \partial_0 M \cup \partial_1 M$ *with* $W = \partial_0 M \cap \partial_1 M$ *as above. Then there exists a bc collar of* $(\partial_0 M, W, p|\partial_0 M)$ *in* $(M, \partial_1 M, p)$.

The proof is temporarily deferred.

Let $(M,p) \in |\mathcal{PL}^c/Z|$, $\partial M = \partial_0 M \cup \partial_1 M$ with $W = \partial_0 M \cap \partial_1 M$, $c: (\partial_0 M\times I, W\times I, (p|\partial_0 M)\pi) \longrightarrow (N, N_1, p|N)$, be a bc collar and identify $\partial_0 M\times I$ with N via c.

Proposition 1.2 *Define* $p': M \longrightarrow Z$ *by setting* $p'|\partial_0 M\times I = (p|\partial_0 M\times 1)\pi$ *and* $p' = p$ *outside* $\partial_0 M\times I$. *Then the identity map* $1: (M,p') \longrightarrow (M,p)$ *is a bc homeomorphism.*

The proof is temporarily deferred.

The proof of 1.1 is based on the following lemma:

Lemma 1.3 *Let* (X,A,p) *and* (Y,B,q) *be bc pairs over Z and suppose* $g: (X,A) \longrightarrow (Y,B)$ *is a (not necessarily bc) homeomorphism of pairs such that* $p|A = qg|A$. *Then there exist neighborhoods* U *and* V *of* A *and* B *respectively such that* $g|U: (U,A,p|U) \longrightarrow (V,B,q|V)$ *is a bc homeomorphism.*

Proof: We first construct neighborhoods U_1 and V_1 of A and B respectively such that $g|U_1: (U_1, A, p|U_1) \longrightarrow (V_1, B, q|V_1)$ is bc and a homeomorphism. To do this, let $d \geq 0$ be such that for all $a \in A$, there exists a minimal element $K_a \in |\mathcal{P}|$ such that $a \in p^{-1}(\text{Int } C^d K_a)$. Such a d exists since the boundedness control structure on Z is tame. Let $U_a = {}^{-1}(\text{Int } C^d K_a) \cap g^{-1}q^{-1}(\text{Int } C^d K_a)$ and note that U_a is open and $a \in U_a$ since $p(a) = qf(a)$. Let $U_1 = \bigcup_{a \in A} U_a$ and $V_1 = g(U_1)$. Clearly $g|U_1: U_1 \longrightarrow V_1$ is a homeomorphism.

We claim $g|U_1: (U_1, A, p|U_1) \longrightarrow (V_1, B, q|V_1)$ is bc; for if $x \in U_1$ and $x \in p^{-1}(K)$ for $K \in |\mathcal{P}|$ then for some $a \in A$, $x \in U_a \subseteq p^{-1}(C^d K_a)$. Hence, $p(x) \in K \cap C^d K_a$; and by II.1.1.(iv), $K_a \subseteq C^{\theta(d)} K$ and $C^d K_a \subseteq C^{d+\theta(d)} K$. But also since $x \in U_a$, $g(x) \in q^{-1}(C^d K_a) \subseteq q^{-1}(C^{d+\theta(d)})$. It now follows that $g|U_1$ is bc.

We now apply the argument above to the map $g^{-1}: (V_1, B) \longrightarrow (U_1, A)$ to find neighborhoods V and U of B and A respectively such that $g^{-1}|V: (V, B, q) \longrightarrow (U, A, p)$ is bc and a homeomorphism. Since $g|U = (g|U_1)|U$ is the restriction of a bc map, it is also bc. The lemma now follows.

Proof of 1.1: We give the proof of 1.1 only in the absolute case where $\partial_0 M = \partial M$ and $W = \partial_1 M = \emptyset$. The straightforward extension to the relative case is left to the reader.

Let $c_1: \partial M \times I \longrightarrow M$ be an ordinary collar of ∂M in M. In particular, $c_1(x, 0) = x$ for $x \in \partial M$ and 1.3 may be applied to the bc pairs $(\partial M \times I, \partial M \times 0, (p|\partial M)\pi)$ and $(c_1(\partial M \times I), \partial M, q|c_1(\partial M \times I))$ to give a bc homeomorphism $c_1|U: (U, \partial M \times 0, (p|\partial M)(\pi|U)) \longrightarrow (V, \partial M, q|V)$ where U and V are neighborhoods of $\partial M \times 0$ and ∂M respectively in $\partial M \times I$ and M respectively.

Let $t: \partial M \longrightarrow (0,1]$ be a PL map such that whenever $0 \leq s \leq t(x)$, $(x, s) \in U$. (Notice that such a map is easy to construct.) Define $F: \partial M \times I \longrightarrow \partial M \times I$ by $F(x, s) = (x, st(x))$ and notice that F is a bc homeomorphism of $(\partial M\times I,(p|\partial M)\pi)$ onto its image in $(\partial M\times I,(p|\partial M)\pi)$. The composite $c = c_1F: (\partial M\times I,(p|\partial M)\pi) \longrightarrow (N,p|N)$, where $N = c_1F(\partial M\times I)$, is the desired bc collar of $(\partial M,p|\partial M)$ in (M,p).

Let $p: \partial M\times I \longrightarrow Z$ and set $p_i = p|\partial M\times\{i\}$ $(i=0,1)$. The proof of 1.2 is based on the following lemma:

Lemma 1.4 *If the identity map* $1: (\partial M\times I,p_0\pi) \longrightarrow (\partial M\times I,p)$ *is a bc homeomorphism, then so is* $1: (\partial M\times I,p_1\pi) \longrightarrow (\partial M\times I,p)$.

Proof: If $1: (\partial M\times I,p_0\pi) \longrightarrow (\partial M\times I,p)$ is a bc homeomorphism, so is $1: (\partial M,p_0) \longrightarrow (\partial M,p_1)$ since the latter map is just the restriction of the former to $\partial M\times 1$. But then $1: (\partial M\times I,p_0\pi) \longrightarrow (\partial M\times I,p_1\pi)$ is also a bc homeomorphism and the lemma follows easily.

Proof of 1.2: Let $c: (\partial M\times I,(p|\partial M)\pi) \longrightarrow (N,p|N)$ be a bc collar and identify $\partial M\times I$ with N via c. Then $p|\partial M = p_0$ and

$$1: (\partial M\times I,p_0\pi) \longrightarrow (\partial M\times I,p|\partial M\times I)$$

is a bc homeomorphism. By 1.4 so is $1: (\partial M\times I,p_1\pi) \longrightarrow (\partial M\times I,p|\partial M\times I)$. Since $p'=p$ outside $\partial M\times I$, $1: (Cl(M-\partial M\times I),p') \longrightarrow (Cl(M-\partial M\times I),p)$ is also a bc homeomorphism and 1.2 is now obvious.

2. Attaching boundedly controlled handles and disks.

In this section we define boundedly controlled handles and disks and establish bc versions of some basic results concerning them.

A *boundedly controlled handle of index r* (or simply an *r-handle)* is a pair $(T\times(D^r,S^{r-1})\times D^{n-r},q)$ in $\mathcal{PL}^c/Z$ where T is a discrete space and $\{q(\alpha\times D^r\times D^{n-r}) \mid \alpha\in T\}$ is a bounded family of subsets of Z. We note that for every $x \in D^r \times D^{n-r}$, $q_x = q|T\times\{x\}: T \longrightarrow Z$ is proper and that (T,q_x) is bc homeomorphic to (T,q_y) for all x,y. We shall denote this homeomorphism class by (T,ρ) and call it the *type* of the r-handle.

Let $(M,p) \in |\mathscr{PL}^c/Z|$ and $N \subseteq M$ be a codimension 0, PL submanifold which is closed as a subspace of M. Then $s = p|N:N \longrightarrow Z$ is proper and $(N,s) \in |\mathscr{PL}^c/Z|$. We say that (M,p) is obtained from (N,s) by *attaching an r-handle* if there exists a bc homeomorphism

$$\Phi: (T\times(D^r,S^{r-1})\times D^{n-r},q) \longrightarrow (W,W\cap \mathrm{Int}\partial N,p|W)$$

of a bc r-handle onto $W = \mathrm{Cl}(M-N)$. In this case, Φ is called a *characteristic homeomorphism* for the r-handle and $\varphi = \Phi|T\times S^{r-1}\times D^{n-r}$ is called its *attaching homeomorphism*. Note that φ is an embedding.

The following lemmas, whose proofs are given later in this section, contain the main results on handle attaching that we will need.

Lemma 2.1 (*Extension of Homeomorphisms*). *Let* (M_i,p_i) *be obtained from* (N_i,s_i) (i=0,1) *by attaching the bc r-handle* $(T\times(D^r,S^{r-1})\times D^{n-r},q)$ *via the embedding* $\varphi_i: T\times S^{r-1}\times D^{n-r} \longrightarrow \partial N_i$. *Let* $h: (N_0,s_0) \longrightarrow (N_1,s_1)$ *be a bc homeomorphism and suppose there exists a bc isotopy* ψ_u $(0\leq u\leq 1)$ *of* ∂N_1 *such that* $\psi_0(h\varphi_0) = \varphi_1$ *and* $\psi_1 = 1$. *Then there exists a bc homeomorphism* $H: (M_0,p_0) \longrightarrow (M_1,p_1)$ *such that* $H|N_0$ *agrees with* h *outside a collar of* ∂N_0.

Lemma 2.2 (*Extension of Structures*) *Let* (M_0,p_0) *be obtained from* (N,s) *by attaching the bc r-handle* $(T\times(D^r,S^{r-1})\times D^{n-r},q)$ *via the embedding* $\varphi_0: T\times S^{r-1}\times D^{n-r} \longrightarrow \partial N$. *Let* $\{D_t^{n-1} \mid t\in T\}$ *be a bounded family of mutually disjoint disks in* ∂N *and suppose there exists an isotopy* $\psi_u: (T\times S^{r-1}\times D^{n-r},q) \longrightarrow (\coprod_t D_t^{n-1},s|)$ $(0\leq u\leq 1)$ *such that for every* $t\in T$, $\psi_u(t\times S^{r-1}\times D^{n-r}) \subseteq D_t^{n-1}$ *and* $\psi_0 = \varphi_0$. *Let* $\varphi_1 = \psi_1$. *Then there exists a map* $p_1: M_1 = N \cup_{\varphi_1} (T\times D^r\times D^{n-r}) \longrightarrow Z$ *extending* s *such that* (M_1,p_1) *is obtained from* (N,s) *by attaching a bc r-handle.*

Corollary 2.3 *In lemma 2.2, there is a bc homeomorphism* $h: (M_0,p_0) \longrightarrow (M_1,p_1)$ *such that* $h|N$ *is the identity outside a collar of* ∂N.

Lemma 2.4 (*Special Pushouts*) *Every diagram in* $\mathcal{PL}^c/Z$ *of the form*

$$\begin{array}{ccc} (N_0,s_0) & \xrightarrow{h} & (N_1,s_1) \\ \cap & & \downarrow \\ (M_0,p_0) & \overset{H}{\dashrightarrow} & (M_1,p_1) \end{array}$$

in which (M_0,p_0) *is obtained from* (N_0,s_0) *by attaching an r-handle and h is a bc homeomorphism has a pushout (dotted) in which* (M_1,p_1) *is obtained from* $(N_1,p_1|N_1)$ *by attaching an r-handle and H is a bc homeomorphism such that* $p_0 = p_1H$.

We remark that in 2.4, $p_1|N_1 = s_0h^{-1}$ and $1: (N_1,s_1) \longrightarrow (N_1,s_0h^{-1})$ is a bc homeomorphism.

A *boundedly controlled* n-*disk* is a pair $(T\times(D^n,D^{n-1}),q)$ such that $D^{n-1} \subseteq \partial D^n$ and $\{q(\alpha\times D^n) \mid \alpha\in T\}$ is a bounded family of subsets of Z. As before, the homeomorphism class of (T,q_x), where $q_x = q|T\times x$ $(x\in D^n)$, is well defined and is called the *type* of the n-disk. It is denoted (T,p).

Let $(M,p) \subseteq |\mathcal{PL}^c/Z|$ and $N\subseteq M$ be a codimension 0, PL submanifold which is closed as a subset of M. Then $(N,s) \in |\mathcal{PL}^c/Z|$ where $s = p|N$. We say (M,p) is obtained from (N,s) *by attaching a bc* n-*disk* if there exists a bc homeomorphism $\Psi: (T\times(D^n,D^{n-1}),q) \longrightarrow (W,W\cap\partial N,p|W)$ where $W = Cl(M-N)$.

Lemma 2.5 *Let* (M,p) *be obtained from* (N,s) *by attaching a bc* n-*disk. Then there is a bc homeomorphism* $h: (N,s) \longrightarrow (M,p)$. *Furthermore,* h *can be chosen to be the identity outside an arbitrarily small neighborhood of* $\Psi(T\times D^{n-1})$.

We turn now to the proofs of these lemmas.

Proof of 2.1: Let $g: (N_1,s_1) \longrightarrow (N_1,s_1)$ be the homeomorphism such that g=1 outside a collar of $(\partial N_1,s_1|)$ and $g=\psi$ on the collar. Then $gh: (N_0,s_0) \longrightarrow (N_1,s_1)$ agrees with h outside a collar of $(\partial N_0,s_0|)$ and $gh\varphi_0 = \varphi_1$. Let $H: (M_0,p_0) \longrightarrow (M_1,p_1)$ agree with gh on N_0 and be the identity on the r-handle. Then H is the desired homeomorphism.

Proof of 2.2: We identify D^r with $D^r \cup S^{r-1}\times I$ via the homeomorphism that expands the disk of radius 1/2, $1/2D^r \subseteq D^r$, radially onto D^r and that maps the annular region $Cl(D^r-1/2D^r)$ onto the collar $S^{r-1}\times I$ in such a way that $\partial(1/2D^r)$ maps to $S^{r-1}\times 0$. Using this identification, we now identify $T\times D^r\times D^{n-r}$ with $T\times\{D^r\times D^{n-r}\cup S^{r-1}\times D^{n-r}\times I\}$. Let

$$p_1(\alpha,x,y,u) = s\psi_u(\alpha,x,y) \text{ for } (\alpha,x,y,u) \in T\times S^{r-1}\times D^{n-r}\times I,$$

$$p_1(\alpha,v,w) = p_0\Phi(\alpha,v,w) \text{ for } (\alpha,v,w) \in T\times D^r\times D^{n-r},$$

where $\Phi: (T\times D^r\times D^{n-r},q) \longrightarrow (M_0,p)$ is a characteristic homeomorphism for the attached handle, and $p_1|N=s$. Clearly p_1 is continuous and (M_1,p_1) is obtained from (N_0,p_0) by attaching a bc r handle. Lemma 2.2 follows.

Proof of 2.3: We note first that $\psi_u: t\times S^{r-1}\times D^{n-r} \longrightarrow D_t^{n-1}$ can be covered by an isotopy (cf. [Hu, p. 147]) $\psi'_{u,t}$ of D_t^{n-1} such that $\psi'_{0,t}=1$ and $\psi'_{u,t}|\partial D_t^{n-1}=1$ for all u $(0\le u\le 1)$. Since the disks $\{D_t^{n-1} \mid t\in T\}$ are mutually disjoint, $\amalg_t\psi'_{u,t}$ extends via the identity to an isotopy ψ'_u $(0\le u\le 1)$ of ∂N. Since $\{D_t^{n-1} \mid t\in T\}$ is bounded, ψ'_u is bc. Corollary 2.3 now follows from 2.1 by setting h=1.

Proof of 2.4: Since h is a bc homeomorphism there is a commutative diagram of bc homeomorphisms

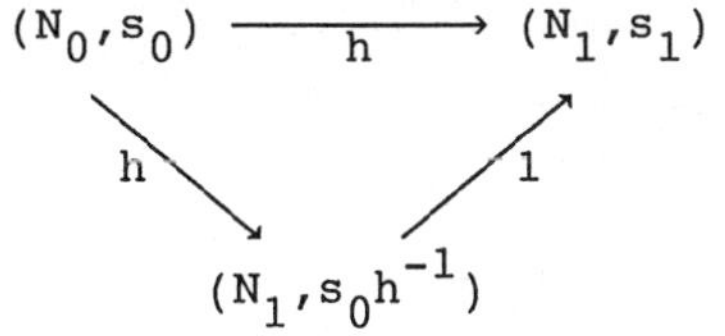

Thus, the angle diagram in 2.4 is isomorphic to one in which the controlling map on N_1 is s_0h^{-1} and it suffices to prove 2.4 in this case. To do this, one simply lets M_1, p_1, and H be determined by the usual pushout (dotted) of the diagram

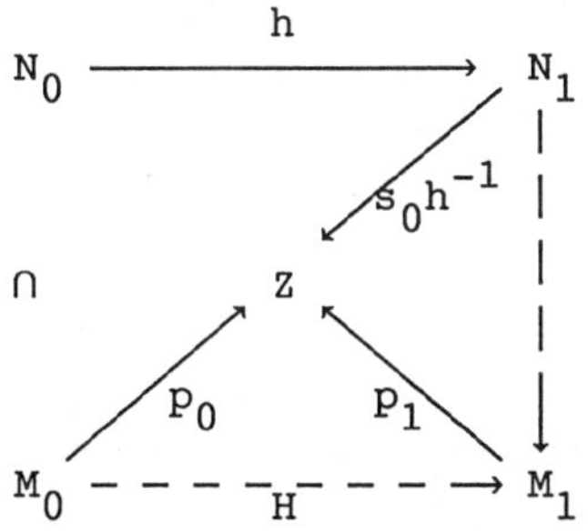

Proof of 2.5: We prove 2.5 first in a special case, where we assume

(i) $(T\times(D^n, D^{n-1}), q) = (T\times(D^{n-1}\times I, D^{n-1}\times 0), q_0\pi)$, in particular $q = q_0\pi$ where $q_0 = q|T\times D^{n-1}\times 0$ and $\pi\colon T\times D^{n-1}\times I \longrightarrow T\times D^{n-1}$ is the projection;

(ii) $q = p\Psi$ where Ψ is a characteristic homeomorphism for the n-disk; and

(iii) $p|\partial N\times I = p_1\pi$ where $\partial N\times I$ is a bc collar of ∂N in N with $\partial N\times 0 = \partial N$, $p_1 = p|\partial N\times 1$, and $\pi\colon \partial N\times I \longrightarrow \partial N$ is the projection.

Thus we have normalized the situation so that the following diagram commutes:

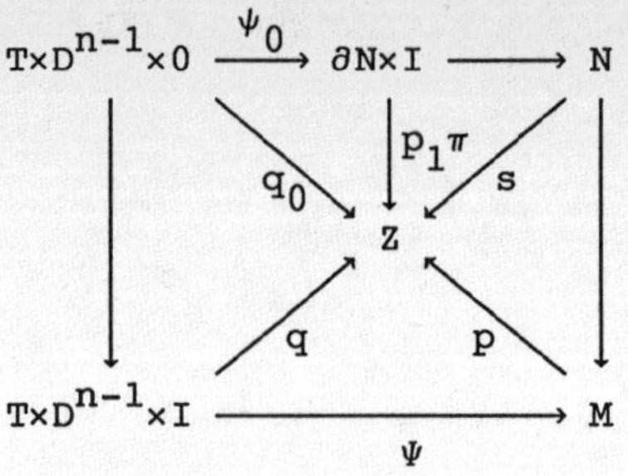

Let $\{K_\alpha \mid \alpha\in T\}$ be a bounded family of sets in $|\mathcal{P}|$ such that $q(\alpha\times D^n) \subseteq C^d K_\alpha$ for each $\alpha\in T$. Then $p\psi_0(\alpha\times D^{n-1}) = q_0(\alpha\times D^{n-1}) \subseteq q(\alpha\times D^n) \subseteq \mathrm{Cl}(K_\alpha)$ and by property (ii) of a tame control structure, $\psi_0(\alpha\times D^{n-1}) \subseteq p^{-1}(\mathrm{Int}\ C^d K_\alpha) \cap \partial N$. Thus, there exists a disk $E_\alpha^{n-1} \subseteq \partial N$ with $\psi_0(\alpha\times D^{n-1}) \subseteq \mathrm{Int}\ E_\alpha^{n-1}$ and $p(E_\alpha^{n-1}) \subseteq \mathrm{Int}\ C^d K_\alpha$. We may assume $E_\alpha^{n-1} \cap E_\beta^{n-1} = \emptyset$ for $\alpha\neq\beta$ since then $(\alpha\times D^{n-1}) \cap (\beta\times D^{n-1}) = \emptyset$. Consider $E_\alpha^{n-1}\times I \subseteq \partial N\times I$ and observe that it is easy to construct a homeomorphism $h_\alpha\colon E_\alpha^{n-1}\times I \longrightarrow E_\alpha^{n-1}\times I \cup \Psi(D_\alpha^{n-1}\times I)$ such that $h_\alpha|\partial E_\alpha^{n-1}\times I \cup E_\alpha^{n-1}\times 1$ is the identity. Hence $\amalg_\alpha h_\alpha$ extends via the identity to a homeomorphism $h\colon N \longrightarrow M$.

It remains to prove that h is a bc homeomorphism. To see that h is bc we show that for all $K\in|\mathcal{P}|$, if $x \in s^{-1}(K)$, then $h(x) \in p^{-1}(C^m K)$ for $m = d+\theta(d)$. If $x \notin E_\alpha^{n-1}\times I$ for any α, this is clear since $h(x) = x$ and $s = p|N$. If $x \in E_\alpha^{n-1}\times I$, then by the normalization of p, $p(x) \in C^d K_\alpha$. Hence $K \cap C^d K_\alpha \neq \emptyset$ and $C^d K_\alpha \subseteq C^m K$ for $m = d+\theta(d)$ by a now standard argument. On the other hand, by the construction of h_α and the normalization of p, $ph_\alpha(x) \in C^d K_\alpha$. Hence, $h(x) \in p^{-1}(C^m K)$ as claimed and h is bc. A similar argument shows that h^{-1} is bc. Hence, h is a bc homeomorphism.

Since E_α^{n-1} can be chosen to lie in an arbitrarily small neighborhood of $\psi_0(\alpha\times D^{n-1})$, the last clause of 2.5 is obvious from the construction of h and the proof of the special case of 2.5 is completed.

To prove 2.5 in general, we first note that by 1.2 there is a map $s': N \longrightarrow Z$ that satisfies condition (iii) above and for which $1: (N,s) \longrightarrow (N,s')$ is a bc homeomorphism. Extend s' over $\Psi(\alpha \times D^n) = \Psi(\alpha \times D^{n-1} \times I)$ by setting $s'\Psi(x,t) = s'\Psi(x,0)$ and call the extension p'. Clearly p' satisfies the normalization conditions.

It now suffices to show that $1: (M,p) \longrightarrow (M,p')$ is a bc homeomorphism and to do this, it is enough to show that $1: (\Psi(T \times D^n),p) \longrightarrow (\Psi(T \times D^n),p')$ is a bc homeomorphism. Now notice that since $(T \times D^n,q)$ is a bc n-disk and Ψ is bc, there exists a bounded family $\{K_\alpha \mid \alpha \in T\}$ with $p\Psi(\alpha \times D^n) \subseteq K_\alpha$ for each $\alpha \in T$. Then $s\Psi(\alpha \times D^{n-1}) = p\Psi(\alpha \times D^{n-1}) \subseteq K_\alpha$ and since $1: (N,s) \longrightarrow (N,s')$ is bc, there exists an integer $m \geq 0$ such that $s'\Psi(\alpha \times D^{n-1}) \subseteq C^m K_\alpha$. But then also $p'\Psi(\alpha \times D^n) \subseteq C^m K_\alpha$. An argument similar to the one above showing $h = \amalg_\alpha h_\alpha$ is bc now shows that $1: (\Psi(T \times D^n),p|) \longrightarrow (\Psi(T \times D^n),p'|)$ is a bc homeomorphism and completes the proof of 2.5.

3. Handle decompositions exist.

In this section we show that manifolds (M,p) in $\mathcal{PL}^c/Z$ have handle decompositions. We begin with two definitions.

Let $(M^n,p) \in |\mathcal{PL}^c/Z|$. A *handle decomposition* of (M,p) is an increasing filtration

$$M^{(-1)} \subseteq M^{(0)} \subseteq \cdots \subseteq M^{(r)} \subseteq \cdots \subseteq M^{(n)} = M$$

of closed subspaces of M for which

(i) $M^{(-1)} = \emptyset$;

(ii) For $0 \leq r \leq n-1$, $M^{(r)}$ is a codimension 0 submanifold of M with boundary;

(iii) For $0 \leq r \leq n$, $(M^{(r)}, p_r)$ is obtained from $(M^{(r-1)}, p_{r-1})$ by attaching a (possibly empty) handle of index r where $p_k = p|M^{(k)}$.

Let $(M^n, p) \in |\mathcal{PL}^c/Z|$ be such that $\emptyset \neq \partial M = \partial_0 M \cup \partial_1 M$ where $W = \partial_0 M \cap \partial_1 M$ is the boundary of $\partial_i M$ $(i=0,1)$. A *handle decomposition* of (M,p) relative to $(\partial_0 M, p_0)$, where $p_0 = p|\partial_0 M$, is an increasing filtration as above except one now requires that the following conditions hold:

(i') $(M^{(-1)}, M^{(-1)} \cap \partial_1 M, p_{-1})$ is a bc collar neighborhood of $(\partial_0 M, W, p_0)$ in $(M, \partial_1 M, p)$ where $p_{-1} = p|M^{(-1)}$ (that is, there exists a bc homeomorphism $c\colon (\partial_0 M \times I, W \times I, p_0 \pi) \longrightarrow (M^{(-1)}, M^{(-1)} \cap \partial_1 M, p_{-1}))$;

(ii') Condition (ii) above holds and for $0 \leq r \leq n-1$, $\partial M^{(r)} = \partial_0 M \cup \partial_1 M^{(r)}$ and $W = \partial_0 M \cap \partial_1 M^{(r)}$;

(iii') Condition (iii) above holds and for $1 \leq r \leq n$, the handle of index r is attached to $M^{(r-1)}$ along $\mathrm{Int}(\partial_1 M^{(r-1)} - c(W \times I))$.

Proposition 3.1 *Let* $(M,p) \in |\mathcal{PL}^c/Z|$ *and suppose* M *is connected. Then there exists a handle decomposition of* (M,p). *If* $\partial M = \partial_0 M \cup \partial_1 M$ *as above; then there exists a handle decomposition of* (M,p) *relative to* $(\partial_0 M, p_0)$.

The proof of 3.1 requires some preliminary results.

Let X be polyhedron (i.e. the underlying space of a finite dimensional, locally finite simplicial complex) and $\mathcal{U} = \{U_\alpha \mid \alpha \in A\}$ be an open cover of X. A triangulation K of X is *subordinate* to $\mathcal{U}$ if for every simplex $\sigma \in K$, there exists a $U_{\alpha(\sigma)} \in \mathcal{U}$ such that $\sigma \subseteq U_{\alpha(\sigma)}$.

Lemma 3.2 *Let X be a connected polyhedron and $\mathcal{U}$ be an open cover of X. Then there exists a triangulation of X subordinate to $\mathcal{U}$.*

Proof: If X is compact, i.e. is triangulated by a finite complex K, this is well known. One lets ε be a Lebesgue number for the cover $\mathcal{U}$ and takes an iterated barycentric subdivision of K with mesh $< \varepsilon$.

Suppose K is a locally finite simplicial complex triangulating X and v_0 is a vertex in K. For any vertex $v \in K$, let $f(v)$ be the minimum number of edges in an edge path from v_0 to v. Clearly, if $v_1,\dots,v_n$ are the vertices of a simplex, then $|f(v_i)-f(v_j)| \le 1$ for any i,j with $1 \le i,j \le n$. Hence the vertex map $f\colon K^{(0)} \longrightarrow \mathbb{Z}_+$ extends to a simplicial map $f\colon K \longrightarrow \mathbb{R}_+$ where $\mathbb{R}_+ = [0,\infty)$ is triangulated so that $\mathbb{R}_+^{(0)} = \mathbb{Z}_+$.

Let $L_j = \bigcup_{k=0}^{\infty} f^{-1}[3k+j,\ 3k+j+1]$ and notice that L_j is full in K (j=0,1,2). Furthermore, $K = L_0 \cup L_1 \cup L_2$. On the other hand, for any k,j ($k \in \mathbb{Z}_+$, j=0,1,2), $J_{3k+j} = f^{-1}[3k+j,3k+j+1]$ is a finite subcomplex of K. By the remarks above, then there is a subdivision J'_{3k+j} of J_{3k+j} subordinate to $\mathcal{U}$. Since J_{3k+j} is disjoint from $J_{3\ell+j}$ for $\ell \ne k$, $L'_j = \bigcup_{k=0}^{\infty} J'_{3k+j}$ is a subdivision of L_j subordinate to $\mathcal{U}$. By [RS, Exercise 3.4, p.32], there is a subdivision K'_j of K (j=0,1,2) such that L'_j is a subcomplex of K'_j and there are no new vertices in $K'_j - L'_j$. Let K' be a common subdivision of K'_1, K'_2, and K'_3. Clearly, K' is a triangulation of X subordinate to $\mathcal{U}$.

Corollary 3.3 *Let* $(M,p) \in |\mathcal{P}\mathcal{L}^c/Z|$. *Then there exists a triangulation J of M such that (J,p) is a bc CW complex.*

Proof: Since the boundedness control structure on Z is tame, there exists an integer $d \ge 0$ such that

$$\mathcal{U} = \{p^{-1}(\mathrm{Int}\ c^d K_0) \mid K_0 \in |\mathcal{P}| \text{ is a minimal element}\}$$

is an open cover of M. Let J be a triangulation of M subordinate to $\mathcal{U}$. Then (J,p) is obviously a bc CW complex.

Proof of 3.1: We give only the proof of the relative case of 3.1 since the proof of the absolute case can easily be obtained from it.

Let $(N^{(-1)}, N^{(-1)} \cap \partial_1 M, p|)$ be a bc collar neighborhood of $(\partial_0 M, W, p_0)$ in $(M, \partial_1 M, p)$ and set $\partial_1 N^{(-1)} = Cl(\partial N^{(-1)} - (\partial_0 M \cup (\partial_1 M \cap N^{(-1)})))$. Let $(J, J_{-1}, \partial_1 J_{-1})$ be a triangulation of $(M, N^{(-1)}, \partial_1 N^{(-1)})$ such that (J,p) is a bc CW complex as in 3.3. Let $J_i = J_{-1} \cup$(i-skeleton of J) and set $M^{(i)} = |N(J_i'', J'')|$ $(-i \leq i \leq n)$ where $|N(J_i'', J'')|$ is the underlying space of the simplicial neighborhood of J_i'' in a second derived subdivision J" of J. It is well known (cf. [Hu, pp. 224-225]) that, as a space $M^{(i+1)}$ is obtained from $M^{(i)}$ by attaching (ordinary) handles of index i+1. In fact, for each (i+1)-simplex $\alpha \in J - J_{-1}$,

$$(H_\alpha, \partial_- H_\alpha) = (|st(\hat{\alpha}, J'')|, \ |st(\hat{\alpha}, J'')| \cap M^{(i)})$$

is a handle of index i+1 where $\hat{\alpha}$ is the barycenter of α in J. Thus, there is a homeomorphism $\varphi_\alpha: (D^{i+1}, S^i) \times D^{n-i} \longrightarrow (H_\alpha, \partial_- H_\alpha)$.

Let $T = \{\hat{\alpha} \mid \alpha$ is an (i+1) simplex of $J - J_{-1}\}$. Then T is a discrete space and certainly

$$\Phi = \amalg \ \varphi_\alpha: (T \times (D^{i+1}, S^i) \times D^{n-i}, (p|)\Phi) \longrightarrow (\amalg (H_\alpha, \partial_- H_\alpha), p|)$$

is a bc homeomorphism where the disjoint union runs over $\hat{\alpha} \in T$. Thus, to prove that $(M^{(i+1)}, p_{i+1})$ is obtained from $(M^{(i)}, p_i)$ by attaching an (i+1)-handle, it suffices to prove $\{p\Phi(\hat{\alpha} \times D^i \times D^{n-i}) \mid \hat{\alpha} \in T\}$ (or, equivalently, $\{p(H_\alpha) \mid \hat{\alpha} \in T\}$) is a bounded family of subsets of Z. Since $H_\alpha \subseteq |st(\alpha, J)|$, it now suffices to prove the latter sets are bounded.

For each simplex $\beta \in J$, let $K_\beta \in |\mathcal{S}|$ be a minimal element with $\beta \subseteq p^{-1}(C^d K_\beta)$ for some d independent of β. Such a K_β exists by the choice of J. Now let τ be a simplex in $st(\alpha, J)$. Then there is a simplex $\rho \in J$

with α and τ faces of ρ. Then $\emptyset\neq\alpha \subseteq C^dK_\alpha \cap C^dK_\rho$. Hence $K_\rho \subseteq C^{d+\theta(d)}K_\alpha$, $C^dK_\rho \subseteq C^eK_\alpha$, and $\rho \subseteq p^{-1}(C^eK_\alpha)$ where $e = 2d+\theta(d)$. Thus $|st(\alpha,J)| \subseteq C^eK_\alpha$ and $\{|st(\alpha,J)| \mid \hat{\alpha}\in T\}$ is a bounded family of sets.

It remains to show that $(M^{(-1)},M^{(-1)}\cap\partial_1M,p_{-1})$ is a bc collar of (∂_0M,W,p_0) in (M,∂_1M,p). Since $(N^{(-1)},N^{(-1)}\cap\partial_1M,p|)$ was chosen to be such a collar, it suffices to show that there is a bc homeomorphism $h\colon (N^{(-1)},N^{(-1)}\cap\partial_1M,p|) \longrightarrow (M^{(-1)},M^{(-1)}\cap\partial_1M,p_{-1})$ with $h|\partial_0M = 1$. Such a homeomorphism is constructed as follows: let $V^{(i)} = N^{(-1)}\cup(\cup_\alpha|st(\hat{\alpha},J'')|)$ $(-1\leq i\leq n-1)$ where the union runs over all simplices α in ∂_1J_{-1} with $\dim\alpha \leq i$. By Lemma 3.4 below, there is a bc homeomorphism $h_i\colon (V^{(i)},V^{(i)}\cap\partial_1M,p|) \longrightarrow (V^{(i+1)},V^{(i+1)}\cap\partial_1M,p|)$ with $h_i|\partial_0M = 1$. Since $V^{(-1)} = N^{(-1)}$ and $V^{(n-1)} = M^{(-1)}$, $h = h_{n-2}\cdots h_0h_{-1}$ is the desired homeomorphism.

Lemma 3.4 *There is a bc homeomorphism* $h_i\colon (V^{(i)},V^{(i)}\cap\partial_1M,p|) \longrightarrow (V^{(i+1)},V^{(i+1)}\cap\partial_1M,p|)$ $(-1\leq i\leq n-2)$ *with* $h_i|\partial_0M = 1$.

Proof: Let $\partial_0V^{(i)} = \partial_0M\cup(\partial_1M\cap V^{(i)})$, $\partial_1V^{(i)} = Cl(\partial V^{(i)}-\partial_0V^{(i)})$ and set $W^{(i)} = \partial_0V^{(i)}\cap\partial_1V^{(i)}$. Let α be an $(i+1)$-simplex in ∂_1J_{-1}. It is well known that $(|st(\hat{\alpha},J'')|\cap Cl(M-V^{(i)}), |st(\hat{\alpha},J'')|\cap\partial_1V^{(i)})$ is a ball pair $(B^n_\alpha,B^{n-1}_\alpha)$. Let $\psi_\alpha\colon (D^n,D^{n-1}) \longrightarrow (B^n_\alpha,B^{n-1}_\alpha)$ be a characteristic homeomorphism for this ball pair. An argument similar to one above shows that $V^{(i+1)}$ is obtained from $V^{(i)}$ by attaching the n-disk $(S\times(D^n,D^{n-1}),(p|)\Psi)$ where $S = \{\hat{\alpha} \mid \alpha$ is an $(i+1)$-simplex of $\partial_1J_{-1}\}$ and $\Psi = \amalg\,\psi_\alpha$. It now follows from 2.5 that there is a bc homeomorphism $h_i'\colon (V^{(i)},p|) \longrightarrow (V^{(i+1)},p|)$ with $h_i'|\partial_0M = 1$.

To see that h_i' can be chosen to carry $V^{(i)}\cap\partial_1M$ into $V^{(i+1)}\cap\partial_1M$ requires a little more care. In particular, notice that if α is not in $W^{(i)}$, then $B^n_\alpha\cap\partial_1M = \emptyset$; while if α is in $W^{(i)}$, then $B^n_\alpha\cap\partial_1M = B^n_\alpha \cap Cl(\partial_1M-\partial_0V^{(i)}) = C^{n-1}_\alpha$ is an (n-1) ball, $B^{n-1}_\alpha \cap Cl(\partial_1M-\partial_0V^{(i)}) = B^{n-1}_\alpha\cap W^{(i)} = C^{n-1}_\alpha\cap W^{(i)} = C^{n-2}_\alpha$ is an (n-2)-ball, and $(B^n_\alpha,B^{n-1}_\alpha,C^{n-1}_\alpha,C^{n-2}_\alpha)$

is a triad of disks. It is now straightforward to modify the proof of 2.5 to obtain the desired bc homeomorphism h_i: $(V^{(i)}, V^{(i)} \cap \partial_1 M, p|) \longrightarrow (V^{(i+1)}, V^{(i+1)} \cap \partial_1 M, p|)$ with $h_i|\partial_0 M = 1$. This completes the proofs of 3.4 and 3.1.

4. Modifying handle presentations.

In this section we prove some basic lemmas concerning the modification of a handle presentation of a manifold. The context for the first two lemmas is the following:

Let $(M,p) \in |\mathcal{PL}^c/Z|$ be obtained from $(N,p|)$ by attaching an r-handle to the interior of a submanifold $\partial_1 N$ of ∂N; (thus, $N \subseteq M$, $p| = p|N$, and $\partial_1 N$ has codimension zero in ∂N). Let Φ: $(S \times (D^r, S^{r-1}) \times D^{n-r}, q) \longrightarrow (M, \partial_1 N, p)$ be a characteristic homeomorphism and $\{K_s \mid s \in S\}$ be a bounded family of subsets of Z for which there exists an integer $m \geq 0$ such that for every $s \in S$, $\Phi(s \times D^r \times D^{n-r}) \subseteq p^{-1}(C^m K_s)$. Such a family exists by the definition of an r-handle. Let $(x_0, y_0) \in S^{r-1} \times S^{n-r-1}$ and if $1 \leq r$ define σ: $S \longrightarrow |\mathcal{P}G_1(\partial_1 N, p|)|$ by $\sigma(s) = (x_s, K_s)$ where $x_s = \Phi(s, x_0, y_0)$. In the sequel, where no confusion will arise, we identify σ with $i_*\sigma$ where i_*: $\mathcal{P}G_1(\partial_1 N, p|) \longrightarrow \mathcal{P}G_1(N, p|)$ is induced by the inclusion i: $(\partial_1 N, p|) \longrightarrow (N, p|)$.

Lemma 4.1 *In the situation described above, if* $1 \leq r$ *or* $r=0$ *and* $(S \times 0 \times D^n, q)$ *is coextensive with* $(N, p|)$, *then* $(N, p|)$ *is coextensive with* (M, p). *If* $1 \leq r$, *then* $\pi_i^c(M, N, p) = 0$ *for* $i < r$. *If either* $3 \leq r$ *or* $r=2$ *and* $\Phi|$: $(S \times S^1 \times y_0, q|) \longrightarrow (\partial N, p|)$ *is bc null homotopic, then the morphism* φ: $F(\sigma) \longrightarrow \pi_r^c(M, N, p)$ *defined by setting*

$$\varphi(x_s, K_s)(1, s) = [\Phi(s \times (D^r, S^{r-1}) \times y_0] \in \pi_r(p^{-1}(C^m K_s), p^{-1}(C^m K_s) \cap N, x_s)$$

is an isomorphism of $\mathbb{Z}\mathcal{G}_1(N,p|)$*-modules. Moreover,* $\pi_r^c(M,N,p)$ *is boundedly finitely generated.*

Lemma 4.2 *Let* $\rho\colon T \longrightarrow |\mathcal{G}C_1(\partial_1 N,p|)|$ *be another basis and* $(\alpha,\nu)\colon (T,\rho) \longrightarrow (S,\sigma)$ *be an isomorphism of bases in* $\mathcal{G}G_1(\partial_1 N,p|)$*. Then there is a manifold* $(\bar{M},\bar{p})$ *obtained from* $(N,p|)$ *by attaching an r-handle such that* (T,ρ) *is a basis for* $\pi_r^c(\bar{M},N,\bar{p})$ *and such that the following diagram commutes*

$$\begin{array}{ccccc} F(\rho) & \xrightarrow{\ \bar{\varphi}\ } & \pi_r^c(\bar{M},N,\bar{p}) & & \\ {\scriptstyle F(\alpha,\nu)}\Big\downarrow & & & \searrow^{\bar{\partial}} & \\ & & & & \pi_{r-1}^c(N,p|) \\ & & & \nearrow_{\partial} & \\ F(\sigma) & \xrightarrow{\ \varphi\ } & \pi_r^c(M,N,p) & & \end{array}$$

Furthermore, there is a bc homeomorphism $h\colon (M,N,p) \longrightarrow (\bar{M},N,\bar{p})$.

In 4.2 we have again identified ρ with $i_*\rho$.

In these lemmas, $F(\sigma)$ is the free $\mathbb{Z}\mathcal{G}G_1(N,p|)$-module with basis (S,σ) defined in IV.1 and we have used IV.1.3, to define φ. We refer the reader to IV.2.2 for the definition of boundedly finitely generated (bfg).

Proof of 4.1: The part about coextensivity is obvious. Let $X = N\cup\Phi(S\times D^r\times y_0)$. Clearly the inclusion $(X,N,p|) \longrightarrow (M,N,p)$ is a homotopy equivalence of pairs. The assertion that $\pi_i^c(M,N,p) = 0$ for $i < r$ when $1 \leq r$ follows from V.1.10; while the fact that φ is an isomorphism follows from V.2.5.(ii). The last sentence of 4.1 follows easily from the facts that $\{K_s \mid s\in S\}$ is a bounded family of subsets of Z, that the boundedness control structure is tame (hence ClK is compact for all $K\in|\mathcal{G}|$), and that $q\colon S\times D^r\times D^{n-r} \longrightarrow Z$ is proper.

The proof of 4.2 is temporarily deferred.

The context for the next lemma is the following: We suppose that the index set for the r-handle of M-N is decomposed into the disjoint union $S_1 \amalg S_2$. We now regard (M,p) as being obtained from (N,p|) by attaching the two r-handles $\Phi_i(S_i\times(D^r,S^{r-1})\times D^{n-r},q_i)$ where Φ_i and q_i (i=1,2) are the restrictions of Φ and q respectively. Let $F(\sigma_1)\oplus F(\sigma_2)$ be the corresponding decomposition of $F(\sigma)$ and $P = N\cup\Phi_1(S_1\times D^r\times D^{n-r}) \subseteq M$. Finally suppose that $\Phi(S\times S^{r-1}\times D^{n-r}) \subseteq \partial_1 N$ where $\partial_1 N$ is a codimension zero submanifold of ∂N, closed as a subspace of ∂N, for which $\pi_1^c(N,\partial_1 N,p|) = 0$.

Lemma 4.3 (*Handle Rearrangement*) *In the situation described above assume* $r\geq 3$ *or that* $r=2$ *and* $\Phi|: (S\times S^1\times y_0,q|) \longrightarrow (\partial_1 N,p|)$ *is bc null homotopic. Let* $\rho: F(\sigma_2) \longrightarrow F(\sigma_1)$ *be a morphism of* $\mathbb{Z}\mathcal{G}_1(N,p|)$*-modules. Then there exists a manifold* $(\bar{M},\bar{p})$ *obtained from* (P,p|) *by attaching an r-handle with characteristic homeomorphism* $\bar{\Phi}_2: (S_2\times(D^r,S^{r-1})\times D^{n-r},q_2) \longrightarrow (\bar{M},N,\bar{p})$ *such that*

(i) $\bar{\Phi} = \Phi_1\amalg\bar{\Phi}_2: ((S_1\amalg S_2)\times(D^r,S^{r-1})\times D^{n-r},q) \longrightarrow (\bar{M},N,\bar{p})$ *is a characteristic homeomorphism for an r-handle; in particular, the images of* Φ_1 *and* $\bar{\Phi}_2$ *are disjoint;*

(ii) *There is a bc homeomorphism* $h: (M,P,p) \longrightarrow (\bar{M},P,\bar{p})$.

(iii) *The following diagram commutes*

$$\begin{array}{ccccc}
F(\sigma_1)\oplus F(\sigma_2) & \xrightarrow{\bar{\varphi}} & \pi_r^c(\bar{M},N,\bar{p}) & \xrightarrow{\bar{\partial}} & \pi_{r-1}^c(N,p|) \\
\eta\downarrow & & & & \| \\
F(\sigma_1)\oplus F(\sigma_2) & \xrightarrow[\varphi]{} & \pi_r^c(M,N,p) & \xrightarrow[\partial]{} & \pi_{r-1}^c(N,p|)
\end{array}$$

where η *has matrix* $\begin{bmatrix} 1 & \rho \\ 0 & 1 \end{bmatrix}$ *and* φ *and* $\bar{\varphi}$ *are the isomorphisms of* 4.1.

The final results on modifying handle presentations concern the following situation: Let (M_2,p_2) be obtained from (M_0,p_0) by successively attaching handles of indices r and r+1 of the same type (S,σ). Then there is a submanifold $M_1 \subseteq M_2$ such that (M_{i+1}, p_{i+1}) is obtained from (M_i, p_i) by attaching an (r+i)-handle (i=0,1) of type (S,σ). (Here $p_i = p_2|M_i$ for $i=0,1$.) Let $\Phi_i\colon (S\times(D^{r+i}, S^{r+i-1})\times D^{n-r-i}, q_i) \longrightarrow (M_{i+1}, \partial M_i, p_{i+1})$ be a characteristic homeomorphism for the (r+i)-handle (i=0,1) and suppose $\{K_{i,s} \mid s\in S\}$ (i=0,1) is a bounded family of subsets of $\mathbb{Z}$ for which there exists an integer $m_i \geq 0$ such that for all $s\in S$, $\Phi_i(s\times D^{r+i}\times D^{n-r-i}) \subseteq p_{i+1}^{-1}(C^{m_i}K_{i,s})$.

Lemma 4.4 (*Handle Cancellation*) *In the situation described above suppose that for all* $s,s' \in S$, $\Phi_1(s\times S^r\times 0) \cap \Phi_0(s'\times 0\times S^{n-r-1}) = \emptyset$ *if* $s\neq s'$ *and that* $\Phi_1(s\times S^r\times 0)$ *meets* $\Phi_0(s\times 0\times S^{n-r-1})$ *transversally in a single point. Then there is a bc homeomorphism* h: $(M_0,p_0) \longrightarrow (M_2,p_2)$ *such that* h *is the identity outside an arbitrarily small neighborhood of* $(\mathrm{Im}\Phi_0 \cup \mathrm{Im}\Phi_1) \cap \partial M_0$.

The proof is temporarily deferred.

Suppose now that either $r\geq 3$ or that $r=2$ and the attaching map for the 2-handle is null-homotopic. Then by II.5.3, the functor $k\colon \mathcal{P}G_1(M_0,p_0) \longrightarrow \mathcal{P}G_1(M_1,p_1)$ induced by the inclusion induces a equivalence of categories $k_!\colon \mathbb{Z}\mathcal{P}G_1(M_0,p_0)\text{-}mod \longrightarrow \mathbb{Z}\mathcal{P}G_1(M_1,p_1)\text{-}mod$ with pseudo-inverse $k^!$. To simplify our notation, we identify these categories via these functors and omit further mention of $k_!$ and $k^!$.

Lemma 4.5 *In the situation described before 4.4, suppose there is a change of basis isomorphism* $(\alpha,\upsilon)\colon (S,\sigma_1) \longrightarrow (S,\sigma_0)$ *such that the diagram*

$$\begin{array}{ccc} F(\sigma_1) & \xrightarrow{\varphi_1} & \pi^c_{r+1}(M_2,M_1,p_2) \\ {\scriptstyle F(\alpha,\upsilon)}\downarrow & & \downarrow{\scriptstyle \partial} \\ F(\sigma_0) & \xrightarrow{\varphi_0} & \pi^c_r(M_1,M_0,p_1) \end{array}$$

commutes. Then there exists a manifold $(\bar{M}_2,\bar{p}_2)$ *obtained from* (M_1,p_1) *by attaching an* $(r+1)$*-handle with characteristic homeomorphism* $\bar{\Phi}_1$ *such that* $(\bar{M}_2,M_1,\bar{p}_2)$ *is bc homeomorphic to* (M_2,M_1,p_2) *and* Φ_0 *and* $\bar{\Phi}_1$ *satisfy the hypothesis of 4.4.*

In this lemma $\sigma_i\colon S \longrightarrow |\mathcal{G}G_1(M_i,p_i)|$ is given by $\sigma_i(s) = (x_s,K_{i,s})$ $(i=0,1)$ as described earlier in this section.

The proof is temporarily deferred.

We now supply the proofs which have been deferred.

Proof of 4.2: Let $\rho(t) = (y_t,L_t)$. Since (α,ν) is an isomorphism of bases, there exists an $\ell \geq 0$ such that $\nu\colon \sigma\alpha \longrightarrow C^{\ell}\rho$. In particular, for all $t \in T$, $\nu(t) = (\omega_t,i_t)\colon (x_{\alpha(t)},K_{\alpha(t)}) \longrightarrow (y_t,C^{\ell}L_t)$. Since the boundedness control structure is tame, we may assume ℓ chosen so that the path ω_t lies in $p^{-1}(\text{Int } C^{\ell}L_t) \cap \partial_1 N$. Similarly, let $m \geq 0$ be chosen so that for every $s \in S$, $\Phi(s \times D^r \times D^{n-r}) \subseteq p^{-1}(\text{Int } C^m K_s)$. Let $\Psi_u\colon (S \times S^{r-1} \times D^{n-r},q) \longrightarrow (\partial_1 N,p|)$ $(0 \leq u \leq 1)$ be an isotopy such that $\Psi_0 = \Phi$, $\Psi_u|\alpha(t) \times S^{r-1} \times y_0$ pulls a small disk about x_0 along $\omega_{\alpha(t)}$ so that $\Psi_1(\alpha(t),x_0,y_0) = y_t$, and $\Psi_u|\alpha(t) \times S^{r-1} \times D^{n-r}$ is supported on the interior of an $(n-1)$-disk in $p^{-1}(\text{Int } C^{\ell+m}L_t) \cap \partial_1 N$. Let $(\bar{M},\bar{p})$ be the manifold obtained from $(N,p|N)$ and Ψ_1 by applying 2.2. Then by 2.3 there is a homeomorphism $h\colon (M,N,p) \longrightarrow (\bar{M},N,\bar{p})$. Since it follows immediately from the construction of $\bar{M}$ that the diagram given in 4.2 commutes, the proof of 4.2 is completed.

Let $\partial_1 P = \text{Cl}(\partial P - (\partial N - \partial_1 N))$. The main ingredient in the proof of 4.3 is the following lemma:

Lemma 4.6 *There exists a bounded family of mutually disjoint disks* $\{D_s^{n-1} \mid s\in S_2\}$ *contained in* $\partial_1 P$ *and an isotopy*

$$\Phi_{2,t}\colon (S_2\times S^{r-1}\times D^{n-r},q_2|) \longrightarrow (\partial_1 P,p|) \quad (0\le t\le 1)$$

such that

(i) *There exists an integer* $u\ge m$ *such that for every* $s\in S_2$, $D_s^{n-1} \subseteq C^u K_s$;

(ii) $\Phi_{2,t}(s\times S^{r-1}\times D^{n-r}) \subseteq D_s^{n-1}$ $(0\le t\le 1)$;

(iii) $\Phi_{2,0} = \Phi_2$ *and* $\Phi_{2,1}(S_2\times S^{n-1}\times D^{n-r}) \subseteq \partial_1 N - \Phi_1(S_1\times S^{n-1}\times D^{n-r})$;

(iv) *For every* $s\in S_2$ *and every* t $(0\le t\le 1)$, $\Phi_{2,t}(x_s) = x_s$; *and*

(v) *If* $\psi\colon F(\sigma_2) \longrightarrow \pi_{r-1}^c(N,p|)$ *is represented by the elements*

$$\Phi_{2,1}(s\times S^{r-1}\times y_0, s\times x_0\times y_0) \in \pi_{r-1}(p^{-1}(C^u K_s)\cap N, x_s) \quad (s\in S_2)$$

in the sense of IV.1.3, *then* $\psi = \partial\varphi(\rho,1)^T$ *where* T *denotes the transpose and* 1 *is the identity of* $F(\sigma_2)$.

The proof of 4.6 is temporarily deferred.

Proof of 4.3 assuming 4.6: Let $\bar{M} = P\cup\Phi_{2,1}(S_2\times D^r\times D^{n-r})$. Then by 2.2 and 2.3, there is a map $\bar{p}\colon \bar{M} \longrightarrow Z$ and a bc homeomorphism $h\colon (M,P,p) \longrightarrow (\bar{M},P,\bar{p})$. Thus (ii) of 4.3 follows; while (i) follows easily from (iii) of 4.6.

Since $\bar{\varphi}|F(\sigma_2)$ is represented by the natural transformation that sends $(1,s) \in F(\sigma_2)(x_s,K_s)$ to

$$\Phi_{2,1}(s\times(D^r,S^{r-1},x_0)\times y_0) \in \pi_r(\bar{p}^{-1}(C^u K_s),\ p^{-1}(C^u K_s)\cap N, x_s)$$
$$= \pi_r^c(\bar{M},N,\bar{p})(x_s,C^u K_s),$$

clearly $\bar{\partial}\bar{\varphi} = \psi$ and (iii) of 4.3 now follows easily from (v) of 4.6.

Proof of 4.6: Let $\rho\colon F(\sigma_2) \longrightarrow F(\sigma_1)$ be represented by the natural transformation $\rho_{u_1}\colon F(\sigma_2) \longrightarrow F(\sigma_1)C^{u_1}$ and recall that ρ_{u_1} is completely determined by its values on the generators $(1,s) \in F(\sigma_2)(x_s,K_s)$ $(s\in S_2)$. Since

$$\rho_{u_1}(x_s,K_s)\colon F(\sigma_2)(x_s,K_s) \longrightarrow F(\sigma_1)(x_s,C^{u_1}K_s)$$

and the latter group is free abelian on the elements (β,s') where $s'\in S_1$ and $\beta\colon (x_{s'},K_{s'}) \longrightarrow (x_s,C^{u_1}K_s)$ is a morphism in $\mathcal{G}G_1(N,p|)$ (cf. IV.1.3 and IV.1.1), we may write

$$\rho_{u_1}(x_s,K_s)(1,s) = \sum n(s,\beta,s')(\beta,s')$$

where $n(s,\beta,s') \in \mathbb{Z}$, and β and s' are as above.

For fixed s, let $U_s = \{\beta \mid n(s,\beta,s') \neq 0 \text{ for some } s'\}$ and for fixed s', let $V_{s'} = \{\beta \mid n(s,\beta,s') \neq 0 \text{ for some } s\}$. Since $F(\sigma_1)$ and $F(\sigma_2)$ are bfg, U_s and $V_{s'}$ are finite sets. Let $U = \coprod_s U_s$ $(s\in S_2)$ be given the discrete topology and $r\colon U \longrightarrow Z$ be given by $r(\beta) = p(x_s)$ if $\beta \in U_s$. Then r is proper. For each $\beta\colon (x_{s'},K_{s'}) \longrightarrow (x_s,C^{u_1}K_s)$ in U, we write $\beta = (\omega_\beta,i_\beta)$ where $i_\beta\colon K_{s'} \longrightarrow C^{u_1}K_s$ is an inclusion and ω_β is a path class in $p^{-1}(C^{u_1}K_s)$ from x_s to $x_{s'}$. Finally, let $f_\beta\colon I \longrightarrow p^{-1}(C^{u_1}K_s)$ be a map representing ω_β and let $f\colon (U\times I, U\times\{0,1\},r\pi) \longrightarrow (N,\partial_1 N,p|)$ be given by $f|\beta\times I = f_\beta$. Then f is a bc map.

Since $\pi_1^c(N,\partial_1 N,p|) = 0$, by II.10.6, there exists a map $g\colon (U\times I,r\pi) \longrightarrow (\partial_1 N,p|)$ bc homotopic to f relative to $U\times\{0,1\}$. In particular, there exist an integer $u_2\geq 0$ and paths $g_\beta\colon I \longrightarrow p^{-1}(C^{u_1+u_2}K_s)\cap\partial_1 N$ $(\beta\in U)$ from x_s to $x_{s'}$ representing ω_β. By general position, we may assume that $g_\beta(I) \cap \Phi(S\times S^{r-1}\times \mathrm{Int}D^{n-r}) = \emptyset$ and that g_β is an embedding.

Let $C \subseteq S^{r-1}\times S^{n-r-1}$ be an $(n-2)$-disk neighborhood of (x_0,y_0) and set $E_t = \Phi(t\times C)$ $(t\in S)$. For every $(s,\beta,s') \in S_2\times U_s\times S_1$ for which $n(s,\beta,s') \neq 0$, let $P(s,\beta,s')$ (or $Q(s,\beta,s')$, respectively) be a collection of $|n_{(s,\beta,s')}|$ points in E_s (or $E_{s'}$, respectively) chosen so that

$P(s,\beta_1,s_1')\cap P(s,\beta_2,s_2') = \emptyset$ for $(\beta_1,s_1')\neq(\beta_2,s_2')$ (or $Q(s_1,\beta_1,s') \cap Q(s_2,\beta_2,s') = \emptyset$ if $(s_1,\beta_1) \neq (s_2,\beta_2)$, respectively). Note that since U_s is finite, $\Sigma_{(\beta,s')}|n(s,\beta,s')|$ is finite and such a choice for the sets $P(s,\beta,s')$ is possible. Similarly, such a choice for the sets $Q(s,\beta,s')$ is possible. Now for each $\beta\in U_s$, let

$$g_{\beta,i}\colon I \longrightarrow p^{-1}(C^{u_1+u_2}K_s) \cap Cl(\partial_1 N-\Phi(S\times S^{r-1}\times D^{n-r})$$

$(1\leq i\leq|n(s,\beta,s')|)$ be a family of mutually disjoint embeddings such that $g_{\beta,i}(0) \in P(s,\beta,s')$, $g_{\beta,i}(1) \in Q(s,\beta,s')$, and such that $g_{\beta,i}$ is homotopic to $\lambda g_\beta \mu$ relative to the endpoints where λ is a path in E_s from $g_{\beta,i}(0)$ to x_s and μ is a path in $E_{s'}$ from $x_{s'}$ to $g_{\beta,i}(1)$. By general position, we may assume that $\{g_{\beta,i}(I) \mid \beta\in B,\ 1\leq i\leq|n(s,\beta,s')|\}$ is a family of mutually disjoint arcs.

For $s\in S_2$, let $F_s = E_s \cup \{g_{\beta,i}(I)\}\cup\{\Phi(s'\times D^r\times y_{\beta,i}\}$ where $\beta\in U_s$, $1\leq i\leq|n(s,\beta,s')|$ and $\Phi(s',x_0,y_{\beta,i}) = g_{\beta,i}(1)$. We claim that $\{F_s \mid s\in S_2\}$ is a bounded family of subsets of $\partial_1 P$. To see this, note first that since $\{K_{s'} \mid s'\in S_1\}$ is a bounded family of subsets of Z, there exists an integer u_3 such that for every $s'\in S_1$ there exists a minimal element $K_{0,s'}$ with $K_{s'} \subseteq C^{u_3}K_{0,s'}$. Since $\Phi(s'\times D^r\times D^{n-r}) \subseteq p^{-1}(C^m K_{s'})$, $E_s \subseteq p^{-1}(C^m K_s)$, and $\{g_{\beta,i}(I)\} \subseteq p^{-1}(C^{u_1+u_2}K_s)$ by construction, if s' is such that $n(s,\beta,s') \neq 0$, $pg_{\beta,i}(1) \in C^{u_3}K_{0,s'} \cap C^{m+u_1+u_2}K_s$. Hence by II.1.1.(iv), $C^{u_3}K_{0,s'} \subseteq C^v K_s$ for $v = m+u_1+u_2+u_3+\theta(u_3)$. Thus $F_s \subseteq p^{-1}(C^v K_s)$. Since $\{K_s \mid s\in S_2\}$ is a bounded family of subsets of Z, the claim follows.

We now observe that by construction F_s is collapsible and $F_{s_1} \cap F_{s_2} = \emptyset$ for $s_1\neq s_2$. Let D_s^{n-1} be a regular neighborhood of F_s in $\partial_1 P$. Then D_s^{n-1} is a disk and we may assume that $\{D_s^{n-1} \mid s\in S_2\}$ is a family of mutually disjoint disks. Finally, we may assume $D_s^{n-1} \subseteq p^{-1}(C^u K_s)$ where $u = v+d$ and d is the integer such that for all $K\in|\mathcal{F}|$, $Cl\ K \subseteq Int\ C^d K$ given by the tameness of the boundedness control structure. Condition (i) of 4.6 is now established.

The desired isotopy $\Phi_{2,t}$ of $S_2\times S^{r-1}\times D^{n-r}$ is now constructed by the usual argument (cf. [Hu, p.228 ff]) of pulling $\Phi_2(s\times S^{r-1}\times y_0)$ along the paths $g_{\beta,i}(I)$ and over the disks $\Phi_1(s'\times D^r\times y_{\beta,i})$ inside D^{n-1}_s(with the orientation reversed if $n(s,\beta,s')<0)$. Properties (ii) to (v) then follow easily from the construction and the proof of 4.6 is complete.

Proof of 4.4: Clearly the families of subspaces of M_2 $\{\Phi_0(s\times D^r\times D^{n-r}) \mid s\in S\}$ and $\{\Phi_1(s\times D^{r+1}\times D^{n-r-1}) \mid s\in S\}$ are bounded. Since $\Phi_0(s\times D^r\times D^{n-r}) \cap \Phi_1(s\times D^{r+1}\times D^{n-r-1}) \neq \emptyset$ for every $s\in S$, it is easy to see that $\{\bigcup_{k=0}^{1} \Phi_k(s\times D^{r+k}\times D^{n-r-k}) \mid s\in S\}$ is also a bounded family of subspaces of M_2. Now standard arguments (cf. [RS, Lemma 6.4]) show that $\Phi_0(s\times D^r\times D^{n-r}) \cup \Phi_1(s\times D^{r+1}\times D^{n-r-1})$ is an n-disk $s\times D^n$ and that M_2 is obtained from M_0 by attaching $S\times D^n$ to M_0 along $S\times D^{n-1}$ where $s\times D^{n-1} \subseteq s\times\partial D^n \subseteq \partial M_0$. Since the remarks above show that $(S\times D^n,p|)$ is a bc n-disk, 4.4 now follows from 2.5.

Proof of 4.5: By applying 4.2 to the change of basis isomorphism $(\alpha,\upsilon)^{-1}$ inverse to (α,υ), we may replace (M_2,M_1,p_2) by a bc homeomorphic pair $(\bar{M}_2,M_1,\bar{p}_2)$, if needed, and assume that $(S,\sigma_1) = (S,\sigma_0)$, that (α,υ) is the identity, and that

$$\begin{array}{ccc} F(\sigma_0) & \xrightarrow{\varphi_1} & \pi^c_{r+1}(M_2,M_1,p_2) \\ \downarrow{\scriptstyle 1} & & \downarrow{\scriptstyle \partial} \\ F(\sigma_0) & \xrightarrow{\varphi_0} & \pi^c_r(M_1,M_0,p_1) \end{array}$$

commutes.

Let $H^{r+i}_s = \Phi_i(s\times D^{r+i}\times D^{n-r-i})$ $(i=0,1)$ and choose $m\geq m_1$ large enough so that for all $s\in S$, $H^{r+1}_s \subseteq p^{-1}(\text{Int } C^m K_{1,s}) = M_s$ and if $H^r_{s'} \cap H^{r+1}_s \neq \emptyset$, then $H^r_{s'} \subseteq M_s$. (Such an m can easily be found by using the tameness and property (iv) of the boundedness control structure.) Then

$$\partial_s\colon \pi_{r+1}(M_s,M_s\cap M_1,x_s) \longrightarrow \pi_r(M_s\cap M_1,M_s\cap M_0,x_s)$$

is essentially one of the homomorphisms in the natural transformation representing ∂. Since $\varphi_0 = \partial\varphi_1$, by increasing m, if necessary, $\partial_s\zeta_s^{r+1} = \zeta_s^r$ where $\zeta_s^{r+i} = \Phi_i[s\times(D^{r+i},S^{r+i-1})\times z_i] \in \pi_{r+i}(M_s\cap M_{r+i+1},M_s\cap M_{r+i},x_s)$ $(i=0,1)$. But this means that the core sphere $\Phi_1(s\times S^r\times 0)$ has algebraic intersection 1 with the belt sphere $\Phi_0(s\times 0\times S^{n-r-1})$ and zero algebraic intersection with $\Phi_0(s'\times 0\times S^{n-r-1})$ if $s\neq s'$ (but $H_{s'}^r\cap H_s^{r+1}\neq\emptyset$). The standard arguments (cf. [RS, pp. 69 ff.]), using the Whitney trick, may be applied in $M_s\cap\partial M_1$ to isotope $\Phi_1(s\times S^r\times 0)$ off $\Phi_0(s'\times 0\times S^{n-r-1})$ for $s\neq s'$ and to make sure that $\Phi_1(s\times S^r\times 0)$ meets $\Phi_0(s\times 0\times S^{n-r-1})$ transversally in a single point.

By using general position in $M_s \cap \partial M_1$ to ensure that the Whitney disks are disjoint, we may assume the isotopy of $\Phi_1|s\times S^r\times D^{n-r-1}$ is supported on an $(n-1)$-disk $E_s \subseteq M_s \cap \partial M_1$ and that the $\{E_s \mid s\in S\}$ are mutually disjoint. Since $E_s\subseteq M_s$ and $\{M_s \mid s\in S\}$ is a bounded family of sets, so is $\{E_s \mid s\in S\}$. It follows that the isotopy of $\Phi_1|S\times S^r\times D^{n-r-1}$ made up of the isotopies of $\Phi_1|s\times S^r\times D^{n-r-1}$ is bc and 4.5 now follows from 2.2 and 2.3.

5. Geometric connectivity of manifolds.

It is the purpose of this section to prove the following theorem:

Theorem 5.1 *Let* $n\geq 6$ *and* (M,p) *be an n-manifold in* $\mathcal{PL}^c/Z$ *with* $\partial M = \partial_0 M\cup\partial_1 M$. *Suppose that* $(\partial_0 M,p_0)$ *is coextensive with* (M,p) *and that for some* r $(0\leq r\leq n-3)$, $\pi_i^c(M,\partial_0 M,p) = 0$ *for* $i\leq r$. *Then* M *has a handle decomposition relative to* $\partial_0 M$ *with no handles of index* i *for* $i\leq r$.

The proof of 5.1 is by induction on r with the inductive hypothesis being that M has a handle decomposition with no handles of index i for $i<r$. Thus, it suffices to prove the following lemma:

Lemma 5.2 *Let $n\geq 6$ and (M,p) be an n-manifold in $\mathcal{PL}^c/Z$ with $\partial M = \partial_0 M\cup\partial_1 M$. For some r $(0\leq r\leq n-3)$ suppose M has a handle decomposition relative to $\partial_0 M$ with no handles of index i for $i<r$*

(i) *If $r=0$, suppose that $(\partial_0 M,p_0)$ is coextensive with (M,p) and that $\pi_0^c(M,\partial_0 M,p) = 0$.*

(ii) *If $r\geq 1$, suppose also that $\pi_i^c(M,\partial_0 M,p) = 0$ for $i\leq r$.*

Then M has a handle decomposition relative to $\partial_0 M$ with no handles of index i for $i\leq r$ and the same handles of index i for $r+2<i$.

Suppose that $M^{(-1)} = M^{(r-1)} \subseteq M^{(r)} \subseteq\ldots\subseteq M^{(n)} = M$ is a handle decomposition of (M,p) relative to $\partial_0 M$ with no handles of index i for $i\leq r-1$ and that the type of the j-handle is (T_j,ρ_j). We remark that the proof of 5.2 to be given below shows that (M,p) has a handle decomposition relative to $\partial_0 M$, $M^{(-1)} = \overline{M}^{(r)} \subseteq \overline{M}^{(r+1)} \subseteq \ldots \subseteq \overline{M}^{(n)} = M$, with no handles of indices i for $i \leq r$ for which the type of the j-handle is (T_j,ρ_j) for $j \neq r+2$ and is $(T_r \amalg T_{r+2},\rho_r \amalg \rho_{r+2})$ for $j = r+2$.

Before proving 5.2, we establish two results that will be useful in its proof. The first result is the following:

Lemma 5.3 *Let $M^{(-1)} \subseteq M^{(0)} \subseteq\cdots\subseteq M^{(r)} \subseteq\cdots\subseteq M^{(n)}$ be a handle decomposition of the n-manifold (M,p) in $\mathcal{PL}^c/Z$ (relative to $\partial_0 M$ if $\partial_0 M\neq\emptyset$). Then*

(i) *$\pi_i^c(M^{(s)},M^{(r)},p_s) = 0$ for $0\leq i\leq r<s$ where $p_s = p|M^{(s)}$. In particular, $\pi_i^c(M,M^{(r)},p) = 0$ for $i\leq r$.*

(ii) *If $N\subseteq M$ is a submanifold of $M^{(r)}$ and $\pi_i^c(M,N,p) = 0$ for $i<r$, then $\pi_i^c(M^{(r)},N,p_r) = 0$ for $i<r$.*

Proof: It follows from 4.1, that $\pi_i^c(M^{(r+1)},M^{(r)},p_{r+1}) = 0$ for $i \leq r$. Part (i) now follows from a simple induction argument using the homotopy exact sequence of the triple $(M^{(s)},M^{(s-1)},M^{(r)},p_s)$ and 4.1. In a similar spirit, part (ii) follows from (i) by using the homotopy exact sequence of $(M,M^{(r)},N,p)$.

The context for the second result, which is quite similar to that of the proof of 5.2, is the following: Let (M,p) be a manifold with $\partial M = \partial_0 M \cup \partial_1 M$ where $\partial_0 M \cap \partial_1 M = W$ is the boundary of $\partial_i M$ $(i=0,1)$. Let $M^{(-1)} \subseteq \cdots \subseteq M^{(r-1)} \subseteq \cdots \subseteq M^{(n)} = M$ be a handle decomposition of M relative to $\partial_0 M$ and set $p_i = p|M^{(i)}$ $(-1 \leq i \leq n)$. Let $\partial_1 M^{(r)} = \mathrm{Cl}(\partial M^{(r)} - \partial_0 M)$. Let $\Phi_i\colon (S_i \times (D^i,S^{i-1}) \times D^{n-i},q_i) \longrightarrow (M^{(i)},M^{(i-1)},p_i)$ be a characteristic homeomorphism for the i-handle $(0 \leq i \leq n)$ and suppose $\{K_s^i \mid s \in S_i\}$ is a bounded family of subsets of Z such that $\Phi_i(s \times D^i \times D^{n-i}) \subseteq p^{-1}(K_s^i)$ for every $s \in S_i$. We assume the handle decomposition satisfies the following technical condition

TC: For every $i \leq n-3$ and every $s \in S_i$, there exists a point $y_s \in \Phi_i(s \times S^{i-1} \times S^{n-i-1}) - \mathrm{Im}\,\Phi_{i+1} - \mathrm{Im}\,\Phi_{i+2}$.

(As we shall soon see, TC imposes no real restriction.)

Let $c\colon (\partial_0 M \times I,(p|\partial_0 M)\pi) \longrightarrow (M^{(-1)},p_{-1})$ be a bc homeomorphism with $c|\partial_0 M \times 0$ the identity, where $\pi\colon \partial_0 M \times I \longrightarrow \partial_0 M$ is the projection and set $W_+ = c(W \times 1)$ and $\partial_+ M^{(r-1)} = \mathrm{Cl}(\partial M^{(r-1)} - c(\partial_0 M \times 0 \cup W \times I))$. Note that $\partial_+ M^{(-1)} = c(\partial_0 M \times 1)$.

Let $(T,\bar{\rho})$ be a bfg basis in $\mathcal{P}G_1(\partial_+ M^{(r-1)} - \bigcup_{i=r}^{r+2} \mathrm{Im}\,\Phi_i - W_+,p|)$ and set $\bar{\rho}(t) = (\bar{x}_t,\bar{K}_t)$. We say that $(T,\bar{\rho})$ is *geometric* if the map $t \longrightarrow \bar{x}_t$ is injective.

Lemma 5.4 *Let* $\bar{\rho}: T \longrightarrow \mathscr{P}G_1(\partial_+ M^{(r-1)} - \bigcup_{i=r}^{r+2} \mathrm{Im}\,\Phi_i - W_+, p|)$ *be a geometric bfg basis. Then for* s *either* r *or* r+1 *there exists a PL manifold* $(\bar{M},\bar{p})$ *and a handle decomposition* $\{(\bar{M}^{(i)},\bar{p}_i) \mid 0 \le i \le n\}$ *of* $(\bar{M},\bar{p})$ *relative to* $\partial_0 M$ *such that*

$$\bar{M} = M \cup H^s \cup H^{s+1} \quad \text{and} \quad \bar{M}^{(i)} = \begin{cases} M^{(i)} & i<s \\ M^{(s)} \cup H^s & i=s \\ M^{(i)} \cup H^s \cup H^{s+1} & i>s \end{cases}$$

where H^s *and* H^{s+1} *form a pair of cancelling handles of type* $(T,\bar{\rho})$. *In particular,* $(\bar{M},\bar{p})$ *is bc PL homeomorphic to* (M,p) *and the attaching map for the* s *handle is bc null homotopic.*

In the sequel we will identify $(\bar{M},\bar{p})$ with (M,p) and will regard $\{(\bar{M}^{(i)},\bar{p}_i)\}$ as a handle decomposition of (M,p).

The reader should note that if $s \ge 3$, the following diagram commutes

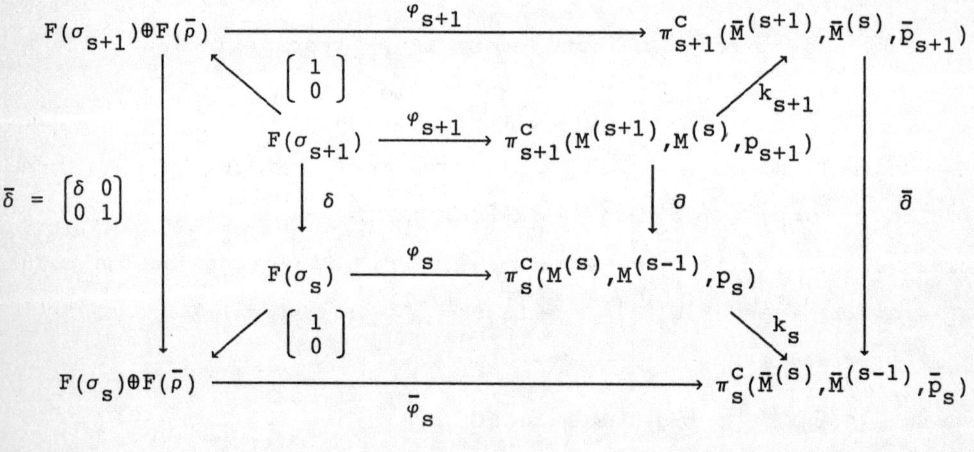

$F(\sigma_{s+1}) \oplus F(\bar{\rho})$
$\bar{\varphi}_{s+1}$
$\pi^c_{s+1}(\bar{M}^{(s+1)},\bar{M}^{(s)},\bar{p}_{s+1})$
$\begin{bmatrix}1\\0\end{bmatrix}$
k_{s+1}
$F(\sigma_{s+1})$
φ_{s+1}
$\pi^c_{s+1}(M^{(s+1)},M^{(s)},p_{s+1})$
$\bar{\delta} = \begin{bmatrix}\delta & 0\\0 & 1\end{bmatrix}$
δ
∂
$\bar{\partial}$
$F(\sigma_s)$
φ_s
$\pi^c_s(M^{(s)},M^{(s-1)},p_s)$
$\begin{bmatrix}1\\0\end{bmatrix}$
k_s
$F(\sigma_s) \oplus F(\bar{\rho})$
$\pi^c_s(\bar{M}^{(s)},\bar{M}^{(s-1)},\bar{p}_s)$
$\bar{\varphi}_s$

In the diagram k_j is induced by the inclusion $(M^{(j)},(M^{j-1}),p_j) \longrightarrow (\bar{M}^{(j)},\bar{M}^{(j-1)},\bar{p}_j)$ $(j=s,s+1)$ and δ and $\bar{\delta}$ are defined to make the inner and outer rectangles, respectively, commute. In addition, to simplify notation, we have not distinguished between $\sigma_j: S_j \longrightarrow |\mathscr{P}G_1(\partial_+ M^{(j)},p|)|$ and its images in $|\mathscr{P}G_1(M^{(j)},p_j)|$ and $|\mathscr{P}G_1(\bar{M}^{(j)},\bar{p}_j)|$. A similar comment applies to $\bar{\rho}$. Furthermore, since $s \geq 3$, each of the inclusions in the diagram

$$\begin{array}{ccc}
(M^{(s-1)},p_{s-1}) & = & (\bar{M}^{(s-1)},\bar{p}_{s-1}) \\
{\scriptstyle i}\downarrow & & \downarrow{\scriptstyle \bar{i}} \\
(M^{(s)},p_s) & \longrightarrow & (\bar{M}^{(s)},\bar{p}_s)
\end{array}$$

induces an equivalence of categories of $\mathbb{Z}\mathscr{G}_1(\ )$-modules. We have identified all of these categories and suppressed all mention of the maps $i^!$, $\bar{i}^!$, etc.

Similarly, if $s=2$, there is a commutative diagram

$$\begin{array}{ccccc}
F(\sigma_3)\oplus F(\bar{\rho}) & & \xrightarrow{\quad\bar{\varphi}_3\quad} & & \pi_3^c(\bar{M}^{(3)},\bar{M}^{(2)},\bar{p}_3) \\
 & \nwarrow{\scriptstyle \begin{bmatrix}1\\0\end{bmatrix}} & & \nearrow{\scriptstyle k_{s+1}} & \\
 & F(\sigma_3) & \xrightarrow{\quad\varphi_3\quad} & \pi_3^c(M^{(3)},M^{(2)},p_3) & \\
\bar{\delta} = \begin{bmatrix}\delta & 0\\0 & 1\end{bmatrix}\Big\downarrow & & {\scriptstyle\delta}\searrow \quad \swarrow{\scriptstyle\partial} & & \Big\downarrow\bar{\partial} \\
 & & \pi_2^c(M^{(2)},M^{(1)},p_2) & & \\
 & \swarrow{\scriptstyle \begin{bmatrix}1\\0\end{bmatrix}} & & \searrow{\scriptstyle k_s} & \\
\pi_2^c(M^{(2)},M^{(1)},p_2)\oplus F(\bar{\rho}) & & \xrightarrow[\quad(k_s,\bar{\varphi}_2)\quad]{} & & \pi_2^c(\bar{M}^{(2)},\bar{M}^{(1)},\bar{p}_2)
\end{array}$$

in which $(k_s,\bar{\varphi}_2)$ is an isomorphism and we follow the notations and conventions described above.

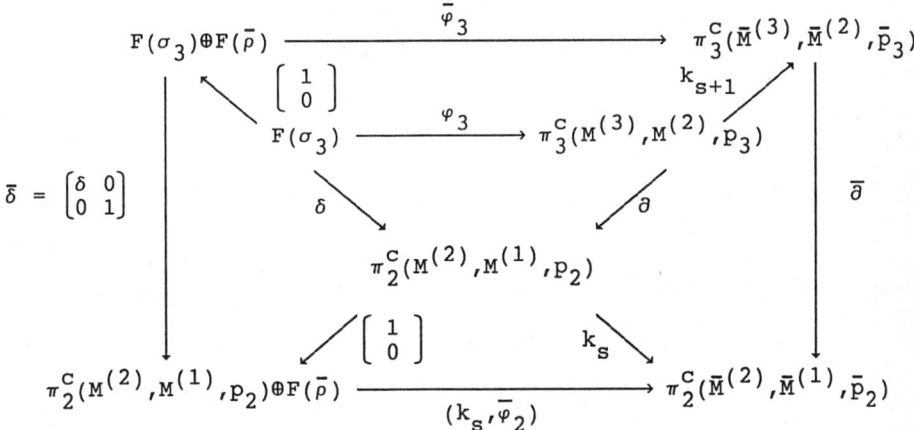

Proof: Since $\bar{\rho}$ is geometric and bfg, there exists a bounded family $\{E_t \mid t \in T\}$ of mutually disjoint $(n-1)$ disks in $\partial_+ M^{(r-1)} - \bigcup_{i=r}^{r+2} \operatorname{Im}\Phi_i - W_+$ with $\bar{x}_t \in \operatorname{Int} E_t$. Define $q\colon T \times D^n \longrightarrow Z$ by $q(t \times D^n) = p(\bar{x}_t)$ and attach the n-disk $(T \times (D^n,D^{n-1}),q)$ to $(M^{(r-1)},p_{r-1})$ in such a way that $t \times D^{n-1}$ is attached to the interior of E_t. Present the n-disk as a pair of cancelling handles (cf. [Hu, Theorem 5.1, p. 235]) of indices s and s+1 with characteristic homeomorphisms Ψ_s and Ψ_{s+1} respectively. We may clearly assume that $\Psi_s(t,x_0,y_0) = \bar{x}_t$. The results of section 2 now show that this construction yields a manifold $(\bar{M},\bar{p})$ and a handle decomposition $\{\bar{M}^{(i)},\bar{p}_i) \mid 0\le i\le n\}$ of $(\bar{M},\bar{p})$ with all the properties required in 5.4.

If $\rho\colon T \longrightarrow |\mathcal{F}G_1(\partial_+M^{(r-1)},p|)$ is a basis, we set $\rho(t) = (x_t,K_t)$ and if $(1,\nu)\colon (T,\bar{\rho}) \longrightarrow (T,\rho)$ is a map of bases, we write

$$\nu(t) = (\omega_t,i_t)\colon (x_t,K_t) \longrightarrow (\bar{x}_t,C^m\bar{K}_t).$$

Lemma 5.5 *Let* $1\le r\le n-3$ *and if* $r=1$, *suppose also that* $\pi_1^c(M^{(3)},M^{(0)},p_3) = 0$. *Let* (T,ρ) *be a bfg basis in* $\mathcal{F}G_1(\partial_+M^{(r-1)},p|)$. *Then there exist a geometric basis* $(T,\bar{\rho})$ where

$$\bar{\rho}\colon T \longrightarrow |\mathcal{F}G_1(\partial_+M^{(r-1)} - \bigcup_{i=r}^{r+2} \operatorname{Im}\Phi_i - W_+,p|)|$$

and an isomorphism $(1,\nu)\colon (T,\bar{\rho}) \longrightarrow (T,\rho)$ *of bases in* $\mathcal{F}G_1(\partial_+M^{(r-1)},p|)$.

Proof: Recall that $\{K_t \mid t\in T\}$ is a bounded family of subsets of Z. It suffices to show that there is an integer $m\ge 0$ and an embedding $T \longrightarrow \partial_+M^{(r-1)} - \bigcup_{i=r}^{r+2} \operatorname{Im}\Phi_i - W_+$ given by $t \longrightarrow \bar{x}_t$ such that for every $t\in T$ there is a path ω_t in $\partial_+M^{(r-1)} \cap p^{-1}(\operatorname{Int} C^mK_t)$ from x_t to $\bar{x}_t$. Indeed, we could then set $\bar{\rho}(t) = (\bar{x}_t,K_t)$ and take $\nu(t) = (\omega_t,i_t)\colon (\bar{x}_t,K_t) \longrightarrow (x_t,C^mK_t)$. In fact, let u_j be such that for every $s\in S_j$, there is a minimal element $K_{0,s}$ with $H_s^j \subseteq p^{-1}(C^{u_j}K_{0,s})$ $(j=r,r+1,r+2)$ where $H_s^j = \Phi_j(s\times D^j\times D^{n-j})$; and let d, coming from the tameness of the boundedness control structure, be such that $\operatorname{Cl} K \subseteq \operatorname{Int} C^dK$ for every $K\in|\mathcal{F}|$. Then we claim that $m_1 = d + \sum_{j=r}^{r+2} (u_j + \theta(u_j))$ will almost work, where θ is the function of II.1.1.(iv).

To see this, notice that if $x_t \in W_+$, then there is a disk neighborhood of x_t contained in $\partial_+M^{(r-1)} \cap p^{-1}(\text{Int } C^dK_t) \subseteq \partial_+M^{(r-1)} \cap p^{-1}(\text{Int } C^{m_1}K_t)$, and it is easy to find the required point $\bar{x}_t$ and path ω_t by pushing along a collar of W_+ in $\partial_+M^{(r-1)}$. On the other hand, if $x_t \in \bigcup_{i=r}^{r+2} \text{Im } \Phi_i$, then there is a chain of handles H_j $(1 \leq j \leq n_t \leq 3)$, of length at most 3, such that

(i) $x_t \in H_1$;

(ii) $r \leq \text{index } H_{j+1} < \text{index } H_j \leq r+2$;

(iii) $H_j \cap H_{j+1} \neq \emptyset$;

(iv) If $H_{n_t} = H_s^k$, then $\Phi_s(s \times S^{k-1} \times S^{n-k-1}) \subseteq \partial_+M^{(r-1)}$.

A straightforward argument, using II.1.1.(iv), shows that $\bigcup_{j=1}^{n_t} H_j \subseteq p^{-1}(\text{Int } C^{m_1}K_t)$. Since each H_j is path con-nected, it is easy to find a path α_t from x_t to a point $y_t \in H_{n_t} \cap \partial_+M^{(r-1)}$. In fact, by (iv) and (TC), we may suppose that $y_t \notin \text{Im } \Phi_i$ for $k<i \leq r+2$. Now let β_t be a path in a small disk in $\partial_+M^{(r-1)} \cap p^{-1}(\text{Int } C^{m_1}K_t)$ from y_t to a point $\bar{x}_t \in \partial_+M^{(r-1)} - \bigcup_{i=r}^{r+2} \text{Im } \Phi_i - W_+$. Then the inverse of the concatenation γ_t of α_t and β_t is almost the desired path ω_t. Clearly, we may assume $t \longrightarrow \bar{x}_t$ is injective.

The problem is that $\gamma_t \subseteq M^{(r+2)}$ and may not be in $\partial_+M^{(r-1)}$. But if $r \geq 2$, $\pi_1^c(M^{(r+2)}, M^{(r-1)}; p_{r+2}) = 0$ by 4.1 and if $r=1$, $\pi_1^c(M^{(3)}, M^{(0)}; p_3) = 0$ by hypothesis. It now follows from II.10.6 that there is an integer $m_2 \geq m_1$ and a path δ_t in $M^{(r-1)} \cap p^{-1}(\text{Int } C^{m_2}K_t)$ from $\bar{x}_t$ to x_t. Since $r \leq n-3$, $\pi_1^c(M^{(r-1)}, \partial_+M^{(r-1)}, p_{r-1})$ vanishes by applying 4.1 to the dual handle decomposition. Thus there is an integer $m \geq m_2$ and a path ω_t from $\bar{x}_t$ to x_t in $\partial_+M^{(r-1)} \cap p^{-1}(\text{Int } C^mK_t)$ by the same argument as above. This completes the proof of 5.5.

Proof of 5.2: We follow the context and notation set preceding 5.4 with the extra observation that, by the hypothesis of 5.2, we may assume that $M^{(-1)} = M^{(r-1)}$. Furthermore, if it is necessary to do so, we may alter the attaching homeomorphisms for the (r+1)- and (r+2)-handles by an isotopy that shrinks them down onto a small neighborhood of their cores $\Phi_i(S_i \times D^i \times 0)$ $(i=r+1,r+2)$ to insure that the handle decomposition given by the hypothesis of 5.2 satisfies the technical condition TC.

There are now three cases to consider depending on whether $r \geq 2$, $r=0$, or $r=1$.

Case I. Suppose $r \geq 2$. In this case, we note first that the inclusion $i_{r-1}\colon (M^{(r-1)},p_{r-1}) \longrightarrow (M^{(r)},p_r)$ induces an equivalence of categories $i_{r-1!}\colon \mathbb{Z}\mathcal{G}_1(M^{(r-1)},p_{r-1})\text{-}mod \longrightarrow \mathbb{Z}\mathcal{G}_1(M^{(r)},p_r)\text{-}mod$ with pseudo-inverse $i_{r-1}^{!}$. This is immediate from 5.3 and II.5.3 if $r>3$ or if $r=2$ and $\Phi_2|S_2 \times S^1 \times D^{n-2}$ is bc null homotopic. Now the composite

$$(S_2 \times S^1 \times D^{n-2},q_2) \xrightarrow{\Phi_2|} (\partial_+ M^{(1)},p_1) = (\partial_+ M^{(-1)},p|) \longrightarrow (M,p)$$

is clearly bc null homotopic. Since the inclusions

$$(\partial_0 M,p|\partial_0 M) \longrightarrow (M^{(-1)},p_{-1}) \longleftarrow (\partial_+ M^{(-1)},p_+) = (\partial_+ M^{(1)},p_+)$$

are both homotopy equivalences and $\pi_i^c(M,\partial_0 M,p) = 0$ for $i \leq 2$, $\pi_i^c(M,\partial_+ M^{(1)},p) = 0$ for $i \leq 2$. It now follows from II.10.6 that $\Phi_2|S_2 \times S^1 \times D^{n-2}$ is bc null homotopic as required.

Since $i_{r-1!}$ is an equivalence of categories, to avoid obscuring the essentially simple (and classical) line of the remainder of the proof with excessive notation, we shall identify $\mathbb{Z}\mathcal{G}_1(M^{(r-1)},p_{r-1})\text{-}mod$ with $\mathbb{Z}\mathcal{G}_1(M^{(r)},p_r)\text{-}mod$ via $i_{r-1!}$ and will suppress mention of the maps $i_{r-1!}$ and $i_{r-1}^{!}$.

Since $\pi_i^c(M,\partial_0 M,p) = 0$ for $i \leq r$, $\pi_i^c(M^{(r+1)},\partial_0 M,p_{r+1}) = 0$ for $i \leq r$ by 5.3 and since $(\partial_0 M,p|\partial_0 M) \longrightarrow (M^{(-1)},p_{-1}) = (M^{(r-1)},p_{r-1})$ is a homotopy equivalence, $\pi_i^c(M^{(r+1)},M^{(r-1)},p_{r+1}) = 0$ for $i \leq r$. An examination of the exact sequence of the triple $(M^{(r+1)},M^{(r)},M^{(r-1)},p_{r+1})$ now shows that $\partial\colon \pi^c_{r+1}(M^{(r+1)},M^{(r)},p_{r+1}) \longrightarrow \pi^c_r(M^{(r)},M^{(r-1)},p_r)$ is epic. Since $\pi^c_r(M^{(r)},M^{(r-1)},p_r)$ is free by 4.1, there is a morphism $\rho\colon \pi^c_r(M^{(r)},M^{(r-1)},p_r) \longrightarrow \pi^c_{r+1}(M^{(r+1)},M^{(r)},p_{r+1})$ such that $\partial\rho = 1$. Let ∂_0 and ρ_0 be chosen so that the following diagram commutes

$$\begin{array}{ccc}
F(\sigma_{r+1}) & \xrightarrow{\varphi_{r+1}} & \pi^c_{r+1}(M^{(r+1)},M^{(r)},p_{r+1}) \\
{\scriptstyle\partial_0}\big\downarrow\ \big\uparrow{\scriptstyle\rho_0} & & {\scriptstyle\partial}\big\downarrow\ \big\uparrow{\scriptstyle\rho} \\
F(\sigma_r) & \xrightarrow{\varphi_r} & \pi^c_r(M^{(r)},M^{(r-1)},p_r)
\end{array}$$

where φ_i $(i=r,r+1)$ is the isomorphism of 4.1. (In particular, $\sigma_i\colon S_i \longrightarrow |\mathcal{G}_1(\partial_+M^{(i-1)},p|)|$ is given by $\sigma_i(s) = (x_{i,s},K_{i,s})$ where $x_{i,s} = \Phi_i(s,x_0,y_0)$ and $(x_0,y_0) \in S^{i-1}\times S^{n-i-1}$ $(i=r,r+1)$.)

By 5.4 and 5.5, there are a basis $(S_r,\bar\sigma_r)$, an isomorphism of bases $(1,\nu)\colon (S_r,\bar\sigma_r) \longrightarrow (S_r,\sigma_r)$, and a handle decomposition $\{(\bar M^{(i)},\bar p_i)\,|\,0\le i\le n\}$ of (M,p) relative to ∂_0M obtained from $\{(M^{(i)},p_i)\}$ by attaching a pair of cancelling handles Ψ^{r+1} and Ψ^{r+2} of type $(S_r,\bar\sigma_r)$. In particular, $\bar M^{(i)} = M^{(i)}$ for $i\le r$, $\bar M^{(r+1)} = M^{(r+1)} \cup \Psi^{r+1}$, and $\pi^c_{r+1}(\bar M^{(r+1)},\bar M^{(r)},\bar p_{r+1})$ is isomorphic to $F(\sigma_{r+1}) \oplus F(\bar\sigma_r)$.

Let the morphism $\rho_0F(1,\nu)\colon F(\bar\sigma_r) \longrightarrow F(\sigma_{r+1})$ play the role of ρ in 4.3 and use it to rearrange the handles of $\{(\bar M^{(i)},\bar p_i)\}$. The result is a new handle decomposition $\{(\bar{\bar M}^{(i)},\bar{\bar p}_i)\,|\,0\le i\le n\}$ of (M,p) relative to ∂_0M with $\bar{\bar M}^{(i)} = M^{(i)}$ for $i\le r$, $\bar{\bar M}^{(r+1)} = M^{(r+1)} \cup \bar{\bar\Psi}^{r+1}$, and such that the outer polygon of the following diagram commutes

$$\begin{array}{cccccc}
F(\sigma_{r+1})\oplus F(\bar\sigma_r) & & \xrightarrow{\bar{\bar\varphi}_{r+1}} & \pi^c_{r+1}(\bar{\bar M}^{(r+1)},\bar{\bar M}^{(r)},\bar{\bar p}) & \xrightarrow{\partial''} & \pi^c_r(\bar{\bar M}^{(r)},\bar{\bar p}) \\
 & & & \big\downarrow{\scriptstyle\bar{\bar\partial}} & & \big\| \\
\begin{bmatrix}1 & \rho_0F(1,\nu)\\ 0 & 1\end{bmatrix}\Big\downarrow & F(\sigma_r) & \xrightarrow{\varphi_r} & \pi^c_r(M^{(r)},M^{(r-1)},p) & \xleftarrow{j_*} & \pi^c_r(M^{(r)},p) \\
 & \big\uparrow{\scriptstyle(\partial_0,0)} & & \big\uparrow{\scriptstyle\bar\partial} & & \big\| \\
F(\sigma_{r+1})\oplus F(\bar\sigma_r) & & \xrightarrow{\bar\varphi_{r+1}} & \pi^c_{r+1}(\bar M^{(r+1)},\bar M^{(r)},\bar p) & \xrightarrow{\partial'} & \pi^c_r(\bar M^{(r)},\bar p)
\end{array}$$

Since $\bar{\partial}\bar{\varphi}_{r+1} = \varphi_r(\partial_0,0)$ by 5.4, $\bar{\bar{\partial}}\bar{\bar{\varphi}}_{r+1}$ has matrix form

$$(\partial_0,0)\begin{bmatrix} 1 & \rho_0 F(1,\nu) \\ 0 & 1 \end{bmatrix} = (\partial_0,\partial_0\rho_0 F(1,\nu)) = (\partial_0,F(1,\nu)) .$$

The calculation of $\bar{\bar{\partial}}$ shows that we may now apply 4.6 and 4.5 (with $M^{(r-1)}, M^{(r)}$, and $M^{(r)} \cup \bar{\bar{\Psi}}^{r+1}$ playing the roles of M_0, M_1, and M_2 respectively) to cancel the r handle with $\bar{\bar{\Psi}}^{r+1}$. The end result is a new handle presentation of (M,p) with no handles of index i for $i \leq r$, the same handles of index i as in the original handle decomposition for $i=r+1$ and $r+3 \leq i$, and a new (r+2)-handle of type (S_r,σ_r). Thus, this completes the proof of Case I of 5.2.

Case II. Suppose r=0. We continue to follow the notation set preceeding 5.4. In this case, then $\Phi_0(s\times D^0\times D^n)$ is an n-disk disjoint from $M^{(-1)}$. Let x_s $(s\in S_0)$ be a point in the boundary of $\Phi_0(s\times D^0\times D^{n-1})$. Without loss of generality, we may assume $x_s \notin \operatorname{Im}\Phi_1 \cup \operatorname{Im}\Phi_2$. Thus, $x_s \in \partial_+M^{(1)}$. Define $\sigma_0\colon S_0 \longrightarrow |\mathcal{G}G_1(\partial_+M^{(1)},p|)|$ by $\sigma_0(s) = (x_s,K_s^0)$. The main step in the proof of this case is the following lemma whose proof is postponed:

Lemma 5.6 *There exist a basis* $(S_0,\bar{\sigma}_0)$ *in*

$$\mathcal{G}G_1(\partial_+M^{(-1)} - \bigcup_{i=0}^{2} \operatorname{Im}\Phi_i - W_+,\ p|)$$

and an isomorphism $(1,\nu)\colon (S_0,\bar{\sigma}_0) \longrightarrow (S_0,\sigma_0)$ *of bases in* $\mathcal{G}G_1(\partial_+M^{(1)},p)$.

By 5.4, we may attach a pair of cancelling 1 and 2 handles to $M^{(-1)}$ to obtain a new handle decomposition of (M,p) relative to $\partial_0 M$ such that the characteristic homeomorphism Ψ_1 for the new 1-handle has $\Psi_1(s,x_0,y_0) = \bar{x}_s$ where $\bar{\sigma}_0(s) = (\bar{x}_s,L_s)$. Let $\nu(s) = (\omega_s,i_s)\colon (x_s,K_s^0) \longrightarrow (\bar{x}_s,C^mL_s)$. By the tameness of the boundedness control structure, we may assume m chosen so that ω_s is represented by a path f_s in $p^{-1}(\mathrm{Int}\ C^m L_s)$ from $\bar{x}_s$ to x_s. By general position, we may assume that $\{f_s(I) \mid s\in S_0\}$ is a family of mutually disjoint embedded arcs.

It is now easy to find a bounded family $\{E_s \mid s\in S_0\}$ of mutually disjoint $(n-1)$ disks and to pull $\Psi_1(s\times\{-1\}\times D^{n-1})$ along f_s via an isotopy supported on $E_s \subseteq \partial_1 M^{(1)} \cap p^{-1}(\mathrm{Int}\ C^m L_s)$ to obtain a new attaching homeomorphism $\Psi_1'\colon (S_0\times S^0\times D^{n-1},q) \longrightarrow (\partial_1 M^{(1)},p_1)$ such that Ψ_1' agrees with Ψ_1 on $S^0\times\{1\}\times D^{n-1}$ and $\Psi_1'(s\times\{-1\}\times D^{n-1})$ is a small disk in $\Phi_0(s\times D^0\times S^{n-1})-(\mathrm{Im}\ \Phi_1 \cup \mathrm{Im}\ \Phi_2)$ containing x_s. The results of section 2 now show that (M,p) can be written in the form

$$M = \bar{M}^{(-1)} \cup \mathrm{Im}\ \Phi_0 \cup \mathrm{Im}\ \Psi_1' \cup \mathrm{Im}\ \Phi_1 \cup \text{handles of index} \geq 2.$$

The construction above, however, shows that Φ_0 and Ψ_1' are cancelling handles. Hence, M can also be written in the form

$$M = \bar{\bar{M}}^{(-1)} \cup \mathrm{Im}\ \Phi_1 \cup \text{handles of index} \geq 2$$

and Case II of 5.2 is established.

Proof of 5.6: Since $(\partial_0 M,p_0)$ is coextensive with (M,p) and there is a bc homeomorphism $c\colon (\partial_0 M\times I,(p|\partial_0 M)\pi) \longrightarrow (M^{(-1)},p_{-1})$, we may choose an integer m_0 large enough so that if $K\in|\mathcal{P}|$ has $p^{-1}(K) \neq \emptyset$, then $p^{-1}(C^{m_0}K) \cap \partial_+ M^{(-1)} \neq \emptyset$. In addition, since $\pi_0^c(M,\partial_0 M,p) = 0$ by hypothesis, $\pi_0^c(M^{(1)},\partial_0 M,p_1) = 0$ by 5.3 and since c is a homotopy equivalence, $\pi_0^c(M^{(1)},\partial_+ M^{(-1)},p_1) = 0$.

Let $q_0' = q_0|S_0\times D^0\times 0\colon S_0 \longrightarrow Z$ and define a bc map $f\colon (S_0,q_0') \longrightarrow (M^{(1)},p_1)$ by $f(s) = x_s$. By II.10.1, there is a bc homotopy $\omega\colon (S_0\times I,q_0'\pi) \longrightarrow (M^{(1)},p_1)$ such that $\omega(s,0) = x_s$ and $\omega(s,1) \in \partial_+ M^{(-1)}$. Thus, there exists an integer m such that for all $s\in S_0$, $\mathrm{Im}\ \omega_s \subseteq p^{-1}(C^m K_s^0)$. A simple argument using property (iv) of a boundedness control structure and its tameness shows that we may assume m is large enough so that $\mathrm{Im}\ \omega_s \subseteq p^{-1}(\mathrm{Int}\ C^m K_s^0)$ and so that if $\mathrm{Im}\ \omega_s$ intersects the i-handle $H_{s'}^i = \Phi_i(s'\times D^i\times D^{n-i})$ then $H_{s'}^i \subseteq p^{-1}(\mathrm{Int}\ C^m K_s^0)$.

Let z_s' be the first point on $\mathrm{Im}\ \omega_s$ that is also in $\partial_+ M^{(-1)}$ and let

$\omega_{s'}$ be the path from x_s to z'_s determined by ω_s. Then there is a function $\alpha\colon S_0 \longrightarrow S_1$ such that $z_{s'}$ is in the 1-handle $H^1_{\alpha(s)}$. We may assume the characteristic homeomorphism chosen so that $\Phi_1(\alpha(s)\times\{+1\}\times D^{n-1}) \subseteq \partial_+M^{-1}$ and that z'_s is in this disk. By using TC, let z''_s be a point in the boundary of this disk not in Im Φ_2 and ω''_s be a path in this disk from z'_s to z''_s. Since $H^1_{\alpha(s)} \subseteq p^{-1}(\mathrm{Int}\ C^mK^0_s)$, it is now easy to find a point $\bar{x}_s \in [\partial_+M^{(-1)} - \bigcup_{i=0}^{2} \mathrm{Im}\ \Phi_i - W_+] \cap p^{-1}(\mathrm{Int}\ C^mK^0_s)$ and a path ω'''_s in $\partial_+M^{(-1)} \cap p^{-1}(\mathrm{Int}\ C^mK^0_s)$ from z''_s to $\bar{x}_s$. By general position in $M^{(1)} \cap p^{-1}(\mathrm{Int}\ C^mK^0_s)$ it is possible to homotope the concatenation of ω'_s, ω''_s, and ω'''_s to a path υ_s in $\partial_+M^{(1)} \cap p^{-1}(\mathrm{Int}\ C^mK^0_s)$ from x_s to $\bar{x}_s$.

Let $\bar{\sigma}_0\colon S_0 \longrightarrow |\mathcal{G}G_1(\partial_+M^{(-1)} - \bigcup_{i=0}^{2} \mathrm{Im}\ \Phi_i - W_+)|$ be given by $\bar{\sigma}_0(s) = (\bar{x}_s, C^mK^0_s)$. If $\upsilon(s) = (\upsilon_s, \mathrm{inc})\colon (x_s, K^0_s) \longrightarrow (\bar{x}_s, C^mK^0_s)$, then $(1,\upsilon)\colon (S_0, \bar{\sigma}_o) \longrightarrow (S_0, \sigma_0)$ is the desired isomorphism of bases. This completes the proof of 5.6.

Case III. Suppose r=1. We again follow the notation set preceeding 5.4. The main result needed in this case is the following lemma:

Lemma 5.7 *There is an integer* $m \geq 0$ *such that for every* $s \in S_1$*, there is an embedding* $f_s\colon D^2 \longrightarrow [\partial_1 M^{(2)} - W_+] \cap p^{-1}(\mathrm{Int}\ C^mK^1_s)$ *with*

(i) $f_s(S^1) \subseteq \partial_1 M^{(1)} - \bigcup_{i=2}^{3} \mathrm{Im}\ \Phi_i - W_+$ *and*

$f_s(S^1) \cap (\partial_+M^{(0)} - \bigcup_{i=1}^{3} \mathrm{Im}\ \Phi_i - W_+) \neq \emptyset$;

(ii) $f_s(S^1)$ *meeting* $\Phi_1(s\times 0\times S^{n-2})$ *transversally in a single point; and*

(iii) $f_s(S^1) \cap \Phi_1(s'\times 0\times S^{n-2}) = \emptyset$ *for* $s \neq s'$.

The proof of 5.7 is given later in this section.

For every $s \in S_1$, let $x_s \in f_s(S^1) \cap [\partial_+ M^{(0)} - \bigcup_{i=1}^{3} \operatorname{Im} \Phi_i - W_+]$, choose a ball D_s in $(\partial_+ M^{(0)} - \bigcup_{i=1}^{3} \operatorname{Im} \Phi_i - W_+) \cap p^{-1}(\operatorname{Int} C^m K_s^1)$ around x_s, and attach a pair of cancelling handles of indices 2 and 3 to $\partial_+ M^{(0)}$ inside $D_s - f_s(S^1)$. Let Ψ_2 and Ψ_3 be the characteristic homeomorphisms for the 2 and 3 handles respectively and set $y_s = \Psi_2(s, x_0, 0)$ for some $x_0 \in S^1$. It is now easy to construct an isotopy of $\Psi_2 | s \times S^1 \times 0$ that pulls a small arc around y_s in $\Psi_2(s \times S^1 \times 0)$ along a path in D_s first to a small arc in $f_s(S^1) \cap D_s$ around x_s and then pulls this arc across $f_s(D^2)$ to the complementary arc in $f_s(S^1)$. Clearly this isotopy is supported on a disk $E_s \subseteq p^{-1}(\operatorname{Int} C^m K_s^1)$ and at time 1 yields an embedding Ψ_2' whose image meets $\Phi_1(s \times 0 \times S^{n-2})$ transversally in a single point and has empty intersection with $\Phi_1(s' \times 0 \times S^{n-2})$ for $s' \neq s$. We apply first 2.2 to give (M,p) a handle decomposition of the form

$$M = M^{(0)} \cup \operatorname{Im} \Phi_1 \cup \operatorname{Im} \Psi_2' \cup \operatorname{Im} \Phi_2 \cup \text{handles of index} \geq 3$$

and then 4.5 to cancel $\operatorname{Im} \Phi_1 \cup \operatorname{Im} \Psi_2'$. The result is a handle decomposition of M of the form

$$M = M^{(0)} \cup \operatorname{Im} \Phi_2 \cup \text{handles of index} \geq 3$$

and completes the proof of Case III of 5.2.

Proof of 5.7: Consider the class $\zeta_s = \Phi_1(s \times D^1 \times y_0) \in \pi_1(M^{(1)} \cap p^{-1}(K_s^1), \partial_+ M^{(0)} \cap p^{-1}(K_s^1), x_s) = \pi_1^c(M^{(1)}, \partial_+ M^{(0)}, p_1)(x_s, K_s^1)$ where $x_s = \Phi_1(s, x_0, y_0)$ and $(x_0, y_0) \in S^0 \times S^{n-2}$. We may assume by TC that $x_s \notin \bigcup_{i=2}^{3} \operatorname{Im} \Phi_i$. Since $\pi_1^c(M, \partial_0 M, p) = 0$ by hypothesis, $\pi_1^c(M^{(2)}, \partial_0 M, p_2) = 0$ by 5.3 and since the inclusions

$$(\partial_0 M, p | \partial_0 M) \longrightarrow (M^{(-1)}, p_{-1}) = (M^{(0)}, p_0) \longleftarrow (\partial_+ M^{(0)}, p | \partial_+ M^{(0)})$$

are both homotopy equivalences, $\pi_1^c(M^{(2)}, \partial_+ M^{(0)}, p_2) = 0$. It now follows from an argument using the exact sequence of the triple $(M^{(2)}, M^{(1)}, \partial_+ M^{(0)}, p_2)$ and I.4.7 that

$$\partial\colon i^!\pi_2^c(M^{(2)},M^{(1)},p_2) \longrightarrow \pi_1^c(M^{(1)},\partial_+M^{(0)},p_1)$$

is eventually epimorphic where $i\colon (\partial_+M^{(0)},p|\partial_+M^{(0)}) \longrightarrow (M^{(1)},p_1)$ is the inclusion. In particular, there exists an integer $m'\geq 0$ and for every $s\in S_1$, a class $\xi_s \in \pi_2(M^{(2)}\cap p^{-1}(C^{m'}K_s^1),\ M^{(1)}\cap p^{-1}(C^{m'}K_s^1),x_s)$ such that $\partial\xi_s$ is the image of ζ_s in $\pi_1(M^{(1)}\cap p^{-1}(C^{m'}K_s^1),\partial_+M^{(0)}\cap p^{-1}(C^{m'}K_s^1),x_s)$. Let $m = m'+d$ where d is the integer given by the tameness of the boundedness control structure.

Let $M_s = p^{-1}(\text{Int } C^mK_s^1)$ and note that M_s, $M^{(i)} \cap M_s$, $\partial M^{(i)} \cap M_s$, etc. are manifolds. Let $f_s\colon (D^2,D_+^1,(1,0)) \longrightarrow (M^{(2)}\cap M_s,M^{(1)}\cap M_s,x_s)$ be a map representing ξ_s. A general position argument shows that we may assume that $f_s\colon D^2 \longrightarrow [\partial_1M^{(2)}-W_+] \cap M_s$, that f_s is an embedding, that $f_s(S^1) \subseteq \partial M^{(1)} - \bigcup_{i=2}^{3} \text{Im } \Phi_1 - W_+$, and that for all $s' \in S_1$, $f_s(S^1)$ meets $\Phi_1(s'\times 0\times S^{n-2})$ in a finite set of points. Since $[f_s|D_+^1] = \zeta_s$, $f_s(S^1)$ has zero algebraic intersection with $\Phi_1(s'\times 0\times S^{n-2})$ for $s'\neq s$ and algebraic intersection ± 1 with $\Phi_1(s\times 0\times S^{n-2})$. Thus, the usual Whitney trick argument can be applied inside $(\partial_1M^{(1)} - \bigcup_{i=2}^{3} \text{Im } \Phi_i-W_+)\cap M_s$ to isotope $f_s|S^1$ to an embedding satisfying (ii) and (iii) of 5.7. By adding a collar $S^1\times I$ to D^2, extending f_s via the isotopy, and using general position we may now assume f_s also satisfies (ii) and (iii) of 5.7. By a now-familiar argument, there is a point $y_s \in M_s\cap[\partial_+M^{(0)} - \bigcup_{i=1}^{3} \text{Im } \Phi_i-W_+]$ and a path g_s from x_s to y_s in $(\partial_+M^{(0)}-W_+)\cap M_s$. There is now an isotopy supported on a disk in $(\partial_+M^{(0)}-W_+)\cap M_s$ that alters $f_s|S^1$ by pulling a small arc in $f_s(S^1)$ around x_s along g_s without disturbing the rest of $f_s(S^1)$. Again we may extend f_s over a collar $S^1\times I$ using this isotopy to obtain a new embedding f_s with all the properties required by 5.7.

This completes the proofs of 5.7 and of Case III of 5.2.

6. The boundedly controlled s-cobordism theorem.

Let $(M^n,p) \in |\mathcal{PL}^c/Z|$ be a manifold with $\partial M = \partial_- M \cup V \cup \partial_+ M$ where $\partial_\pm M \cap V = W_\pm$ is the boundary of $\partial_\pm M$ and where there exists a homeomorphism $h_0\colon (W_- \times I, W_- \times 0, W_- \times 1, (p|W_-)\pi) \longrightarrow (V, W_-, W_+, p)$. If each of the inclusions $(\partial_\pm M, p_\pm) \longrightarrow (M,p)$ $(p_\pm = p|\partial_\pm M)$ is a bc homotopy equivalence, $(M, \partial_- M, \partial_+ M, p)$ is called a *(relative) bc h-cobordism* (if $V \neq \emptyset$).

Let $(M,\partial_- M,\partial_+ M,p)$ be a (relative) bc h-cobordism. By 3.3, $(M,\partial_- M,p)$ has a triangulation (J,J_-,p) as a bc CW complex. Let $[J,J_-,p] \in Wh^c(J_-,p_-)$ be the class determined by (J,J_-,p) and let $\tau(M,\partial_- M,p) \in Wh\mathbb{Z}\mathcal{G}_1(\partial_- M,p_-)$ be the image of $[J,J_-,p_-]$ under the isomorphism τ of V.3.1. That $\tau(M,\partial_- M,p)$ is independent of the choice of the triangulation follows from the combinatorial invariance of torsions, see V.4.1.

This section gives the proof of the following theorem:

Theorem 6.1 (*Boundedly Controlled s-Cobordism Theorem*) *Let* $n \geq 6$*; let* $(M^n,\partial_- M,\partial_+ M,p)$ *be a (relative) bc* h-cobordism*; and in the relative case let* $h_0\colon (W_- \times I, W_- \times 0, W_- \times 1, (p|W_-)\pi) \longrightarrow (V, W_-, W_+, p|V)$ *be a bc PL homeomorphism such that* $h_0|W_- \times 0 = 1$*. There exists a bc PL homeomorphism* $h\colon (\partial_- M \times I, \partial_- M \times 0, \partial_- M \times 1, p_-\pi) \longrightarrow (M, \partial_- M, \partial_+ M, p)$ *with* $h|\partial_- M \times 0$ *the identity and extending* h_0*, in the relative case, if and only if* $\tau(M,\partial_- M,p) = 0$.

The proofs of both this theorem and the Realization Theorem require the following lemma:

Lemma 6.2 *Let* (M,p) *be obtained from* $(N,p|N)$ *by attaching handles of indices* r *and* r+1 *where* $r \geq 2$*. If* $r=2$*, assume that the attaching map for the 2-handle is bc null homotopic. If the inclusion of* $(N,p|N)$ *into* (M,p) *is a bc homotopy equivalence, then*

$$\partial\colon \pi^c_{r+1}(M,M^{(r)},p) \longrightarrow \pi^c_r(M^{(r)},N,p_r)$$

is an isomorphism of based $\mathbb{Z}\mathcal{G}_1(N,p|)$ *modules and* $\tau(M,N,p) = (-1)^r\tau(\partial)$.

In this lemma, $M^{(r)} = M \cup$ the r-handle and $p_r = p|M^{(r)}$. We have also identified the different categories of $\mathbb{Z}\mathcal{G}_1(\ )$ modules by the usual device.

Proof: That ∂ is an isomorphism follows from an obvious exact sequence argument. Let $X \subseteq M$ be the CW complex obtained from N by attaching the cores of the r- and the (r+1)-handle. Then there are homotopy equivalences $(N,p|) \longrightarrow (X,p|X) \longrightarrow (M,p)$, the second of which is simple. There is also a commutative diagram of isomorphisms

$$\begin{array}{ccccc}
\pi^c_{r+1}(M,M^{(r)},p) & \xleftarrow{i_{r+1}} & \pi^c_{r+1}(X,X^{(r)},p|) & \xrightarrow{h_{r+1}} & C^{F,\mathrm{cell}}_{r+1}(\tilde{X},\bar{N},p|) \\
\Big\downarrow{\scriptstyle\partial} & & \Big\downarrow & & \Big\downarrow{\scriptstyle\bar{\partial}} \\
\pi^c_r(M^{(r)},N,p_r) & \xleftarrow[i_r]{} & \pi^c_r(X^{(r)},N,p|) & \xrightarrow[h_r]{} & C^{F,\mathrm{cell}}_r(\tilde{X},\bar{N},p|)
\end{array}$$

in which $X^{(r)} = N \cup$ (r skeleton of X), i_r and i_{r+1} are induced by the inclusions, and the right hand square is that of V.2.7. It now follows from V.2.7 that $\tau(M,N,p) = \tau([X,N,p|]) = (-1)^r\tau(\bar{\partial}) = (-1)^r\tau(\partial)$.

Proof of 6.1: Recall that if (X,q) is any bc CW complex, then $(X\times I,p\pi)$ is the mapping cylinder of $1_X\colon (X,q) \longrightarrow (X,q)$. Hence $\tau(X\times I,X\times 0,q\pi) = 0$ by III.5. Thus, if there exists a bc PL homeomorphism h as above, $\tau(M,\partial_-M,p) = 0$ by the combinatorial invariance of torsions.

Suppose now that $\tau(M,\partial_-M,p)=0$. Let $\partial_0M=\partial_-M\cup V$ and $\partial_1M=\partial_+M$. By a standard argument using bc collars, it suffices to prove that there is a bc homeomorphism $g\colon (\partial_0M\times I,\partial_0M\times 0,(p|\partial_0M)\pi) \longrightarrow (M,\partial_0M,p)$ extending the identity on $\partial_0M\times 0$. By applying 5.1 first to the pair (M,∂_0M,p) and then to the dual handle decomposition (M,∂_1M,p), we obtain a handle decomposition of (M,p) relative to $\partial_0 M$ whose only handles are of index r and $r+1$. We may further suppose that $r=2$.

To set notation, we let $M^{(1)} \subseteq M^{(2)} \subseteq M^{(3)} = M$ be this handle decomposition and $p_i = p|M^{(i)}$ $(1 \leq i \leq 3)$. Then there exists a bc PL homeomorphism $c\colon (\partial_0 M\times I,(p|\partial_0 M)\pi) \longrightarrow (M^{(1)},p_1)$ and we let $\partial_+ M^{(1)} = c(\partial_- M\times 1)$ and in the relative case, $W_+ = c(W_-\times 1)$. Finally, let

$$\Phi_i\colon (S_i\times(D^i,S^{i-1})\times D^{n-i},q_i) \longrightarrow (M^{(i)},M^{(i-1)},p_i) \quad (i=2,3)$$

be a characteristic homeomorphism for the i-handle.

It follows immediately from II.5.3 that $k\colon (\partial_0 M,p|\partial_0 M) \longrightarrow (M^{(1)},p_1)$ and $\ell\colon (M^{(1)},p_1) \longrightarrow (M^{(2)},p_2)$ induce equivalences $k_!$ and $\ell_!$ on the categories of $\mathbb{Z}\mathcal{G}_1(\ )$-modules. As is customary, we shall identify these categories under $k_!$ and $\ell_!$ and will drop these maps, and their pseudo-inverses $k^!$ and $\ell^!$ from our notation.

A simple argument using the homotopy exact sequences of the triples $(M,M^{(1)},\partial_0 M,p)$ and $(M,M^{(2)},M^{(1)},p)$ shows that $\partial\colon \pi_3^c(M,M^{(2)},p) \longrightarrow \pi_2^c(M^{(2)},M^{(1)},p_2)$ is an isomorphism. Thus, we obtain a commutative diagram of isomorphisms

$$\begin{array}{ccc} F(\sigma_3) & \xrightarrow{\varphi_3} & \pi_3^c(M,M^{(2)},p) \\ \delta\downarrow & & \downarrow\partial \\ F(\sigma_2) & \xrightarrow{\varphi_2} & \pi_2^c(M^{(2)},M^{(1)},p_2) \end{array}$$

in which φ_i $(i=2,3)$ is the isomorphism of 4.1 and $\delta = \varphi_2^{-1}\partial\varphi_3$. Since δ is an isomorphism, by IV.2.6 there is an isomorphism of bases $(\alpha,\nu)\colon (S_2,\sigma_2) \longrightarrow (S_3,\sigma_3)$. By 4.2, we may now replace $(M,M^{(2)},p)$ by a bc homeomorphic pair, if necessary, and assume that $(S_3,\sigma_3) = (S_2,\sigma_2)$. In particular, the diagram above becomes

$$\begin{array}{ccc} F(\sigma_2) & \xrightarrow{\varphi_3} & \pi_3^c(M,M^{(2)},p) \\ \delta\downarrow & & \downarrow\partial \\ F(\sigma_2) & \xrightarrow{\varphi_2} & \pi_2^c(M^{(2)},M^{(1)},p_2) \end{array}$$

and $[F(\sigma_2),\delta] = \tau(M,M^{(1)},p) \in \mathrm{Wh}\mathbb{Z}\mathcal{P}G_1(M^{(1)},p_1)$ by 6.2. Since each of the inclusions $(\partial_-M,p|\partial_-M) \xrightarrow{k} (M^{(1)},p_1) \longrightarrow (M,p)$ is a homotopy equivalence, $k_*\tau(M,\partial_-M,p) = \tau(M,M^{(1)},p) + k_*\tau(M^{(1)},\partial_-M,p_1)$ by III.5.4. Since $\tau(M,\partial_-M,p) = 0$ by hypothesis and $\tau(M^{(1)},\partial_-M,p_2)$ obviously vanishes, it follows from this equation that $0 = \tau(M,M^{(1)},p) = [F(\sigma_2),\delta]$.

Lemma 6.3 *There exist a geometric bfg basis* (T,τ) *in* $\mathcal{P}G_1(\partial_+M^{(1)}-\mathrm{Im}\Phi_1-\mathrm{Im}\,\Phi_2-W_+,p|)$, *an isomorphism* $(\alpha,1)\colon (R_1\amalg R_2,\rho_1\amalg\rho_2) \longrightarrow (S_2\amalg T,\sigma_2\amalg\tau)$ *of bases, an automorphism* ξ_j *of* $F(\rho_j)$ $(j=1,2)$, *and elementary automorphisms* ε_k *of* $F(\sigma_2)\oplus F(\tau)$ $(1\leq k\leq 16)$ *such that the following diagram commutes*

$$\begin{array}{ccc}
F(\rho_1)\oplus F(\rho_2) & \xrightarrow{F(\alpha,1)} & F(\sigma_2)\oplus F(\tau) \\
\Big\downarrow{\scriptstyle \mathrm{diag}(\xi_1,\xi_2)} & & \Big\downarrow{\scriptstyle (\delta+1_{F(\tau)})\prod_{k=1}^{16}\varepsilon_k} \\
F(\rho_1)\oplus F(\rho_2) & \xrightarrow{F(\alpha,1)} & F(\sigma_2)\oplus F(\tau)
\end{array}$$

Furthermore, ξ_j $(j=1,2)$ *is an automorphism of type* j *below:*

(1) *A change of basis automorphism;*

(2) *The negative of the identity automorphism;*

Proof: This follows immediately from IV.3.9.

The proof of 6.1 now continues as follows: Apply 5.4 with s=2 to add a pair of cancelling handles of indices 2 and 3 to (M,p) and to stabilize $F(\sigma_2)$ by adding $F(\tau)$ and δ by adding $1_{F(\tau)}$. Now apply the Handle Rearrangement Lemma 4.3 successively 16 times to obtain a new handle presentation $\bar{M}^{(1)}\subseteq\bar{M}^{(2)}\subseteq\bar{M}^{(3)} = M$ of (M,p) relative to ∂_0M such that the outer square in the following diagram commutes

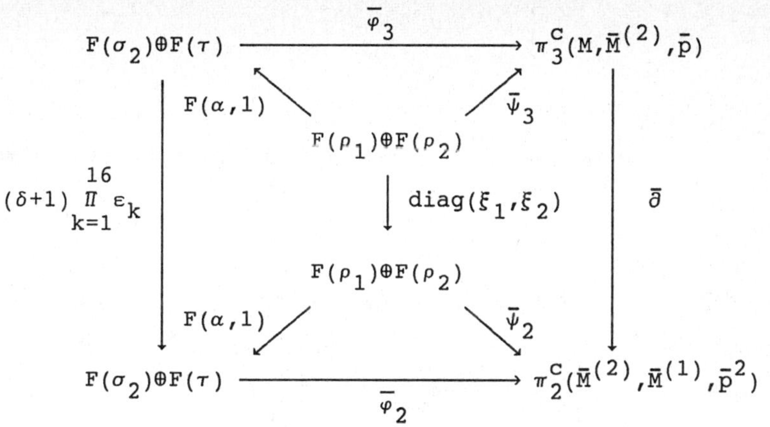

Let $\bar{\psi}_i = \bar{\varphi}_i\, F(\alpha,1)$ (i=2,3). Then the entire diagram commutes.

Since $\alpha\colon R_1 \amalg R_2 \longrightarrow S_2 \amalg T$ is a bijection, there is a decomposition of the i-handle of $\bar{M}$ (i=2,3) as the disjoint union of 2 handles of index i of the form $(R_j\times(D^i,S^{i-1})\times D^{n-i},q_{i,j})$ (i=2,3; j=1,2). Furthermore, $\bar{\psi}_i$ (i=2,3) is just the isomorphism arising from this decomposition by the construction in 4.2. We will now show how to alter this handle presentation of (M,p) relative to $\partial_0 M$ to replace $\operatorname{diag}(\xi_1,\xi_2)$ by $\operatorname{diag}(1,1)$.

To do this, we note that since ξ_1 is a change of basis isomorphism, so are ξ_1^{-1} and $\operatorname{diag}(\xi_1^{-1},1)$. Then by 4.6 we may modify the handle presentation of M to obtain a new handle presentation in which the boundary map from π_3^c to π_2^c corresponds to $\operatorname{diag}(1,\xi_2)$. Since $\xi_2 = -1_{F(\rho_2)}$, it can be changed to the identity by replacing the characteristic homeomorphism Φ for the handle $(R_2\times(D^3,S^2)\times D^{n-3},q_{3,2})$ by $\Phi(1\times g\times 1)$ where g is an orientation reversing homeomorphism of (D^3,S^2). Although this does not change the handle presentation of M, it does change the isomorphism of $F(\rho_1)\oplus F(\rho_2)$ with π_3^c so that the boundary map from π_3^c to π_2^c corresponds to $\operatorname{diag}(1,1)=1$.

Theorem 6.1 now follows easily from 4.5 and 4.4.

7. The realization and duality theorems.

This section contains the proofs of the Realization and Duality Theorems. We begin by stating the main results and defer their proofs to the end.

Theorem 7.1 (*Realization Theorem*). *Let* $n \geq 6$, $(N^{n-1},q) \in |\mathcal{P}\mathcal{L}^c/Z|$ *and* $\tau_0 \in Wh\mathbb{Z}\mathcal{P}G_1(N,q)$. *Then there exists a bc h-cobordism* $(M,\partial_-M,\partial_+M,p)$ *in* $|\mathcal{P}\mathcal{L}^c/Z|$ *such that* $(\partial_-M,p_-) = (N,q)$ *and* $\tau(M,\partial_-M,p) = \tau_0$.

To set the context for the Duality Theorem we observe first that for any bc PL manifold (M,p) (in fact, for any bc space), there is a forgetful functor $F\colon \mathcal{P}G_1(M,p) \longrightarrow G_1(M)$, the (honest) fundamental groupoid of M, that sends (x,K) to x and (ω,i) to the path class in M determined by ω. If M is connected, $x \in M$, and we view $\pi_1(M,x)$ as a category with one object x whose (auto)morphisms are $\pi_1(M,x)$, then there is an inclusion of categories $J\colon \pi_1(M,x) \longrightarrow G_1(M)$.

A Stiefel-Whitney class $w_1\colon \mathcal{P}G_1(M,p) \longrightarrow \mathbb{Z}_2$ is *geometric* if it fits into a commutative diagram of functors

$$\begin{array}{ccc}
\mathcal{P}G_1(M,p) & & \\
F\downarrow & \searrow^{w_1} & \\
G_1(M) & \xrightarrow{w_1''} & \mathbb{Z}_2 \\
J\uparrow & \nearrow_{w_1'} & \\
\pi_1(M,x) & &
\end{array}$$

in which w_1' is the usual Stiefel-Whitney class for M. (We recall from IV.5.1 that w_1 is a functor from $\mathcal{P}G_1(M,p)$ to $\mathbb{Z}_2$, regarded as a category, for which $w_1(\tau^n) = 1$ for all τ-morphisms in $\mathcal{P}G_1(M,p)$.)

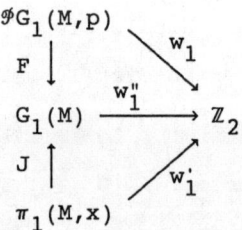

Proposition 7.2 *Let* M *be connected. Then there exists a geometric Stiefel-Whitney class for* (M,p). *If* $w_1, \bar{w}_1\colon \mathcal{G}G_1(M,p) \longrightarrow \mathbb{Z}_2$ *are two geometric Stiefel-Whitney classes for* (M,p), *then there exists a natural equivalence* $\upsilon\colon w_1 \longrightarrow \bar{w}_1$. *In particular, the involutions on* $Wh\mathbb{Z}\mathcal{G}G_1(M,p)$ *associated with* w_1 *and* $\bar{w}_1$ *coincide.*

Theorem 7.3 (*Duality Theorem*) *Let* $(M^n, \partial_-M, \partial_+M, p)$ *be a bc PL h-cobordism. Then*

$$i_{+_*}\tau(M,\partial_+M,p) = (-1)^{n-1}\,(i_{-_*}\tau(M,\partial_-M,p))^*$$

where the homomorphisms $i_{\pm_*}\colon Wh\mathbb{Z}\mathcal{G}G_1(\partial_\pm M, p_\pm) \longrightarrow Wh\mathbb{Z}\mathcal{G}G_1(M,p)$ *are induced by the inclusions and* $(\ )^*$ *is the duality involution on* (M,p), *defined in* V.5.

Proof of 7.1: Let $(M^{(1)},p_1) = (N\times I, q\pi)$ and $N_+ = N\times 1$. Suppose $\tau_0 = [F(\sigma),\delta]$ where $\delta\colon F(\sigma) \longrightarrow F(\sigma)$ is an automorphism and $\sigma\colon S \longrightarrow |\mathcal{G}G_1(N,q)| = |\mathcal{G}G_1(N_+,p_1|)|$. By 5.5 we may assume σ is geometric and we may also find another basis $\bar{\sigma}\colon S \longrightarrow |\mathcal{G}G_1(N_+,p_+)|$ and an isomorphism of bases $(1,\alpha)\colon (S,\bar{\sigma}) \longrightarrow (S,\sigma)$ such that $(S \amalg S, \sigma \amalg \bar{\sigma})$ is also geometric. Now use 5.4 to attach to $(M^{(1)},p_1)$ a pair of cancelling 2- and 3-handles of type (S,σ) and a 3-handle of type $(S,\bar{\sigma})$ with null homotopic attaching homeomorphism. The resulting manifold (Q,q) has a handle decomposition $M^{(1)} = Q^{(1)} \subseteq Q^{(2)} \subseteq Q^{(3)} = Q$ for which the following diagram commutes

$$\begin{array}{ccc}
F(\sigma)\oplus F(\bar{\sigma}) & \xrightarrow{\ \varphi_3\ } & \pi_3^c(Q^{(3)},Q^{(2)},q) \\
\Big\downarrow{\scriptstyle (1\ 0)} & & \Big\downarrow{\scriptstyle \partial} \\
F(\sigma) & \xrightarrow{\ \varphi_2\ } & \pi_2^c(Q^{(2)},Q^{(1)},q_2)
\end{array}$$

(We have suppressed the $i^!$ maps by the now standard device.) By 4.3 there is another handle presentation $\{(\bar{Q}^{(i)},\bar{q}_i)\}$ of (Q,q) with $\bar{Q}^{(i)} = Q^{(i)}$ for $i\leq 2$ such that the following diagram commutes

$$
\begin{array}{ccccc}
F(\sigma)\oplus F(\bar{\sigma}) & \xrightarrow{\bar{\varphi}_3} & \pi_3^c(\bar{Q},Q^{(2)},\bar{q}) & & \\
\left[\begin{matrix} 1 & \delta F(1,\alpha) \\ 0 & 1 \end{matrix}\right]\Bigg\downarrow & \searrow^{(1\ \delta F(1,\alpha))} & & \searrow^{\bar{\partial}} & \\
 & & F(\sigma) & \xrightarrow{\varphi_2} & \pi_2^c(Q^{(2)},Q^{(1)},q_2) \\
 & \nearrow_{(1\ 0)} & & \nearrow_{\partial} & \\
F(\sigma)\oplus F(\bar{\sigma}) & \xrightarrow{\varphi_3} & \pi_3^c(Q,Q^{(2)},q) & &
\end{array}
$$

Let (M,p) be the manifold obtained from (Q,q) by deleting the 3-handle corresponding to (S,σ). Then (M,p) has a handle presentation $M^{(1)} \subseteq M^{(2)} \subseteq M^{(3)}$ for which the diagram

$$
\begin{array}{ccc}
F(\bar{\sigma}) & \xrightarrow{\varphi_3} & \pi_3^c(M,M^{(2)},p) \\
{\scriptstyle \delta F(1,\alpha)}\Big\downarrow & & \Big\downarrow{\scriptstyle \partial} \\
F(\sigma) & \xrightarrow{\varphi_2} & \pi_2^c(M^{(2)},M^{(1)},p_2)
\end{array}
$$

commutes. Let $\partial_-M = N\times 0$; $V = \partial N\times I$ if $\partial N\neq\emptyset$; and $\partial_+M = Cl(\partial M-(\partial_-M\cup V))$. It follows immediately from 6.2 that $\tau(M,\partial_-M,p) = \tau_0$ provided that $(M,\partial_-M,\partial_+M,p)$ is a bc PL h-cobordism. Thus, we must show that the inclusions $(\partial_-M,p_-) \longrightarrow (M,p)$ and $(\partial_+M,p_+) \longrightarrow (M,p)$ are homotopy equivalences.

To see the first of these, let X be the CW complex obtained from $M^{(1)}$ by attaching the cores of the 2- and the 3-handles. Then there are inclusions

$$(\partial_-M,p_-) \xrightarrow{i} (M^{(1)},p_1) \xrightarrow{j} (X,p|) \xrightarrow{k} (M,p)$$

of which i and k are obviously homotopy equivalences. That j is a homotopy equivalence follows from V.3.8.

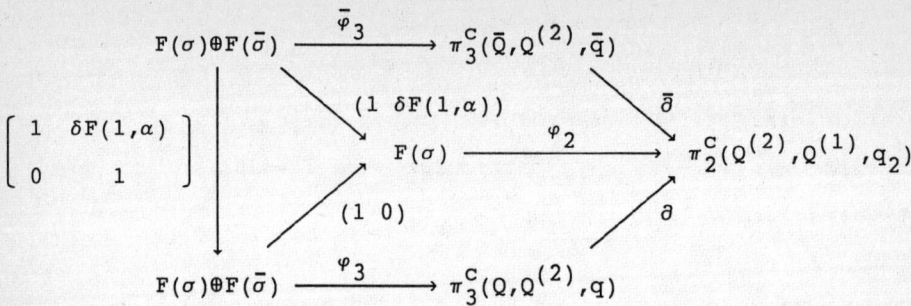

To see that the inclusion $(\partial_+M, p_+) \longrightarrow (M,p)$ is a homotopy equivalence, we note that (∂_+M, p_+) is coextensive with (M,p) and $\pi_i^c(M, \partial_+M, p) = 0$ for $i \leq 2$ since M is obtained from a bc collar on ∂_+M by attaching (dual) handles of indices n-3 and n-2. On the other hand a careful examination of Milnor's proof of the Duality Lemma [Mi2, pp. 395-397], shows that the only boundary operator in the handle chains on the universal cover of (M, ∂_+M, p) is the isomorphism $(\delta F(1,\alpha))^*$. Thus, $(\partial_+M, p_+) \longrightarrow (M,p)$ is a homotopy equivalence by an argument similar to the one above. This completes the proof of 7.1.

Proof of 7.2: Since M is connected, $J: \pi_1(M,x) \longrightarrow G_1(M)$ is an equivalence of categories. In fact, there exists a functor $R: G_1(M) \longrightarrow \pi_1(M,x)$ for which RJ=1 and for which there is a natural equivalence $\eta: 1 \longrightarrow JR$. Clearly $w_1'RF$ is a geometric Stiefel-Whitney class for (M,p). On the other hand, if w_1 is any geometric Stiefel-Whitney class for (M,p), then $w_1 = w_1'' F$ for some $w_1'': G_1(M) \longrightarrow \mathbb{Z}_2$. But then,

$$w_1 = w_1''F = w_1'' \, 1 \, F \xrightarrow{\upsilon} w_1''(JR)F = w_1'' \, RF$$

where $\upsilon = w_1'' \, \eta \, F$ is a natural equivalence and the second sentence of 7.2 follows. The last sentence of 7.2 follows immediately from IV.5.2. This completes the proof of 7.2.

Proof of 7.3: By the remarks in the proof of 7.1, this follows immediately from Milnor's proof of the classical Duality Theorem [Mi2, pp. 394-397].

CHAPTER VII
TOWARD COMPUTATIONS

Let Z be a boundedness control space and $(X,p) \in |\mathcal{CW}_{\mathcal{L}}^{c}/Z|$. In this chapter we begin the process of describing $K_1R\mathcal{G}G_1(X,p)$ and $WhR\mathcal{G}G_1(X,p)$ in terms of more standard K-theoretic invariants of X. We take two approaches to doing this: one approach is to compare our work with that of other authors, in particular Pedersen, [Pe1], [Pe2], [Pe3], Pedersen and Weibel [PW1], [PW2], and Siebenmann, [Si]; the other approach is to calculate $Wh\mathbb{Z}\mathcal{G}G_1(X,p)$ explicitly in one special, but quite interesting, case.

Most of the results of this chapter deal with the case when $Z = \mathbb{R}^k$ with the box metric boundedness control structure of II.1.3; throughout the chapter we shall assume that $\mathbb{R}^k$ is endowed with this structure. We remark that by IV.4.11 and IV.4.12, the results are equally valid when $\mathbb{R}^k$ is endowed with any one of several other naturally arising boundedness control structures.

This chapter is organized as follows: In section 1 we compare our work with that of Pedersen and Pedersen and Weibel. We first focus on comparing $Wh^c(Y\times\mathbb{R}^k,\text{proj})$ with $\tilde{K}_{1-k}\mathbb{Z}\pi_1(Y)$ (see Theorem 1.2 and Corollaries 1.3 and 1.4), and then give more general results in Theorems 1.6 and 1.7. In section 2, $Wh^c(N\times\mathbb{R}^k,\text{proj})$ is compared with Siebenmann's group, $\mathcal{S}(N\times\mathbb{R}^k)$, of proper simple homotopy types, cf. [Si], in the case when N is a compact PL manifold and k = 1 or 2. The main result is Theorem 2.1. Section 3 contains some algebraic results needed for the proof of Theorem 2.1. Finally, in section 4 we construct a bc CW complex $(X,p) \in |\mathcal{CW}_{\mathcal{L}}^{c}/\mathbb{R}|$ for which $Wh^c(X,p)$ is not countably generated although $\pi_1(X)$ and $\mathcal{S}(X)$ vanish.

1. A comparison of $Wh^c(Y\times\mathbb{R}^k,proj)$ with $\tilde{K}_{1-k}\mathbb{Z}\pi_1(Y)$.

In this section we compare our results with those of Pedersen, [Pe1], [Pe2], [Pe3], and Pedersen and Weibel, [PW1],[PW2]. A comparison of $Wh^c(Y\times\mathbb{R}^k,proj)$ with $\tilde{K}_{1-k}\mathbb{Z}\pi_1(Y)$ emerges as a special case. For the comparison the following definition plays an important role. In it, Z is any boundedness control space, $(X,p) \in |\mathcal{T}op^c/Z|$, and $i\geq 0$ is an integer.

Definition 1.1 *We say that* $\pi_i(X)$ *is* uniformly locally presented, *if there exists an integer* $d \geq 0$ *such that the following conditions hold for all* $K \in |\mathcal{B}|$ *and all* $x \in p^{-1}K$:

(i) $\pi_i(p^{-1}C^dK,x) \longrightarrow \pi_i(X,x)$ *is onto*

(ii) $Ker[\pi_i(p^{-1}K,x) \longrightarrow \pi_i(X,x)] = Ker[\pi_i(p^{-1}K,x) \longrightarrow \pi_i(p^{-1}C^d,x)]$

(iii) $p^{-1}(C^dK) \neq \emptyset$.

In the case when $Z = \mathbb{R}^k$ with the box metric boundedness control structure of II.1.3, we will prove the following 3 results later in this section:

Theorem 1.2 *Let* $k\geq 1$ *and* $(X,p) \in |\mathcal{T}op^c/\mathbb{R}^k|$. *There is a homomorphism*

$$\varepsilon\colon Wh\mathbb{Z}\mathcal{B}G_1(X,p) \longrightarrow \tilde{K}_{1-k}\mathbb{Z}\pi_1(X)$$

which is natural with respect to bc maps $(X,p) \longrightarrow (Y,q)$. *If* $\pi_i(X)$ *is uniformly locally presented for* $i=0,1$, *then* ε *is an isomorphism.*

Corollary 1.3 *Let* $k\geq 1$ *and* Y *be a finite CW complex. Then there is an isomorphism*

$$\varepsilon\colon Wh^c(Y\times\mathbb{R}^k,\mathrm{proj}) \longrightarrow \tilde{K}_{1-k}\mathbb{Z}\pi_1(Y)$$

which is natural with respect to maps of finite CW complexes.

Corollary 1.4 *Let* $k\geq 1$, $n+k\geq 5$, *and* N^n *be a closed PL manifold. Then there is a 1-1 corrraspondance between* $\tilde{K}_{1-k}\mathbb{Z}\pi_1(N)$ *and the set of bc PL homeomorphism classes of bc PL h-cobordisms* $(W^{n+k+1},N\times\mathbb{R}^k,\partial_+W,p)$ *over* $\mathbb{R}^k$ *with* $p|N\times\mathbb{R}^k = \mathrm{proj}$.

Remarks 1.5 (i) In the above results, proj: $Y\times\mathbb{R}^k \longrightarrow \mathbb{R}^k$ is the projection and $\tilde{K}_{1-k}\mathbb{Z}\pi_1(Y) = \Sigma_\alpha\tilde{K}_{1-k}\mathbb{Z}\pi_1(Y,y_\alpha)$ where the direct sum runs over a set $\{y_\alpha\}$ of representatives of the components of Y.

(ii) Corollary 1.3 was announced by the first named author in [An] and was described in his talk at the A.M.S. Summer Meeting in Albany in 1983.

(iii) In section 4, a bc CW complex (X,p) over $\mathbb{R}$ is constructed for which $Wh^c(X,p)$ is not countably generated while $\pi_1(X) = 0$. Hence $\tilde{K}_0\mathbb{Z}\pi_1(X) = 0$ and ε is not in general one to one.

(iv) If f: $(X,p) \longrightarrow (Y,g)$ is a bc homotopy equivalence, then $\varepsilon\tau(f) = \sigma(f)$, the invariant defined by Pedersen in [Pe2, Definition 4.4]. Since ε is not in general one to one, Pedersen's invariant σ is not in general a complete invariant for the bc (or bounded) simple homotopy classification of f, although τ is.

(v) The bounded h-cobordism theorem described by Pedersen in [Pe3] is a special case of 1.2 and the results of Chapter VI. For connected X his hypothesis that $\pi_1(X)$ be t-bounded is equivalent to our hypothesis that $\pi_i(X)$ be uniformly locally presented for i=0,1.

Theorem 1.2 is actually a special case of more general results, viz. Theorems 1.6 and 1.7. In these theorems, we let $G(X): \mathcal{P} \longrightarrow \mathcal{G}poid$ be the constant functor whose value on $K \in |\mathcal{P}|$ is the fundamental groupoid of (all of) X.

Theorem 1.6 *Let* $(X,p) \in |\mathcal{T}op^c/Z|$. *Then there is an additive functor*

$$A: R\mathcal{P}G_1(X,p)\text{-}mod \longrightarrow R\mathcal{P}G(X)\text{-}mod$$

which preserves bfg free modules and which is natural with respect to bc maps $f:(X,p) \longrightarrow (Y,g)$. *If* $\pi_i(X)$ *is uniformly locally presented for* $i=0,1$, *then both* A *and its restriction to the subcategory of bfg free modules are equivalences of categories.*

Before we can state the next theorem we need to recall the category $\mathcal{C}_Z(\mathcal{F}R)$ defined by Pedersen and Weibel in [PW1, Remark (1.2.3)]. Here (Z,ρ) is any metric space, R is a ring, and $\mathcal{F}R$ is the category of finitely generated free R modules. An object F of $\mathcal{C}_Z(\mathcal{F}R)$ is a collection $\{F(z) \mid z \in Z\}$ of objects in $\mathcal{F}R$ subject to the constraint that for every $d > 0$ and every $z \in Z$, $F(w) \neq 0$ for only finitely many $w \in B(z,d)$. A morphism $\xi: F_1 \longrightarrow F_2$ consists of a family in $\mathcal{F}R$ $\{\xi(z,w): F_1(z) \longrightarrow F_2(w)\}$ for which there exists a $d \geq 0$ such that $\xi(z,w) = 0$ if $\rho(z,w) > d$. The number d is called a *bound* for ξ.

We note that if $Z = \mathbb{Z}^k$ with the box metric, then $\mathcal{C}_Z(\mathcal{F}R)$ is the category $\mathcal{C}_k(R)$ introduced by Pedersen in [Pe1, p. 461].

In the following theorem we suppose that the metric ρ is proper and satisfies Condition A of II.1.3. We endow Z with the corresponding metric boundedness control structure.

Theorem 1.7 *Let Z be as above,* $(X,p) \in |\mathcal{T}op^c/Z|$ *and suppose that* X *is path connected. Then there exists an additive functor*

$$B: \mathcal{C}_Z(\mathcal{F}R\pi_1(X)) \longrightarrow \text{bfgfree}R\mathcal{P}G(X)\text{-}mod$$

which is an equivalence of categories and natural with respect to bc maps $f: (X,p) \longrightarrow (Y,g)$ *between path connected spaces.*

We note that B induces an isomorphism $\beta: K_1\mathcal{C}_Z(\mathcal{F}R\pi_1(X)) \longrightarrow K_1R\mathcal{P}G(X)$, which is natural with respect to bc maps.

Proof of 1.6: The inclusions $i_K: p^{-1}K \longrightarrow X$, $K \in |\mathcal{P}|$, induce a natural transformation μ from the functor $G_1(X,p): \mathcal{P} \longrightarrow \mathcal{G}poid$ to the constant functor $G(X): \mathcal{P} \longrightarrow \mathcal{G}poid$. Let $\mathcal{P}\mu: \mathcal{P}G_1(X,p) \longrightarrow \mathcal{P}G(X)$ be the functor of II.5 and let $A = \mathcal{P}\mu_!: R\mathcal{P}G_1(X,p)\text{-}mod \longrightarrow R\mathcal{P}G(X)\text{-}mod$. Clearly, A is additive and natural with respect to bc maps. Also, A preserves bfg free modules by IV.4.1. If $\pi_i(X)$ is uniformly locally presented for $i = 0,1$, then it is easy to check that G(X) is coextensive with $G_1(X,p)$ and that $\mu_*: \pi_i(\mathcal{P}G_1(X,p)) \longrightarrow \mu^!\pi_i G(X)$ is an isomorphism for $i = 0,1$. Hence the rest of 1.6 follows from IV.4.10 and the remark following it.

Proof of 1.7: Choose a base point $x_0 \in X$, and let $\pi: \mathcal{P} \longrightarrow \mathcal{G}poid$ be the constant functor having $\pi(K) = \pi_1(X,x_0)$ viewed as a category with the unique object x_0. We define a functor

$$E: \mathcal{C}_Z(\mathcal{F}R\pi_1(X)) \longrightarrow \text{bfgfree}R\mathcal{P}\pi\text{-}mod$$

as follows: If $M \in |\mathcal{C}_Z(\mathcal{F}R\pi_1(X))|$ and (x_0,K) is an arbitrary object of $\mathcal{P}\pi$, then $E(M)(x_0,K) = \oplus_{z\in K}M(z)$. If $(\gamma,i): (x_0,K) \longrightarrow (x_0,L)$ is an arbitrary morphism of $\mathcal{P}\pi$, then $\gamma \in \pi_1(X,x_0)$ so γ acts on each M(z) and we let $(\gamma,i)_*: E(M)(x_0,K) \longrightarrow E(M)(x_0,L)$ be the composition

$$\oplus_{z\in K}M(z) \subseteq \oplus_{z\in L}M(z) \xrightarrow{\oplus\gamma} \oplus_{z\in L}M(z).$$

Clearly, $E(M) \in |R\mathcal{P}\pi\text{-}mod|$. Moreover, if $S(z)$ is a $\mathbb{Z}\pi_1(X,x_0)$ basis for $M(z)$, $(z \in Z)$, then $E(M)$ is easily seen to be isomorphic to $F(\sigma)$ where the basis (S,σ) has $S = \coprod_z S(z)$ and $\sigma(s) = (x_0,\{z\})$ for $s \in S(z)$ (note that $\{z\} \in |\mathcal{P}|$). Moreover, the conditions imposed on M imply that (S,σ) is bfg. Thus $E(M)$ is bfg free.

Next let $f: M \longrightarrow N$ in $\mathcal{C}_Z(\mathcal{F}R\pi_1(X))$ have components $f(z,w): M(z) \longrightarrow N(w)$ which vanish when $\rho(z,w) > d$. We define $E(f): E(M) \longrightarrow E(N)$ to be represented by the natural transformation $f_d: E(M) \longrightarrow E(N)C^d$ for which

$$f_d(K)|M(z) = \Sigma_w f(z,w): M(z) \longrightarrow \oplus_{w \in C^d K} N(w).$$

It is easily verified that this definition makes E into a functor, which is full and faithful.

To see that E is an equivalence of categories, it now suffices to show that any bfg free $R\mathcal{P}\pi$ module $F(\sigma)$ is isomorphic to $E(M)$ for some M. Let the given bfg basis (S,σ) have $\sigma(s) = (x_0, B(z_s,r_s))$, and define $\sigma': S \longrightarrow \mathcal{P}\pi$ to have $\sigma'(s) = (x_0,\{z_s\})$. There is an obvious morphism $(1,\nu): (S,\sigma') \longrightarrow (S,\sigma)$ which is an iso-morphism by IV.2.11. Hence we may assume that $\sigma = \sigma'$, and we can simply put $M(z) = F(\sigma')(x_0,\{z\})$ for each $z \in Z$ (note that $F(\sigma')(x_0,\{z\}) = 0$, unless $z = z_s$ for some $s \in S$). This finishes the proof that E is an equivalence.

We next observe that the inclusion $\pi_1(X,x_0) \subseteq G(X)$ can be viewed as a natural transformation $\iota: \pi \longrightarrow G(X)$ between functors $\mathcal{P} \longrightarrow \mathcal{G}poid$. Since X is path connected, it is easy to see that ι satisfies the conditions of IV.4.10. Hence

$$\mathcal{P}\iota_! : \text{bfgfree}R\mathcal{P}\pi\text{-}mod \longrightarrow \text{bfgfree}R\mathcal{P}G(X)\text{-}mod$$

is an equivalence of categories.

We now put $B = (\mathcal{P}\iota_!)\circ E$ and leave it to the reader to verify the naturality of this functor.

Proof of 1.2: Suppose first that X is path connected (in the usual sense). Let $\mathcal{C}_k(\mathbb{Z}\pi_1(X))$ be the category defined by Pedersen on p. 461 of [Pe1] (cf. our remark preceding 1.7). Then there is an inclusion of categories

$$D\colon \mathcal{C}_k(\mathbb{Z}\pi_1(X)) \longrightarrow \mathcal{C}_{\mathbb{R}^k}(\mathcal{F}\mathbb{Z}\pi_1(X))$$

which is obviously natural with respect to bc maps f: (X,p) ⟶ (Y,g) and is an equivalence by [PW1; Remark (1.2.3)]. We apply the functor K_1 to the diagram

$$\begin{array}{ccc} \text{bfgfree}\mathbb{Z}\mathcal{G}G_1(X,p)\text{-}mod & & \mathcal{C}_k(\mathbb{Z}\pi_1(X)) \\ A\downarrow & & \downarrow D \\ \text{bfgfree}\mathbb{Z}\mathcal{G}G(X)\text{-}mod & \xleftarrow[B]{} & \mathcal{C}_{\mathbb{R}^k}(\mathcal{F}\mathbb{Z}\pi_1(X)) \end{array}$$

where A and B are the functors of 1.6 and 1.7 and use the fact that B and D induce isomorphisms β and δ, respectively, to define a homomorphism $\bar{\varepsilon}'$ which makes the diagram

$$\begin{array}{ccc} K_1\mathbb{Z}\mathcal{G}G_1(X,p) & \xrightarrow{\bar{\varepsilon}'} & K_1\mathcal{C}_k(\mathbb{Z}\pi_1(X)) \\ \alpha\downarrow & & \downarrow\delta \\ K_1\mathbb{Z}\mathcal{G}G(X) & \xleftarrow[\beta]{} & K_1\mathcal{C}_{\mathbb{R}^k}(\mathcal{F}\mathbb{Z}\pi_1(X)) \end{array}$$

commute. The main theorem of [Pe1] shows that there is an isomorphism $\eta\colon K_1\mathcal{C}_k(\mathbb{Z}\pi_1(X)) \longrightarrow K_{1-k}\mathbb{Z}\pi_1(X)$ and we let $\varepsilon' = \eta\bar{\varepsilon}'$. Since η is, in fact, natural with respect to bc maps, as are α, β and δ, ε' is natural as claimed. Moreover, if $\pi_i(X)$ is uniformly locally presented for i=0,1, then A is an equivalence and α is an isomorphism. Hence so is ε'.

We now consider the following diagram where the columns are exact, p_0 and p_1 are the natural projections, and i_1 is induced by the inclusion $\mathbb{Z} \subseteq \mathbb{Z}\pi_1(X)$:

$$\begin{array}{ccc}
\mathrm{Ker}\,p_0 & \overset{\varepsilon_1'}{\dashrightarrow} & K_{1-k}\mathbb{Z} \\
\downarrow i_0 & & \downarrow i_1 \\
K_1\mathbb{Z}\mathcal{P}G_1(X,p) & \xrightarrow{\varepsilon'} & K_{1-k}\mathbb{Z}\pi_1(X) \\
\downarrow p_0 & & \downarrow p_1 \\
\mathrm{Wh}\mathbb{Z}\mathcal{P}G_1(X,p) & \overset{\varepsilon}{\dashrightarrow} & \tilde{K}_{1-k}\mathbb{Z}\pi_1(X) \\
\downarrow & & \downarrow \\
0 & & 0
\end{array}$$

The following lemma finishes the proof of 1.2 for path connected X.

Lemma 1.8 *In the above diagram* $p_1\varepsilon' i_0 = 0$. *Hence there are homomorphisms* ε_1' *and* ε *(dotted arrows) which make the diagram commute. Moreover, if* $\pi_i(X)$ *is uniformly locally presented for* $i = 0,1$, *then* ε_1' *is onto, and* ε *is an isomorphism.*

Proof: We first show that $p_1\varepsilon' i_0 = 0$. Since Ker p_0 is generated by elements of the form $x = [F(\sigma),-1_{F(\sigma)}]$ and $y = [F(\sigma),F(\alpha,\nu)]$, where $(\alpha,\nu)\colon (S,\sigma) \longrightarrow (S,\sigma)$ is an automorphism of a bfg basis (S,σ) over $\mathcal{P}G_1(X,p)$, it suffices to show that $p_1\varepsilon'(x) = p_1\varepsilon'(y) = 0$. An easy study of the functors A, B and D which enter into the definition of $\bar{\varepsilon}'$ shows that $\bar{\varepsilon}'(x)$ and $\bar{\varepsilon}'(y)$ are of the form $\bar{\varepsilon}'(x) = [M,-1_M]$, respectively, $\bar{\varepsilon}'(y) = [M,f]$, where $M \in |\mathcal{C}_Z(\mathbb{Z}\pi_1(X))|$ has a basis $\{m_s | s\in S\}$ over $\mathbb{Z}\pi_1(X)$ and where $f(m_s)$ is of the form $f(m_s) = g_s m_{\alpha(s)}$, $(s \in S)$, with $g_s \in \pi_1(X)$. It follows from [Pel, Lemma 1.18] that $\bar{\varepsilon}'(x) = 0$. Hence so is $p_1\varepsilon'(x)$. To show that $p_1\varepsilon'(y) = 0$, we note that f can be decomposed as $f = f_1 f_2$ where $f_1(m_s) = m_{\alpha(s)}$, $f_2(m_s) = g_s m_s$. Hence $[M,f] = [M,f_1] + [M,f_2]$. The second summand vanishes by [Pel, Lemma 1.18]. Therefore, it suffices to show that $p_1\eta[M,f_1] = 0$, and this follows from the commutative diagram

$$\begin{array}{ccc} K_1\mathscr{C}_k(\mathbb{Z}) & \xrightarrow{\eta} & K_{1-k}\mathbb{Z} \\ \downarrow i_0 & & \downarrow i_0 \\ K_1\mathscr{C}_k(\mathbb{Z}\pi_1(X)) & \xrightarrow[\eta]{} & K_{1-k}\mathbb{Z}\pi_1(X) \end{array}$$

and the observation that $[M,f_1] \in i_0(K_1\mathscr{C}_k(\mathbb{Z}))$ because f_1 just permutes the basis elements of M.

Next we prove that ε_1' is onto when $\pi_i(X)$ is uniformly locally presented for $i = 0,1$. If $k \geq 2$, $K_{1-k}\mathbb{Z} = 0$, so there is nothing to prove. Therefore, let $k=1$. Since $i_0K_{1-k}\mathbb{Z}$ is generated by $[\mathbb{Z}\pi_1(X)] \in K_0\mathbb{Z}\pi_1(X)$, it suffices to exhibit an element $x \in \operatorname{Ker}p_0$ with $\varepsilon'(x) = [\mathbb{Z}\pi_1(X)]$.

Clearly $\eta^{-1}[\mathbb{Z}\pi_1(X)] = [M,\mu]$ where $M \in |\mathscr{C}_1(\mathscr{F}\mathbb{Z}\pi_1(X))|$ has $M(z) = \mathbb{Z}\pi_1(X)$, $(z \in \mathbb{Z})$, and where μ maps each $M(z)$ by the identity onto $M(z-1)$. To construct an element $[F(\zeta),F(\alpha,\upsilon)] \in \operatorname{Ker}p_0$ which maps to $[M,\mu]$ under $\bar{\varepsilon}'$, we proceed as follows: Since X is connected, so is $p(X) \subseteq \mathbb{R}$. Moreover, by 1.1(iii), $\sup(p(X)) = \infty$ and $\inf(p(X)) = -\infty$. Hence $p\colon X \longrightarrow \mathbb{R}$ is onto. For each $z \in \mathbb{Z}$ we choose a point $x_z \in X$ with $p(x_z) = z$. There is then a bfg basis $(\mathbb{Z},\zeta)$ over $\mathscr{P}G_1(X,p)$ given by the formula $\zeta(z) = (x_z,\{z\})$. Since X is connected and because 1.1(ii) holds for π_0, for each z there is a path class γ_z from x_z to x_{z-1} in $p^{-1}C^{d+1}\{z\}$. Then there is a morphism in $\mathscr{P}G_1(X,p)$

$$\upsilon_z = (\gamma_z,i_z)\colon (x_{z-1},\{z\}) \longrightarrow C^{d+1}(x_z,\{z\})$$

and as z varies over $\mathbb{Z}$, these morphisms form a natural transformation $\upsilon\colon \zeta\alpha \longrightarrow C^{d+1}\zeta$, where $\alpha\colon \mathbb{Z} \longrightarrow \mathbb{Z}$ has $\alpha(z) = z-1$. This defines an automorphism (α,υ) of the basis $(\mathbb{Z},\zeta)$ and hence an element $[F(\zeta),F(\alpha,\upsilon)]$ of $\operatorname{Ker} p_0$. We leave it to the reader to check that $\bar{\varepsilon}'$ maps this element to $[M,\mu]$.

The final statement in 1.8 now follows from the 5-lemma, since ε' is already known to be an isomorphism when $\pi_i(X)$ is uniformly locally presented for $i = 0,1$.

This finishes the proof of 1.8 and hence the proof of 1.2 for path connected X. If X is not path connected, then (X,p) is the disjoint union, $\coprod_\nu (X_\nu, p_\nu)$, of its usual path components, and the general case of 1.2 follows easily. In fact, α is already defined in the general case, and the remaining groups involved in the definition of $\bar{\varepsilon}'$ decompose into direct sums over the components $\{X_\nu\}$, so that one can use the special case to define β and δ componentwise. We note that if $\pi_i(X)$ is uniformly locally presented for i = 0,1, then so is $\pi_i(X_\nu)$ for each component X_ν. In that case there is also a decomposition

$$K_1\mathbb{Z}\mathcal{G}_1(X,p) = \coprod_\nu K_1\mathbb{Z}\mathcal{G}_1(X_\nu,p_\nu)$$

and $\alpha = \coprod_\nu \alpha_\nu$ with $\alpha_\nu\colon K_1\mathbb{Z}\mathcal{G}_1(X_\nu,p_\nu) \longrightarrow K_1\mathbb{Z}\mathcal{G}(X_\nu)$. Therefore, by applying the special case to each component, we see that ε is an isomorphism.

Proof of 1.3: The hypothesis on Y easily implies that $(Y\times\mathbb{R}^k,\text{proj})$ meets the conditions of 1.2. The conclusion then follows from 1.2 and V.3.1.

Proof of 1.4: This follows immediately from 1.2, VI.6.1, and VI.7.1.

2. A comparison of $Wh^c(N\times\mathbb{R}^k,\text{proj})$ with $\mathcal{S}(N\times\mathbb{R}^k)$.

Let N be a closed PL manifold and $\text{proj}\colon N\times\mathbb{R}^k \longrightarrow \mathbb{R}^k$ be the projection. In this section we shall compare $Wh^c(N\times\mathbb{R}^k,\text{proj})$ with the group $\mathcal{S}(N\times\mathbb{R}^k)$ of proper simple homotopy types in the proper homotopy type of $N\times\mathbb{R}^k$, defined by Siebenmann in [Si]. The comparison is made via a "forgetful" homomorphism $\Box\colon Wh^c(N\times\mathbb{R}^k,\text{proj}) \longrightarrow \mathcal{S}(N\times\mathbb{R}^k)$ constructed as follows: If $(W,N\times\mathbb{R}^k,\partial_+W,p)$ is a bc PL h-cobordism realizing an element ξ of

$Wh^c(N\times\mathbb{R}^k,\mathrm{proj})$, then $\square(\xi)$ is the class of the proper h-cobordism $(W,N\times\mathbb{R}^k,\partial_+W)$ obtained by forgetting the map p. Siebenmann shows that $\mathcal{S}(N\times\mathbb{R}^k) = 0$ for $k\geq 3$. Hence we shall treat only the cases k=1 and k=2. In these cases Siebenmann defines isomorphisms

$$\sigma_\infty\colon \mathcal{S}(N\times\mathbb{R}) \longrightarrow \tilde{K}_0(\mathbb{Z}\pi);\text{ and}$$

$$\sigma_\infty\colon \mathcal{S}(N\times\mathbb{R}^2) \longrightarrow \mathrm{Ker}(\tilde{K}_0(\mathbb{Z}[\pi\times\mathbb{Z}]) \longrightarrow \tilde{K}_0(\mathbb{Z}\pi)),$$

where $\pi = \pi_1(N)$. On the other hand, by 1.3 we can regard our torsion invariant as an isomorphism $\tau\colon Wh^c(N\times\mathbb{R}^k,\mathrm{proj}) \longrightarrow \tilde{K}_{1-k}(\mathbb{Z}\pi)$.

Theorem 2.1 *Let* dim N = n-k-1 *where* n $\geq$ 6. *If* k=1, *then the diagram*

$$\begin{array}{ccc} Wh^c(N\times\mathbb{R},\mathrm{proj}) & \xrightarrow{\ \tau\ } & \tilde{K}_0(\mathbb{Z}\pi) \\ \big\downarrow \square & & \big\downarrow 1 \\ \mathcal{S}(N\times\mathbb{R}) & \xrightarrow[\sigma_\infty]{} & \tilde{K}_0(\mathbb{Z}\pi) \end{array}$$

commutes. If k=2 *and* i *is the Bass-Heller-Swan inclusion, then the diagram*

$$\begin{array}{ccc} Wh^c(N\times\mathbb{R}^2,\mathrm{proj}) & \xrightarrow{\ \tau\ } & K_{-1}(\mathbb{Z}\pi) \\ \big\downarrow \square & & \big\downarrow i \\ \mathcal{S}(N\times\mathbb{R}^2) & \xrightarrow[\sigma_\infty]{} & \mathrm{Ker}[\tilde{K}_o(\mathbb{Z}[\pi\times\mathbb{Z}]) \longrightarrow \tilde{K}_o(\mathbb{Z}\pi)] \end{array}$$

commutes.

The proof of 2.1 begins later in this section and is completed in section 3. Since σ_∞, respectively τ, vanishes if and only if the h-cobordism in question admits a product structure (in the appropriate category) and since i is known to be a monomorphism, the following corollary is obvious:

Corollary 2.2 *Let* $(W^n, N\times\mathbb{R}^k, \partial_+W, p)$ *be a bc PL h-cobordism with* $n \geq 6$, $k = 1$ *or* 2. *If* $(W, N\times\mathbb{R}^k)$ *admits a PL product structure, then* $(W, N\times\mathbb{R}^k, p)$ *admits a bc PL product structure.*

Remark In section 4 we shall see that the situation (over $\mathbb{R}$) is quite different if $\pi_1(\partial_-W)$ is not uniformly locally presented.

Before we begin the proof of 2.1, we recall the definitions of τ and σ_∞. Let $\xi \in Wh^c(N\times\mathbb{R}^k, proj)$ be represented by the bc PL h-cobordism $(W, N\times\mathbb{R}^k, \partial_+W, p)$. We may assume that (W,p) has a handle structure relative to $(N\times\mathbb{R}^k, proj)$ of the form

$$(W^{(1)}, p_1) \subseteq (W^{(2)}, p_2) \subseteq (W^{(3)}, p_3) = (W, p)$$

where $(W^{(1)}, p_1)$ is a bc collar on $(N\times\mathbb{R}^k, proj)$, $(W^{(i)}, p_i)$ is obtained from $(W^{(i-1)}, p_{i-1})$ by attaching an i-handle (i=2,3), the attaching homeomorphism for the 2-handle is bc null homotopic, and $p_i = p|W^{(i)}$. In the sequel we shall identify $(W^{(1)}, p_1)$ with $(N\times\mathbb{R}^k\times I, proj)$ so that, in particular, $\partial_+W^{(1)} = N\times\mathbb{R}^k\times 1$. Let $\tilde{W}$ be the universal cover of W and $\tilde{W}^{(i)}$ the corresponding universal cover of $W^{(i)}$ (i=1,2,3). Then V.2.7, VI.6.2 and 1.5(iii) combine to show that $\tau(\xi)$ is the torsion of the chain complex

$$(2.3) \qquad C_*: \qquad 0 \longleftarrow H_2(\tilde{W}^{(2)}, \tilde{W}^{(1)}) \xleftarrow{\partial} H_3(\tilde{W}^{(3)}, \tilde{W}^{(2)}) \longleftarrow 0$$

It is explained below how C_* is viewed as a chain complex of based objects of $\mathcal{C}_k(\mathbb{Z}\pi)$.

To define $\sigma_\infty\square(\xi)$ we need to consider the part of W "near ∞". For each integer $K\geq 0$, let $\mathbb{R}^k_K = \{z\in\mathbb{R}^k \mid \|z\| \geq K\}$ and let W_K be the largest subhandlebody of W having $W_K \cap N\times\mathbb{R}^k\times I = N\times\mathbb{R}^k_K\times I$. (DRA: Change OK?)
Thus W_K arises from $N\times\mathbb{R}^k_K\times I$ by first attaching all those 2-handles of W which attach only to $N\times\mathbb{R}^2_K\times I$, and then adding all the 3-handles of W attaching only to the resulting manifold. Let $\tilde{W}_K$ be the universal cover

of W_K and $\tilde{W}_K^{(i)}$ be the corresponding universal cover of $W_K^{(i)} = W_K \cap W^{(i)}$.

If k=1 and K is sufficiently large, then W_K has two components and $\sigma_\infty\square(\xi)$ is the relative finiteness obstruction of the "positive" half W_{K+}, i.e. of the following chain complex D_* of $\mathbb{Z}\pi$ modules:

$$(2.4) \qquad D_*: \qquad 0 \longleftarrow H_2(\tilde{W}_{K+}^{(2)},\tilde{W}_{K+}^{(1)}) \xleftarrow{\overline{\partial}} H_3(\tilde{W}_{K+}^{(3)},\tilde{W}_{K+}^{(2)}) \longleftarrow 0.$$

If k = 2 and K is sufficiently large, then $\sigma_\infty\square(\xi)$ is the (relative) finiteness obstruction of the following chain complex of $\mathbb{Z}[\pi\times\mathbb{Z}]$ modules:

$$(2.5) \qquad D_*: \quad 0 \longleftarrow H_2(\tilde{W}_K^{(2)},\tilde{W}_K^{(1)}) \xleftarrow{\overline{\partial}} H_3(\tilde{W}_K^{(3)},\tilde{W}_K^{(2)}) \longleftarrow 0.$$

The proof of 2.1 consists of a comparison of the chain complexes C_* and D_*. In case k = 1, this is fairly straightforward as we are dealing with chain complexes over the same ring, viz. $\mathbb{Z}\pi$. If k = 2, the proof is complicated by the change of ring involved. In either case we need an explicit description of the chain complex C_* of (2.3) as a based chain complex in $\mathscr{C}_k(\mathbb{Z}\pi)$. To get such a description we write

$$(W^{(i)},p_i) = (W^{(i-1)},p_{i-1}) \cup_{\varphi_i} (S_i\times D^i\times D^{n-i},q_i)$$

(i=2,3), where S_i is an indexing set and φ_i is an attaching homeomorphism for the i-handle, (i=2,3). We put $H_\lambda^i = \lambda\times D^i\times D^{n-i}$ for $\lambda\in S_i$. Choose a base point $n_0\in N$ and let $x_0=(n_0,0,1) \in N\times\mathbb{R}^k\times 1$ and for each (i,λ), $(i=2,3;\lambda\in S_i)$, choose a path $\omega_{i\lambda}$ from x_0 to a point $x_{i\lambda}\in\lambda\times D^i\times 0 \subseteq H^i_\lambda$ and an element $\alpha(i,\lambda) \in \mathbb{Z}^k$ within distance 1 of $p(x_{i\lambda})$. We may assume the handle decomposition chosen so that $x_{i\lambda}$ and $\omega_{i\lambda}$ are in $N\times\mathbb{R}^k\times 1$. Let $\omega_{i\lambda}$ have components $(\nu_{i\lambda},\beta_{i\lambda},1)$ in $N\times\mathbb{R}^k\times 1$. Finally choose a point $\tilde{x}_0 \in \tilde{N}\times\mathbb{R}^k\times 1$ above x_0. As in V.2.5 the paths $\omega_{i\lambda}$ determine lifts $\tilde{D}_\lambda^i\times 0$ of $\lambda\times D^i\times 0$ and $\tilde{H}_\lambda^i$ of H_λ^i and it is well known that $C_i = H_i(\tilde{W}^{(i)},\tilde{W}^{(i-1)})$ is the free $\mathbb{Z}\pi$ module with basis $\{\tilde{D}_\lambda^i\times 0 \mid \lambda\in S_i\}$.

Furthermore, C_i becomes an object of $\mathcal{C}_k(\mathbb{Z}\pi)$ by assigning degree $\alpha(i,\lambda)$ to the basis element $\tilde{D}^i_\lambda\times 0$.

The boundary map $\partial\colon C_3 \longrightarrow C_2$ is given by the formula

$$\partial[\tilde{D}^3_\lambda\times 0] = \Sigma_\mu(\Sigma_x \varepsilon_x g_x[\tilde{D}^2_\mu\times 0]) \tag{2.6}$$

Here μ ranges over S_2 while x ranges over all intersection points of the oriented spheres $\mu\times 0\times S^{n-3}$ and $\lambda\times S^2\times 0$ in the oriented upper boundary $\partial_+ W^{(2)}$; ε_x is the sign of the intersection in question, and $g_x \in \pi$ is given as follows in terms of the paths $\omega_{i\lambda}$ and arbitrarily chosen paths $\delta_{i\lambda x}$ in H^i_λ from $x_{i\lambda}$ to x (cf. [Mil, Corollary 7.3]):

$$g_x = [\omega_{3\lambda}\delta_{3\lambda x}\delta^{-1}_{2\mu x}\omega^{-1}_{2\mu}] \in \pi. \tag{2.7}$$

Since the diameter of $p(H^i_\lambda)$ is bounded, ∂ is a bounded isomorphism of based objects of $\mathcal{C}_k(\mathbb{Z}\pi)$.

We now begin the proof of 2.1.

Proof of 2.1: First suppose that k = 1. Then C_* is a chain complex of based $\mathcal{C}_1(\mathbb{Z}\pi)$ objects and D_* is the subcomplex of C_* obtained by looking at the part of C_* "near ∞" and forgetting the $\mathbb{Z}$-grading on C_2 and C_3. Under the identification of $\tilde{K}_1\mathcal{C}_1(\mathbb{Z}\pi)$ and $\tilde{K}_0(\mathbb{Z}\pi)$, the Whitehead torsion of C_* maps to [coker $\partial\colon {C_3}^{(m+d)} \longrightarrow {C_2}^{(m)}$] where $C_i^{(r)}$ is the submodule of C_i generated by all $\tilde{D}^i_\lambda\times 0$ with $\alpha(i,\lambda) \geq r$ and m and d are both sufficiently large. But when m and d are chosen correctly, $\partial\colon {C_3}^{(m+d)} \longrightarrow {C_2}^{(m)}$ is exactly the chain complex D_*. Thus the above cokernel is the finiteness obstruction of D_* and commutativity of the first diagram in 2.1 has been established.

Next suppose that k=2. In this case we begin by describing explicitly how the chain complex D_* of (2.5) becomes a chain complex of based $\mathbb{Z}[\pi\times\mathbb{Z}]$ modules. As before we choose a base point $\bar{x}_0 = (n_0,K,1) \in N\times\mathbb{R}^2_K\times 1$ and we use paths $\bar{\omega}_{i\lambda}$ in $N\times\mathbb{R}^2_K\times 1 \subseteq W_K$ from $\bar{x}_0$ to $x_{i\lambda} \in H^i_\lambda$ to determine preferred liftings of $\lambda\times D^i\times 0$ which form the basis for D_i (i=2,3). For $\bar{\omega}_{i\lambda}$, we take $(\nu_{i\lambda},\gamma_{i\lambda},1)$ where $\nu_{i\lambda}$ is the first component of $\omega_{i\lambda}$, as above, and $\gamma_{i\lambda}$ is a path in $\mathbb{R}^2_K$, whose inverse is described as follows: If $\alpha(i,\lambda)$ has imaginary part ≥ 0, then $\gamma^{-1}_{i\lambda}$ runs clockwise along the circle of radius $\|\alpha(i,\lambda)\|$ to the positive real axis and linearly back to K. If $\alpha(i,\lambda)$ has imaginary part < 0, then $\gamma^{-1}_{i\lambda}$ runs counterclockwise along the same circle to the positive real axis and back to K along the real axis.

We use the gradings $\alpha(i,\lambda) \in \mathbb{Z}^2$ to make D_i into an object of $\mathscr{C}_2(\mathbb{Z}(\pi\times\mathbb{Z}))$ for i=2,3. The boundary map $\bar{\partial}$ of D_* can be described by formulas similar to (2.6) and (2.7). Of course, only generators corresponding to handles that occur in W_K will appear in the formula for $\bar{\partial}$, and the group elements in the formula will belong to $\pi\times\mathbb{Z}$ rather than π. We denote by t be the generator of $\mathbb{Z} = \pi_1(\mathbb{R}^2_K,K)$ represented by the loop $(K\cos 2\pi\theta,-K\sin 2\pi\theta)$ $(0\leq\theta\leq 1)$. Let x be a point of intersection between $\mu\times 0\times S^{n-3}$ and $\lambda\times S^2\times 0$ where the handles involved are in W_K. Then x contributes a term $\varepsilon_x g_x[\tilde{D}^2_\mu\times 0]$ to $\partial[\tilde{D}^3_\lambda\times 0]$ in the chain complex (2.3), and a term $\bar{\varepsilon}_x\bar{g}_x[\tilde{D}^2_\mu\times 0]$ to $\bar{\partial}[\tilde{D}^3_\lambda\times 0]$ in the chain complex D_* of (2.5). To describe $\bar{\partial}$ in terms of ∂, we must relate these two terms. Obviously $\bar{\varepsilon}_x = \varepsilon_x$, while $\bar{g}_x = [\bar{\omega}_{3\lambda}\delta_{3\lambda x}\delta^{-1}_{2\mu x}\bar{\omega}^{-1}_{2\mu}] \in \pi\times\mathbb{Z}$. If K is sufficiently large, then the paths $\delta_{2\mu x}$ and $\delta_{3\lambda x}$ contribute only trivially to the $\mathbb{Z}$-component of $\bar{g}_x$, and one easily checks that the following formula holds:

$$(2.8)\qquad \bar{g}_x = \begin{cases} g_x t^{-1}, & \text{if } \alpha_1(2,\mu)<0,\ \alpha_2(2,\mu)\geq 0,\ \alpha_1(3,\lambda)<0,\ \alpha_2(3,\lambda)<0, \\ g_x t, & \text{if } \alpha_1(2,\mu)<0,\ \alpha_2(2,\mu)<0,\ \alpha_1(3,\lambda)<0,\ \alpha_2(3,\lambda)\geq 0, \\ g_x, & \text{otherwise} \end{cases}$$

where $\alpha(i,\kappa) = (\alpha_1(i,\kappa),\alpha_2(i,\kappa))$ for $(i,\kappa) = (2,\mu)$ or $(3,\lambda)$. The reader should note that when K is large, $\mu\times 0\times S^{n-3}$ and $\lambda\times S^2\times 0$ can intersect only if $\alpha(2,\mu)$ and $\alpha(3,\lambda)$ are in neighbouring quadrants in $\mathbb{R}^2$.

We summarize the above discussion in the following formula for the boundary operator $\bar{\partial}$: $D_3 \longrightarrow D_2$ in (2.5), valid provided K is sufficiently large

(2.9) $$\partial[\tilde{D}^3_\lambda \times 0] = \Sigma_\mu(\Sigma_x \varepsilon_x \bar{g}_x[\tilde{D}^2_\mu \times 0]).$$

Here x ranges over precisely the same points as in (2.6), λ and μ are restriced to those values for which the corresponding handles belong to W_K, and $\bar{g}_x$ is given by (2.8).

3. The algebra of germs at infinity.

This section completes the proof of 2.1 after developing the algebra needed for the completion.

Let A be a ring with unit and satisfying IV.2.1. We introduce the category $\mathcal{C}_{k,\infty}(A)$ of *germs at* ∞ *of* $\mathcal{C}_k(A)$. The objects are those of $\mathcal{C}_k(A)$ while a morphism from $M = (M_i)_{i\in\mathbb{Z}^k}$ to $N = (N_j)_{j\in\mathbb{Z}^k}$ is an equivalence class α_∞ of morphisms α: $M \longrightarrow N$ in $\mathcal{C}_k(A)$. One has $\alpha_\infty = \beta_\infty$ if there is some $d\in\mathbb{Z}$ such that $(\alpha-\beta)(M_i) = 0$ whenever $\|i\|\geq d$. The composition in $\mathcal{C}_{k,\infty}(A)$ has $\alpha_\infty\gamma_\infty = (\alpha\gamma)_\infty$ so that there is a functor π: $\mathcal{C}_k(A) \longrightarrow \mathcal{C}_{k,\infty}(A)$ with $\pi(\alpha) = \alpha_\infty$. We call $\pi(\alpha)$ the *germ of* α at ∞.

Clearly $\mathcal{C}_{k,\infty}(A)$ is a an additive category under degreewise direct sum and we let $K_1\mathcal{C}_{k,\infty}(A)$ be the group defined by Bass in [Ba; VII.5]. Thus $K_1\mathcal{C}_{k,\infty}(A)$ is the abelian group generated by $[M,\alpha_\infty]$, where M ranges over all objects of $\mathcal{C}_{k,\infty}(A)$ and α_∞ over all automorphisms of M, and where the relations are the usual ones:

(3.1) $[M,\alpha_\infty] = [M',\phi_\infty\alpha_\infty\phi_\infty^{-1}]$, whenever α_∞: $M \longrightarrow M$ and ϕ_∞: $M \longrightarrow M'$ are isomorphisms in $\mathcal{C}_{k,\infty}(A)$.

(3.2) $[M,\beta_\infty\alpha_\infty] = [M,\beta_\infty] + [M,\alpha_\infty]$, whenever $\alpha_\infty,\beta_\infty$: $M \longrightarrow M$ are isomorphisms in $\mathcal{C}_{k,\infty}(A)$.

(3.3) $[M'\oplus M'', \alpha'_\infty \oplus \alpha''_\infty] = [M', \alpha'_\infty] + [M'', \alpha''_\infty]$, whenever α'_∞ and α''_∞ are isomorphisms in $\mathscr{C}_{k,\infty}(A)$.

If $M \in |\mathscr{C}_{k,\infty}(A)|$ and $d \geq 0$, we let $M^{(d)} = \bigoplus_{\|i\| \geq d} M_i$.

Lemma 3.4 *Let $\alpha: M \longrightarrow N$ be a representative for an isomorphism $\alpha_\infty: M \longrightarrow N$ in $\mathscr{C}_{k,\infty}(A)$. For any sufficiently large* d, *α restricts to a split monomorphism $\alpha^{(d)}: M^{(d)} \longrightarrow N^{(0)}$. Furthermore, the cokernel of $\alpha^{(d)}$ is a finitely generated projective* A *module.*

Proof: Let $\beta: N \longrightarrow M$ be a representative for the inverse of α_∞. For d sufficiently large, the diagram

$$\begin{array}{ccc} M^{(d)} & \xrightarrow{\alpha^{(d)}} & N^{(0)} \\ & \searrow^{\text{incl}} & \downarrow{\scriptstyle \beta^{(0)}} \\ & & M^{(0)} \end{array}$$

commutes in $\mathscr{C}_k(A)$. Since the inclusion is split monic so is $\alpha^{(d)}$. Hence $\mathrm{Cok}(\alpha^{(d)})$ is projective. Similarly, for e sufficiently large,

$$\begin{array}{ccc} N^{(e)} & \xrightarrow{\beta^{(e)}} & M^{(d)} \\ & \searrow^{\text{incl}} & \downarrow{\scriptstyle \alpha^{(d)}} \\ & & N^{(0)} \end{array}$$

commutes. It follows that $\mathrm{Im}\ \alpha^{(d)} \supseteq N^{(e)}$ so that the epimorphism $N^{(0)} \longrightarrow \mathrm{Cok}(\alpha^{(d)})$ factors through the finitely generated A-module $N^{(0)}/N^{(e)}$. Thus $\mathrm{Cok}(\alpha^{(d)})$ is finitely generated.

To the isomorphism $\alpha_\infty: M \longrightarrow N$, we associate the element

$$c(\alpha_\infty) = [\mathrm{Cok}(\alpha^{(d)})] - [M^{(0)}/M^{(d)}] \in K_0(A)$$

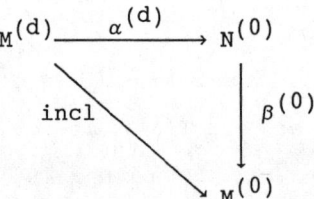

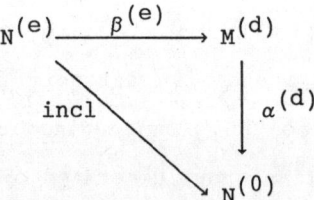

where d is chosen large enough to make $\mathrm{Cok}(\alpha^{(d)})$ finitely generated projective. An application of the snake lemma, see e.g. [Ma, p. 202], to the diagram

$$\begin{array}{ccccccccc} 0 & \longrightarrow & M^{(d+1)} & \subseteq & M^{(d)} & \longrightarrow & M^{(d)}/M^{(d+1)} & \longrightarrow & 0 \\ & & \big\downarrow \alpha^{(d+1)} & & \big\downarrow \alpha^{(d)} & & \big\downarrow & & \\ 0 & \longrightarrow & N^{(0)} & = & N^{(0)} & \longrightarrow & 0 & \longrightarrow & 0 \end{array}$$

shows that $c(\alpha_\infty)$ is independent of the choice of d. In particular, $c(\alpha_\infty)$ is independent of the choice of α in the class α_∞.

Proposition 3.5 *For any* $k \geq 1$ *the formula* $c([M,\alpha_\infty]) = c(\alpha_\infty)$ *defines a homomorphism* c *which fits into an exact sequence*

$$K_1\mathscr{C}_k(A) \xrightarrow[\pi_*]{} K_1\mathscr{C}_{k,\infty}(A) \xrightarrow[c]{} K_0(A) \longrightarrow 0 \ .$$

Proof: We must see that c respects the relations (3.1)-(3.3). First consider isomorphisms $\alpha_\infty\colon M \longrightarrow N$ and $\beta_\infty\colon N \longrightarrow L$. Choose d so that $\beta^{(d)}\colon N^{(d)} \longrightarrow L^{(0)}$ is split monic and choose $e \geq d$ so that $\alpha^{(e)}$ is split monic and factors through $N^{(d)} \subseteq N^{(0)}$ as shown in the diagram

$$\begin{array}{ccccccccc} 0 & \longrightarrow & M^{(e)} & \xrightarrow{1} & M^{(e)} & \longrightarrow & 0 & \longrightarrow & 0 \\ & & \big\downarrow \hat{\alpha}^{(e)} & & \big\downarrow \alpha^{(e)} & & \big\downarrow & & \\ 0 & \longrightarrow & N^{(d)} & \subseteq & N^{(0)} & \longrightarrow & N^{(0)}/N^{(d)} & \longrightarrow & 0 \end{array}$$

The snake lemma then implies that

$$c(\alpha_\infty) = [\mathrm{Cok}\hat{\alpha}^{(e)}] + [N^{(0)}/N^{(d)}] - [M^{(0)}/M^{(e)}].$$

Applying the snake lemma to the diagram

$$\begin{array}{ccccccccc} 0 & \longrightarrow & M^{(e)} & \xrightarrow{\hat{\alpha}^{(e)}} & N^{(d)} & \longrightarrow & \mathrm{Cok}\hat{\alpha}^{(e)} & \longrightarrow & 0 \\ & & \downarrow 1 & & \downarrow \beta^{(d)} & & \downarrow & & \\ 0 & \longrightarrow & M^{(e)} & \xrightarrow[(\beta\alpha)^{(e)}]{} & L^{(0)} & \longrightarrow & \mathrm{Cok}(\beta\alpha)^{e} & \longrightarrow & 0 \end{array}$$

shows that

$$[\mathrm{Cok}(\beta\alpha)^{(e)}] = [\mathrm{Cok}\hat{\alpha}^{(e)}] + [\mathrm{Cok}\beta^{(d)}]$$

It easily follows that

$$(3.6) \qquad c((\beta\alpha)_\infty) = c(\beta_\infty) + c(\alpha_\infty) .$$

and that c respects (3.2). Since $c(\phi_\infty) + c(\phi_\infty^{-1}) = c(1) = 0$,

$$c(\phi_\infty\alpha_\infty\phi_\infty^{-1}) = c(\phi_\infty) + c(\alpha_\infty) + c(\phi_\infty^{-1}) = c(\alpha_\infty)$$

and c respects (3.1). If $\alpha'^{(d)}$ and $\alpha''^{(d)}$ are split monomorphisms, so is $(\alpha'\oplus\alpha'')^{(d)}$ and it follows easily that $c(\alpha'_\infty\oplus\alpha''_\infty) = c(\alpha''_\infty) + c(\alpha''_\infty)$. Hence c respects (3.3) and defines a homomorphism $c\colon K_1\mathscr{C}_{k,\infty}(A) \longrightarrow K_0(A)$.

We proceed to show the exactness of the sequence in 3.5. If $\alpha\colon M \longrightarrow M$ is an automorphism in $\mathscr{C}_k(A)$, then one may choose d=0 when one computes $c([M,\alpha_\infty])$. Hence $c\pi_*$ vanishes.

To check exactness at $K_1\mathscr{C}_{k,\infty}(A)$, let $\alpha\colon M \longrightarrow M$ in $\mathscr{C}_k(A)$ represent an isomorphism α_∞ with $c(\alpha_\infty)=0$. Then, for d sufficiently large, $\mathrm{Cok}\alpha^{(d)}$ and $M^{(0)}/M^{(d)}$ are stably isomorphic finitely generated projective A modules. Replacing α by $\alpha\oplus 1\colon M\oplus F \longrightarrow M\oplus F$ where F is a suitable object of $\mathscr{C}_k(A)$ concentrated in degree zero, we can assume that $\mathrm{Cok}\alpha^{(d)}$ and $M^{(0)}/M^{(d)}$ are isomorphic. Then M splits as a direct sum $N \oplus L$ where $N_i = 0$ for $\|i\|\geq d$, $L_i = 0$ for $\|i\|<d$, and α has the form

$$\alpha = \begin{bmatrix} \alpha_{11} & \alpha_{12} \\ \alpha_{21} & \alpha_{22} \end{bmatrix} : N\oplus L \longrightarrow N\oplus L$$

with

$$\alpha^{(d)} = \begin{bmatrix}\alpha_{12}\\ \alpha_{22}\end{bmatrix}: M^{(d)} = L^{(0)} \longrightarrow (N\oplus L)^{(0)}$$

having cokernel isomorphic to $N^{(0)}$. We view this cokernel as an object of $\mathscr{C}_k(A)$ concentrated in degree zero. It is then isomorphic to N in $\mathscr{C}_k(A)$ so we can choose a splitting

$$\begin{bmatrix}\beta_1\\ \beta_2\end{bmatrix}: N \longrightarrow N\oplus L$$

in $\mathscr{C}_k(A)$. Then

$$\beta = \begin{bmatrix}\beta_1 & \alpha_{12}\\ \beta_2 & \alpha_{22}\end{bmatrix}: N\oplus L \longrightarrow N\oplus L$$

is an isomorphism in $\mathscr{C}_k(A)$ and $\beta_\infty = \alpha_\infty$ (because N is concentrated "near" degree 0) so $[M,\alpha_\infty] = [N\oplus L,\beta_\infty] = \pi_*([N\oplus L,\beta])$ belongs to $\mathrm{Im}\pi_*$ as desired.

To see that c is onto, let P be any finitely generated projective A-module. Choose a complement Q with $A^n = P\oplus Q$ and let $p: A^n \longrightarrow A^n$ be the projection onto P along Q. Now define M by letting $M_i = A^n$ for all i of the form $(i_1,0,\ldots,0) \in \mathbb{Z}^k$, $i_1\geq 0$, while $M_i=0$ for all other values of $i \in \mathbb{Z}^k$; and define $\alpha: M \longrightarrow M$ in $\mathscr{C}_k(A)$ by letting all components $M_i \longrightarrow M_j$ be 0 except that $M_{(i_1,0,\ldots,0)} = A^n$ maps to itself by 1-p for $i_1\geq 0$ and to $M_{(i_1-1,0,\ldots,0)}$ by p for $i_1>0$. Symbolically, α is represented by the following diagram

$$\begin{array}{ccccccccccc}
\ldots 0 & & 0 & & A^n & & A^n & & A^n & & \ldots \\
 & & & & {\scriptstyle 1-p}\downarrow & \swarrow^{p} & {\scriptstyle 1-p}\downarrow & \swarrow^{p} & {\scriptstyle 1-p}\downarrow & \swarrow^{p} & \\
\ldots 0 & & 0 & & A^n & & A^n & & A^n & & \ldots
\end{array}$$

Obviously, α_∞ has an inverse β_∞, where β is represented by the diagram

$$\begin{array}{ccccccccccc}
\cdots\ 0 & & 0 & & A^n & & A^n & & A^n & & \cdots \\
 & & & & \big\downarrow{\scriptstyle 1-p} & \searrow^{p} & \big\downarrow{\scriptstyle 1-p} & \searrow^{p} & \big\downarrow{\scriptstyle 1-p} & \searrow^{p} & \\
\cdots\ 0 & & 0 & & A^n & & A^n & & A^n & & \cdots\ .
\end{array}$$

Thus $[M,\alpha_\infty] \in K_1\mathscr{C}_{k,\infty}(A)$. In computing $c([M,\alpha_\infty])$, we choose d=1 and find that $c([M,\alpha_\infty]) = [Q]-[A^n] = -[P]$. Thus c is onto.

We now consider the Laurent series ring $A[t,t^{-1}]$. We are going to define a functor $f\colon \mathscr{C}_2(A) \longrightarrow \mathscr{C}_{2,\infty}(A[t,t^{-1}])$. We start by writing any object M in $\mathscr{C}_2(A)$ as $M = M^{[1]}\oplus M^{[2]}\oplus M^{[3]}$ where

$$M_\alpha^{[1]} = \begin{cases} M_\alpha & \text{if } \ \alpha = (\alpha_1,\alpha_2) \text{ has } \alpha_1 \geq 0 \\ 0 & \text{otherwise} \end{cases}$$

$$M_\alpha^{[2]} = \begin{cases} M_\alpha & \text{if } \ \alpha = (\alpha_1,\alpha_2) \text{ has } \alpha_1 < 0 \ ,\ \alpha_2 \geq 0 \\ 0 & \text{otherwise} \end{cases}$$

$$M_\alpha^{[3]} = \begin{cases} M_\alpha & \text{if } \ \alpha = (\alpha_1,\alpha_2) \text{ has } \alpha_1 < 0 \ ,\ \alpha_2 < 0 \\ 0 & \text{otherwise} \end{cases}$$

Any morphism $\mu\colon M \longrightarrow N$ in $\mathscr{C}_2(A)$ decomposes accordingly as a matrix

$$\mu = \begin{bmatrix} \mu_{11} & \mu_{12} & \mu_{13} \\ \mu_{21} & \mu_{22} & \mu_{23} \\ \mu_{31} & \mu_{32} & \mu_{33} \end{bmatrix} : M \longrightarrow N$$

where $\mu_{ij}\colon M^{[j]} \longrightarrow N^{[i]}$ in $\mathscr{C}_2(A)$. On objects we let

$$f(M) = M\otimes_A A[t,t^{-1}] = M[t,t^{-1}].$$

This decomposes as

$$f(M) = M^{[1]}[t,t^{-1}]\oplus M^{[2]}[t,t^{-1}]\oplus M^{[3]}[t,t^{-1}]$$

We define f on morphisms by setting

$$(3.7)\qquad f((\mu_{ij})) = \begin{bmatrix} \hat{\mu}_{11} & \hat{\mu}_{12} & \hat{\mu}_{13} \\ \hat{\mu}_{21} & \hat{\mu}_{22} & t^{-1}\hat{\mu}_{23} \\ \hat{\mu}_{31} & t\hat{\mu}_{32} & \hat{\mu}_{33} \end{bmatrix}_{\infty}$$

where

$$\hat{\mu}_{ij} = \mu_{ij} \otimes_A A[t,t^{-1}]: M^{[j]}[t,t^{-1}] \longrightarrow N^{[i]}[t,t^{-1}].$$

Since the morphism μ is bounded and since $f((\mu_{ij}))$ is a germ at ∞, a simple computation shows that f is, indeed, functorial. Also f respects the restrictions and the relations that define the K_1 groups so f induces a homomorphism $f_*: K_1\mathscr{C}_2(A) \longrightarrow K_1\mathscr{C}_{2,\infty}(A[t,t^{-1}])$.

It is the composition cf_* which turns out to be of geometric interest. Algebraically, it is described as follows in terms of homomorphisms defined by Pedersen in [Pe1]:

Proposition 3.8 *The following diagram commutes*

$$\begin{array}{ccc} K_1\mathscr{C}_2(A) & \xrightarrow{\quad f_* \quad} & K_1\mathscr{C}_{2,\infty}(A[t,t^{-1}]) \\ \| & & \downarrow c \\ K'_{-1}(A) & & K_0(A[t,t^{-1}]) \\ \downarrow \lambda_t^1 & & \uparrow \psi \\ K'_0(A[t,t^{-1}]) & \xrightarrow[\quad \phi^1 \quad]{} & K''_0(A[t,t^{-1}]) \end{array}$$

Proof: We start by giving some explicit references to Pedersen's paper [Pe1]. The functors K'_{-i}, K''_{-i} are defined on p. 462, ϕ^1 on p. 464 and ψ is one of the (unchristened) identifications of proposition 2.6. Finally, λ_t^1 is defined on p. 469. By Pedersen's Theorem 2.16 (in which three occurrences of "-i" are misprints for "-i+1"), λ_t^1 is a split mono morphism, and the composition $\psi\phi^1\lambda_t^1$ is the Bass-Heller-Swan injection (which we have labelled i in our section 2). From the formulae below, it is easily seen to map into $\mathrm{Ker}[K_0(A[t,t^{-1}]) \longrightarrow K_0(A)]$, and hence into $\mathrm{Ker}[\tilde{K}_0(A[t,t^{-1}]) \longrightarrow \tilde{K}_0(A)]$ when A is augmented, i.e. when there is a preferred homomorphism of A onto $\mathbb{Z}$.

In the proof we shall actually work with λ_t^2 in place of λ_t^1. By 2.18 of [Pel], $\lambda_t^2 = -\lambda_t^1$ so we have to prove that the diagram with λ_t^2 instead of λ_t^1 commutes up to a minus sign.

Let $x \in K'_{-1}(A)$. We have the isomorphism $\phi^2\colon K'_{-1}(A) \longrightarrow K''_{-1}(A)$ so we may assume that $x = (\phi^2)^{-1}[(M,p)]$ where $p\colon M \longrightarrow M$ is a bounded projection of a $\mathbb{Z}$-graded, free A-module M. From the proof of lemma 1.15 of [Pel], it then follows that x is represented by a bounded automorphism $\alpha\colon N \longrightarrow N$ where the bigraded free A module N has $N_{ij} = M_i$ $(i,j \in \mathbb{Z})$ and where α is given by the diagram

$$\begin{array}{ccccccccccccc}
j & = & \cdots & -2 & & -1 & & 0 & & 1 & & 2 & \cdots \\
N & = & \cdots & M & & M & & M & & M & & M & \cdots \\
\Big\downarrow \alpha & = & \cdots & \Big\downarrow 1-p & \searrow^{p} & \Big\downarrow 1-p & \searrow^{p} & \Big\downarrow 1-p & \searrow^{p} & \Big\downarrow 1-p & \searrow^{p} & \Big\downarrow 1-p & \\
N & = & \cdots & M & & M & & M & & M & & M & \cdots
\end{array}$$

We denote the functor $-\otimes_A A[t,t^{-1}]$ by $\hat{}$ and we split the projection p as

$$p = \begin{bmatrix} p_{--} & p_{-+} \\ p_{+-} & p_{++} \end{bmatrix} : M^-\oplus M^+ \longrightarrow M^-\oplus M^+ = M$$

where

$$M_i^+ = \begin{cases} M_i\,, & i \geq 0 \\ 0\,, & i<0 \end{cases} \quad \text{and} \quad M_i^- = \begin{cases} M_i\,, & i<0 \\ 0\,, & i\geq 0 \end{cases}.$$

Letting

$$q = \begin{bmatrix} p_{--} & 0 \\ 0 & 0 \end{bmatrix} : M^-\oplus M^+ \longrightarrow M^-\oplus M^+,$$

a simple computation shows that $f(\alpha) = \beta_\infty : \hat{N} \longrightarrow \hat{N}$, where β is given by the diagram

$$
\begin{array}{ccccccccccccc}
j & = & \cdots & -2 & & -1 & & 0 & & 1 & & 2 & \cdots \\
\hat{N} & = & \cdots & \hat{M} & & \hat{M} & & \hat{M} & & \hat{M} & & \hat{M} & \cdots \\
\downarrow \beta & = & \cdots & 1-\hat{p}\,\downarrow & \overset{\hat{p}}{\searrow} & 1-\hat{p}\,\downarrow & \overset{Q}{\searrow} & 1-\hat{p}\,\downarrow & \overset{\hat{p}}{\searrow} & 1-\hat{p}\,\downarrow & \overset{\hat{p}}{\searrow} & 1-\hat{p}\,\downarrow & \\
\hat{N} & = & \cdots & \hat{M} & & \hat{M} & & \hat{M} & & \hat{M} & & \hat{M} & \cdots
\end{array}
$$

with $Q = t^{-1}\hat{q} + (\hat{p}-\hat{q})$. Thus $f_*(x) = [\hat{N},\beta_\infty]$. By 3.5, $c\pi_* = 0$. Hence we also have

$$f_*(x) = [\hat{N},(\hat{\alpha})_\infty^{-1}\beta_\infty]. \tag{3.9}$$

It is easy to show that $\hat{\alpha}^{-1}\beta$ is given by the diagram

$$
\begin{array}{ccccccccccccc}
j & = & \cdots & -2 & & -1 & & 0 & & 1 & & 2 & \cdots \\
\hat{N} & = & \cdots & \hat{M} & & \hat{M} & & \hat{M} & & \hat{M} & & \hat{M} & \cdots \\
\downarrow \hat{\alpha}^{-1}\beta & = & \cdots & 1\,\downarrow & & R\,\downarrow & \overset{S}{\searrow} & \downarrow\,1 & & \downarrow\,1 & & \downarrow\,1 & \\
\hat{N} & = & \cdots & \hat{M} & & \hat{M} & & \hat{M} & & \hat{M} & & \hat{M} & \cdots
\end{array}
$$

with $R = (1-\hat{p}) + \hat{p}Q$, $S = (1-\hat{p})Q$. For elements $m \in \hat{M}$, one has

$$\hat{q}(m) = \begin{cases} 0 & , \deg m \gg 0 \\ \hat{p}(m) & , \deg m \ll 0 \end{cases}, \quad \text{so} \quad Q(m) = \begin{cases} \hat{p}(m) & , \deg m \gg 0 \\ t^{-1}\hat{p}(m) & , \deg m \ll 0 \end{cases}$$

Since only the germ at infinity matters, we may change Q to make

$$Q(m) = \begin{cases} \hat{p}(m) & , \deg m \geq 0 \\ t^{-1}\hat{p}(m) & , \deg m < 0 \end{cases}$$

Then (3.9) still holds but now (in the diagram for $\hat{\alpha}^{-1}\beta$) $S=0$ and

$$R(m) = \begin{cases} m & , \deg m \geq 0 \\ (t^{-1}\hat{p}+1-\hat{p})(m) & , \deg m < 0 \end{cases} \tag{3.10}$$

Hence $\hat{\alpha}^{-1}\beta = 1$ on most of $\hat{N}$, and one easily gets

$$(3.11) \qquad cf_*(x) = \left[\operatorname{Cok} t^{-1}\hat{p}+1-\hat{p}\colon \bigoplus_{i\leq -d} \hat{M}_i \longrightarrow \bigoplus_{i<d} \hat{M}_i\right] - \left[\bigoplus_{|i|<d} \hat{M}_i\right]$$

for d sufficiently large.

With x as above, $\lambda_t^2(x)$ is represented by a certain restriction of the commutator $c = [\hat{\alpha}, tp_-^2+1-p_-^2]\colon \hat{N} \longrightarrow \hat{N}$ where $p_-^2\colon \hat{N} \longrightarrow \hat{N}$ is the projection onto $\hat{N}_-^1 = \bigoplus_{j<0} \hat{N}_{*j}$. The commutator in turn is represented as follows

$$\begin{array}{ccccccccc}
j & = & \cdots & -2 & -1 & 0 & 1 & 2 & \cdots \\
\hat{N} & = & \cdots & \hat{M} & \hat{M} & \hat{M} & \hat{M} & \hat{M} & \cdots \\
\big\downarrow c & = & \cdots & 1\big\downarrow & 1\big\downarrow & t^{-1}\hat{p}+1-\hat{p}\big\downarrow & 1\big\downarrow & 1\big\downarrow & \cdots \\
\hat{N} & = & \cdots & \hat{M} & \hat{M} & \hat{M} & \hat{M} & \hat{M} & \cdots
\end{array}$$

so we can restrict to the copy of $\hat{M}$ in degree j=0 and get

$$(3.12) \qquad \lambda_t^2(x) = [\hat{M}, t^{-1}\hat{p}+1-\hat{p}] \in K_0'(A[t,t^{-1}]).$$

Under the augmentation map $A[t,t^{-1}] \longrightarrow A$, this element maps to $[M,1] = 0 \in K_0'(A)$. Hence $\lambda_t^2(x) \in \tilde{K}_0'(A[t,t^{-1}])$.

To finish the proof of Proposition 3.8 we now need to show the following lemma:

Lemma 3.13 *For any augmented ring* R *the composition* $\psi\phi^1\colon K_0'(R) \longrightarrow K_0(R)$ *has*

$$\psi\phi^1([L,\gamma]) = \left[\bigoplus_{i=0}^{d-1} L_i\right] - \left[\operatorname{Cok} \gamma\colon \bigoplus_{i<0} L_i \longrightarrow \bigoplus_{i<d} L_i\right]$$

for d sufficiently large. Here $\gamma\colon L \longrightarrow L$ *is a bounded automorphism of a* $\mathbb{Z}$*-graded, free* R *module* L.

Proof: One has $\phi^1([L,\gamma]) = [\bar{L},\gamma p_-^1\gamma^{-1}] - [\bar{L},p_-^1]$ where $p_-^1\colon L \longrightarrow L$ projects L to $\bigoplus_{i<0} L_i$ and $\bar{L} = \bigoplus_{|i|<d} L_i$ for some sufficiently large d. By Proposition 2.6 of [Pe1], for any (ungraded) free A-module K and any projection $q\colon K \longrightarrow K$, one has $\psi([K,q]) = [\mathrm{Im}\ q] \in K_0(A)$. Equivalently, $\psi([K,q]) = [K] - [\mathrm{Cok}\ q]$. Thus

$$\psi\phi^1([L,\gamma]) = \left[\bigoplus_{i=0}^{d-1} L_i\right] - \left[\mathrm{Cok}\ \gamma p_-\gamma^{-1}\colon \bigoplus_{|i|<d} L_i \longrightarrow \bigoplus_{|i|<d} L_i\right]$$

But d is chosen so large that $\gamma p_-\gamma^{-1} = 1$ on L_i for $i \leq -d$. Also $p_-\gamma^{-1}\left[\bigoplus_{i<d} L_i\right] = \bigoplus_{i<0} L_i$, so

$$\left[\mathrm{Cok}\ \gamma p_-\gamma^{-1}\colon \bigoplus_{|i|<d} L_i \longrightarrow \bigoplus_{|i|<d} L_i\right] = \left[\mathrm{Cok}\ \gamma p_-\gamma^{-1}\colon \bigoplus_{i<d} L_i \longrightarrow \bigoplus_{|i|<d} L_i\right]$$

$$= \left[\mathrm{Cok}\ \gamma\colon \bigoplus_{i<0} L_i \longrightarrow \bigoplus_{i<d} L_i\right]$$

and the proof of 3.13 is complete. We note that this description of $\psi\phi^1$ makes it clear that $\psi\phi^1$ also gives an isomorphism of the reduced groups.

Using (3.12) and 3.13, we get

$$(3.14)\qquad \psi\phi^1\lambda_t^2(x) = \left[\bigoplus_{i=0}^{d-1} \hat{M}_i\right] - \left[\mathrm{Cok}\ t^{-1}\hat{p}+1-\hat{p}\colon \bigoplus_{i<0} \hat{M}_i \longrightarrow \bigoplus_{i<d} \hat{M}_i\right]$$

which must be compared to (3.11). Applying the snake lemma to the diagram

$$\begin{array}{ccccccccc}
0 & \longrightarrow & \bigoplus_{i\leq -d} \hat{M}_i & \xrightarrow{t^{-1}\hat{p}+1-\hat{p}} & \bigoplus_{i<d} \hat{M}_i & \longrightarrow & \mathrm{Cok}_{-d} & \longrightarrow & 0 \\
 & & \big\downarrow{\scriptstyle \mathrm{incl.}} & & \big\downarrow{\scriptstyle 1} & & \big\downarrow & & \\
0 & \longrightarrow & \bigoplus_{i<d} \hat{M}_i & \xrightarrow{t^{-1}\hat{p}+1-\hat{p}} & \bigoplus_{i<d} \hat{M}_i & \longrightarrow & \mathrm{Cok}_{-1} & \longrightarrow & 0
\end{array}$$

one gets the desired conclusion, i.e. $\psi\phi^1\lambda_t^2(x) = -cf_*(x)$.

Completion of the proof of 2.1: We recall the setup from section 2: We consider an element $\xi \in Wh^c(N\times\mathbb{R}^2,\text{proj})$ and want to show that $\sigma_\infty\square(\xi) = i\tau(\xi)$. We have identified $\sigma_\infty\square(\xi)$ as the finiteness obstruction of the chain complex D_* of (2.5), whose boundary operator is given in (2.9). Similarly, $\tau(\xi)$ is the torsion of the chain complex C_* of (2.3), whose boundary operator is given by (2.6).

We can identify the indexing sets S_2 and S_3 for the 2- and the 3-handles of W (cf. IV.2.6). Then λ and μ range over the same set in (2.6), i.e. ∂ is an automorphism of a based object, $C = C_2 = C_3$, in $\mathcal{C}_2(\mathbb{Z}\pi)$, and $\tau(\xi) = [C,\partial]$ under the identification of $K_1\mathcal{C}_2(\mathbb{Z}\pi)$ with $K_{-1}(\mathbb{Z}\pi)$.

The above identification of indexing sets makes $D_3 \subseteq D_2 \subseteq C[t,t^{-1}]$. Moreover, the formulas (2.8) and (2.9) are easily seen to imply that $\bar{\partial}\colon D_3 \longrightarrow D_2$ has $\bar{\partial}_\infty = f(\partial)|D_3\colon D_3 \longrightarrow D_2$ in $\mathcal{C}_{2,\infty}(\mathbb{Z}[\pi\times\mathbb{Z}])$. If K is sufficiently large, it follows that $cf_*\tau(\xi) = [\text{Cok}\bar{\partial}] - [D_2/D_3]$. But this is precisely the finiteness obstruction of D_*, i.e. we have $\sigma_\infty\square(\xi) = cf_*(\tau(\xi))$. Since Pedersen, [Pe 1], shows that $\psi\phi^1\lambda_t^1$ equals the Bass-Heller-Swan inclusion (when one identifies $K_1\mathcal{C}_2(\mathbb{Z}\pi)$ with $K_{-1}(\mathbb{Z}\pi)$), Proposition 3.8 now gives the desired identity, viz. $\sigma_\infty\square(\xi) = i\tau(\xi)$.

4. An example over $\mathbb{R}$.

In this section we prove the following result.

Theorem 4.1 *There exists a finite* bc CW *complex* (X,p) *over* $\mathbb{R}$ *such that* $Wh^c(X,p)$ *is not countably generated whereas Siebenmann's group* $\mathcal{S}(X)$ *vanishes.*

By realizing such a bc homotopy type as a bc PL manifold (∂_-W,p_-) over $\mathbb{R}$ with $\dim \partial_-W \geq 5$, one gets the following corollary:

Corollary 4.2 *Let* $n \geq 5$. *There exists a bc PL* n*-manifold* $(\partial_- W, p_-)$ *over* $\mathbb{R}$, *such that there are uncountably many pairwise non bc PL homeomorphic bc PL h-cobordisms* $(W, \partial_- W, \partial_+ W, p)$ *on* $(\partial_- W, p_-)$. *However, each of these admits a PL (non-bc, in general) product structure.*

The proof consists of a construction of (X,p) and a computation of $\mathcal{S}(X)$ and $Wh^c(X,p)$. For the construction of (X,p), we consider an increasing sequence of integers a_i and a sequence of compact, connected CW complexes X_i $(-\infty<i<\infty)$ with $\{\dim X_i\}$ bounded. Let $x_i^{(0)}$ be a base point in X_i and let $\pi_1(X_i, x_i^{(0)}) = \pi_i$. Subdivide the interval $[a_i, a_{i+1}]$ at all integer points, impose the product CW structure on $\coprod_i X_i \times [a_i, a_{i+1}]$, and take X to be the quotient CW complex

$$X = (\coprod_i X_i \times [a_i, a_{i+1}])/\sim$$

where $\sim$ identifies each $X_i \times \{a_i\} \cup X_{i-1} \times \{a_i\}$ to one point. Points of X will be written as $[x,t]$ with $t \in [a_i, a_{i+1})$ and $x \in X_i$ for some unique i. We let $p([x,t]) = t$. Then (X,p) is a finite bc CW complex over $\mathbb{R}$.

For any $i \in \mathbb{Z}_+$, the subspaces $p^{-1}([a_i,\infty))$ and $p^{-1}((-\infty,a_{-i}])$ are 1-connected. Moreover, as i varies, $p^{-1}((-\infty,a_{-i}] \cup [a_i,\infty))$ form a cofinal subset in the set of all cofinite subcomplexes of X. Therefore, all the inverse systems that are involved in Siebenmann's computation, [Si; p. 483], of $\mathcal{S}(X)$ consist of trivial abelian groups. Hence $\mathcal{S}(X)$ vanishes. To describe $Wh\mathbb{Z}\mathcal{G}_1(X,p)$ we need the following definition:

Definition 4.3 *Let* A_i *or* ℓ_i, *respectively, be a sequence of abelian groups or integers, respectively,* $(-\infty<i<\infty)$. *The element* $(a_i) \in \Pi_i A_i$ *is said to have* bounded life *relative to* (ℓ_i) *(or, simply,* bounded life *when* (ℓ_i) *is clear from the context) if for some integer* $n \geq 0$, $a_i = 0$ *whenever* $\ell_i \geq n$. *The set of elements of bounded life forms a subgroup* B *of* $\Pi_i A_i$. *The quotient group* $\Pi_i A_i / B$ *will be called* *the group of elements of unbounded life and will be denoted* $U(A_i)$, *or* $U(A_i,\ell_i)$ *if the dependence on the sequence* (ℓ_i) *needs to be emphasized.*

Remark 4.4 If the sequence (ℓ_i) is bounded, then $U(A_i)$ vanishes. On the other hand, if $\ell_i \longrightarrow \infty$ as $i \longrightarrow \pm\infty$, then $U(A_i) = (\Pi_i A_i)/(\Sigma_i A_i)$; if, furthermore, each $A_i \neq 0$, then $U(A_i)$ is not countable and therefore is also not countably generated.

Theorem 4.1, and therefore Corollary 4.2, follow immediately from Theorem 4.5 below, since it is easy to choose (a_i) so that $a_{i+1}-a_i \longrightarrow \infty$ as $i \longrightarrow \pm\infty$ and to choose X_i so that $\tilde{K}_0(\mathbb{Z}\pi_i) \neq 0$ for all i (e.g. take $X_i = S^3/Q8$ for all i).

Theorem 4.5 *For the above* (X,p) *there is an isomorphism*

$$\ell c\colon \mathrm{Wh}\mathbb{Z}\mathcal{P}G_1(X,p) \longrightarrow U(\tilde{K}_0(\mathbb{Z}\pi_i),a_{i+1}-a_i)$$

The proof occupies the rest of this section. The first step is to replace $\mathcal{P}G_1(X,p)$ with a more tractable category $\mathcal{P}\hat{G}$. In particular, let $\hat{G}\colon \mathcal{P} \longrightarrow \mathcal{G}poid$ be the functor given by

$$\hat{G}(K) = \begin{cases} \pi_1(X_i,x_i^{(0)}) = \pi_i, & \text{if } K \subseteq (a_i,a_{i+1}) \\ \{1\} & \text{, if some } a_i \in K \end{cases}$$

for $K \in |\mathcal{P}|$. If $i\colon K \longrightarrow L$ is the inclusion, let

$$\hat{G}(i) = \begin{cases} 1, & \text{if } K \subseteq L \subseteq (a_i,a_{i+1}) \\ 0, & \text{otherwise} \end{cases}$$

where 0 is the trivial homomorphism.

Lemma 4.6 *There is an equivalence of categories*

$$E\colon \mathrm{bfgfree}\mathbb{Z}\mathcal{P}G_1(X,p)\text{-}\mathit{mod} \longrightarrow \mathrm{bfgfree}\mathbb{Z}\mathcal{P}\hat{G}\text{-}\mathit{mod}.$$

In particular, there is an isomorphism $E_*: Wh\mathbb{Z}\mathcal{F}G_1(X,p) \longrightarrow Wh\mathbb{Z}\mathcal{F}\hat{G}$.

Proof: The equivalence of categories E is constructed from several other equivalences of categories which we now describe.

Let $s: \mathbb{R} \longrightarrow X$ be defined by setting $s(t) = [x_i^{(0)},t]$ and observe that $ps = 1$. For $K \in |\mathcal{F}|$, let $H(K)$ be the full subcategory (subgroupoid) of $G_1(K)$ with $|H(K)| = \{s(x) \mid x \in K\}$. It is easily seen that $H: \mathcal{F} \longrightarrow \mathcal{G}poid$ is a functor, that the inclusions $i_K: H(K) \subseteq G_1(K)$ define a natural transformation $\mu: H \longrightarrow G_1(X,p)$, and that μ satisfies the hypotheses of II.5.1 and IV.4.10. Hence

$$\mathcal{F}\mu_!: \text{bfgfree}\mathbb{Z}\mathcal{F}H\text{-}mod \longrightarrow \text{bfgfree}\mathbb{Z}\mathcal{F}G_1(X,p)\text{-}mod$$

is an equivalence of categories.

Let $\hat{H}: \mathcal{F} \longrightarrow \mathcal{G}poid$ be the functor defined as follows: If $K \in |\mathcal{F}|$, choose a point $z_K \in K$ and let $\hat{H}(K) = \pi_1(p^{-1}(K),s(z_K))$. If $K \subseteq L$, let λ_K^L be the linear path in $\mathbb{R}$ from z_L to z_K and let $\hat{H}(i)$ be the composite

$$\pi_1(p^{-1}K,s(z_K)) \xrightarrow{i_*} \pi_1(p^{-1}L,s(z_K)) \xrightarrow{s(\lambda)_\#} \pi_1(p^{-1}L,s(z_L))$$

where $s(\lambda)_\#$ is the change of base point isomorphism induced by $s(\lambda_K^L)$. Since $\lambda_K^L\lambda_L^M$ is homotopic to λ_K^M, $\hat{H}$ is a functor.

We now define $\rho_K: H(K) \longrightarrow \hat{H}(K)$ by setting $\rho_K(\omega: s(x) \longrightarrow s(y)) = s(\lambda_K^y)\omega s(\lambda_K^x)^{-1}$ where λ_K^z is the linear path in $\mathbb{R}$ from z_K to z ($z=x,y$). It is easy to check that $\rho = \{\rho_K \mid K \in |\mathcal{F}|\}$ is a natural transformation satisfying the hypotheses of II.5.1 and IV.4.10. Hence

$$\mathcal{F}\rho_!: \text{bfgfree}\mathbb{Z}\mathcal{F}H\text{-}mod \longrightarrow \text{bfgfree}\mathbb{Z}\mathcal{F}\hat{H}\text{-}mod$$

is also an equivalence of categories.

Finally, we note that since $(p^{-1}K,s(K)) = (X_i \times K, x_i^{(0)} \times K)$ for $K \subseteq (a_i,a_{i+1})$, there is an isomorphism $\eta_K: \hat{G}(K) \longrightarrow \hat{H}(K)$ and $\eta = \{\eta_K \mid K \in |\mathcal{F}|\}$ is a natural transformation satisfying the hypotheses of II.5.1 and IV.4.10. Hence

$$\mathcal{F}\eta_!: \text{bfgfree}\mathbb{Z}\mathcal{F}\hat{G}\text{-}mod \longrightarrow \text{bfgfree}\mathbb{Z}\mathcal{F}\hat{H}\text{-}mod$$

is an equivalence of categories.

The functor E is the composite of a pseudo inverse for $\mathcal{F}\mu_!$, $\mathcal{F}\rho_!$, and a pseudo inverse for $\mathcal{F}\eta_!$. This completes the proof of 4.6.

To prove 4.5, it now suffices to construct an isomorphism $\ell c\colon \text{Wh}\mathbb{Z}\mathcal{F}\hat{G} \longrightarrow U(\tilde{K}_0\mathbb{Z}\pi_i, a_{i+1}-a_i)$. We notice first that if M is a $\mathbb{Z}\mathcal{F}\hat{G}$ module and $K \subseteq (a_i, a_{i+1})$ for some i, then M(K) is a $\mathbb{Z}\pi_i$ module, while if $a_i \in K$, then M(K) is simply an abelian group. If $\sigma\colon S \longrightarrow |\mathcal{F}| = |\mathcal{F}\hat{G}|$ is a basis and $M = F(\sigma)$, then M(K) is a free $\mathbb{Z}\pi_i$ module with basis $\{(1,s) \mid \sigma(s) \subseteq K\}$ whenever $K \subseteq (a_i, a_{i+1})$, and a free abelian group with the similar basis when some $a_i \in K$; if $K \subseteq L \subseteq (a_i, a_{i+1})$ or $a_i \in K \subseteq L$, then M(K) is a direct summand in M(L).

Let (S,σ) be a basis for a bfg free $\mathbb{Z}\mathcal{F}\hat{G}$ module $F(\sigma)$, $S_1 \subseteq S$, $\sigma_1 = \sigma|S_1$, and $A \subseteq B \subseteq \mathbb{R}$. We say that $F(\sigma_1)$ is *full over* A, respectively, *supported over* B, (relative to $F(\sigma)$), if (4.7)(i), respectively, (4.7)(ii) holds

(4.7) (i) If $\sigma(s) \subseteq A$, then $s \in S_1$,

(ii) If $s \in S_1$, then $\sigma(s) \subseteq B$.

We call $F(\sigma_1)$ an *essential submodule of* $F(\sigma)$ if there are integers $d_1 \geq d_2 > 0$ such that $F(\sigma_1)$ is full over $\cup_i[a_i+d_1, a_{i+1}-d_1]$ and supported over $\cup_i[a_i+d_2, a_{i+1}-d_2]$.

Put $c_i = (a_i+a_{i+1})/2$. If $F(\sigma_1)$ is an essential submodule of $F(\sigma)$ and $S_\ell \subseteq S_1$, $\sigma_\ell = \sigma_1|S_\ell$, then $F(\sigma_\ell)$ is called a *left fringe* of $F(\sigma_1)$ provided there are integers $d_3 \geq d_4$ such that $F(\sigma_\ell)$ is full over $\cup_i([a_i+d_2, a_i+d_4]\cap[a_i, c_i))$ (relative to $F(\sigma_1)$) and supported over $\cup_i([a_i+d_2, a_i+d_3]\cap[a_i, c_i))$. Similarly define a *right fringe* $F(\sigma_r)$ of $F(\sigma_1)$ by requiring that it be full over $\cup_i([a_{i+1}-d_5, a_{i+1}-d_2]\cap(c_i, a_{i+1}])$ and supported over $\cup_i([a_{i+1}-d_6, a_{i+1}-d_2]\cap(c_i, a_{i+1}])$ for some $d_5 \leq d_6$.

For any based submodule $F(\sigma_2)$ of an essential submodule $F(\sigma_1)$ of $F(\sigma)$, we let

$$(4.8) \qquad |F(\sigma_2)|_i = \lim_{\varepsilon \to 0+} F(\sigma_2)([a_i+\varepsilon, a_{i+1}-\varepsilon]).$$

Note that $|F(\sigma_2)|_i$ is a free $\mathbb{Z}\pi_i$ module with basis (in 1-1 correspondence with) the set of all $s \in S_2$ for which $\sigma_2(s) \subseteq (a_i, a_{i+1})$. The limit construction is needed because $\sigma_2(s) \subseteq (a_i, a_{i+1})$ does not necessarily imply that $\sigma_2(s) \subseteq [a_i+1, a_{i+1}-1]$.

Lemma 4.9 *Let* $\phi\colon F(\sigma) \longrightarrow F(\sigma')$ *be an isomorphism of based* $\mathbb{Z}\mathcal{G}\hat{G}$ *modules. For any sufficiently small, essential submodule* $F(\sigma_1)$ *of* $F(\sigma)$, *for any sufficiently large, essential submodule* $F(\sigma_1')$ *of* $F(\sigma')$, *and for any integer* i, ϕ *restricts to a morphism* $|\phi|_i$ *of* $\mathbb{Z}\pi_i$ *modules which fits into a split short exact sequence*

$$0 \longrightarrow |F(\sigma_1)|_i \xrightarrow{|\phi|_i} |F(\sigma_1')|_i \xrightarrow{\gamma_i} C_i \longrightarrow 0$$

Moreover, for any sufficiently large left and right fringes, $F(\sigma_\ell')$ *and* $F(\sigma_r')$, *of* $F(\sigma_1')$, *and for any* i *with* $a_{i+1}-a_i$ *sufficiently large,* $C_i = P_i \oplus Q_i$ *where* $P_i = \gamma_i(|F(\sigma_\ell')|_i)$, $Q_i = \gamma_i(|F(\sigma_r')|_i)$. *In particular,* C_i, P_i, *and* Q_i *are finitely generated projective* $\mathbb{Z}\pi_i$ *modules.*

The module P_i is called the ith *left cokernel* of ϕ.

Proof: For any representative ϕ_d of ϕ there is a representative, say $\psi_{d'}\colon F(\sigma') \longrightarrow F(\sigma)C^{d'}$, of $\psi = \phi^{-1}$ such that the triangles

$$\begin{array}{ccc} F(\sigma) & \xrightarrow{\phi_d} & F(\sigma')C^{d} \\ & \searrow^{F(\sigma)\tau^{d+d'}} & \downarrow{\scriptstyle \psi_{d'}C^{d}} \\ & & F(\sigma)C^{d+d'} \end{array} \qquad \begin{array}{ccc} F(\sigma') & \xrightarrow{\psi_{d'}} & F(\sigma)C^{d'} \\ & \searrow^{F(\sigma')\tau^{d+d'}} & \downarrow{\scriptstyle \phi_{d}C^{d'}} \\ & & F(\sigma')C^{d+d'} \end{array}$$

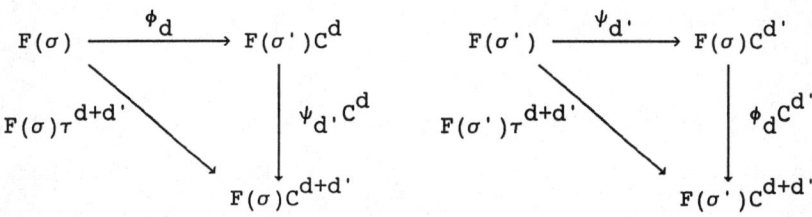

commute. Now choose $F(\sigma_1)$ to be supported over $B = \cup_i[a_i+d_1,a_{i+1}-d_1]$ where $d_1 > d+d'$ and let $F(\sigma_1')$ be full over $A' = \cup_i[a_i+d_2,a_{i+1}-d_2]$ where $d_2 \leq d_1-d$. The first triangle above immediately shows that $\phi_d([a_i+d_1,a_{i+1}-d_1])$ maps $|F(\sigma_1)|_i = F(\sigma_1)[a_i+d_1,a_{i+1}-d_1]$ into $F(\sigma_1')([a_i+d_1-d,a_{i+1}-d_1+d]) \subseteq |F(\sigma_1')|_i$. Since this map is split by means of the composition

$$\begin{array}{ccc} |F(\sigma_1')|_i & \xrightarrow{\text{proj}} & F(\sigma_1')([a_i+d_1-d,a_{i+1}-d_1+d]) \\ & & \Big\downarrow \psi_{d'}([a_i+d_1-d,a_{i+1}-d_1+d]) \\ |F(\sigma_1)|_i & \xleftarrow{\text{proj}} & F(\sigma)([a_i+d_1-d-d',a_{i+1}-d_1+d+d']) \end{array}$$

we have the desired split exact sequence.

Let essential submodules $F(\sigma_1)$ of $F(\sigma)$ and $F(\sigma_1')$ of $F(\sigma)$ be given such that $|\phi_d|_i\colon |F(\sigma_1)|_i \longrightarrow |F(\sigma_1')|_i$ is split monic for each i. The second of the above triangles shows that any sufficiently small essential submodule $F(\sigma_2')$ of $F(\sigma_1')$ will have $|F(\sigma_2')|_i \subseteq \mathrm{Im}|\phi_d|_i$. Also one can pick a left fringe, $F(\sigma_\ell')$, and a right fringe, $F(\sigma_r')$, for $F(\sigma_1')$, such that $F(\sigma_1') = F(\sigma_\ell')\oplus F(\sigma_2')\oplus F(\sigma_r')$. For such a choice one has $C_i = P_i+Q_i$ since $\gamma_i(|F(\sigma_2)|_i) = 0$. We claim that when $a_{i+1}-a_i$ is sufficiently large, this sum is direct. In fact, assume that $x \in |F(\sigma_\ell')|_i$, $y \in |F(\sigma_r')|_i$ and $\gamma_i(x) = \gamma_i(y)$, i.e. $x-y \in |F(\sigma_1')|_i$ is in $|\phi_d|_i(|F(\sigma_1)|_i)$, say $x-y = |\phi_d|_i(z)$. If we write $|\psi_{d'}|_i$ for the splitting of $|\phi_d|_i$ constructed above, then $z = |\psi_{d'}|_i(x)-|\psi_{d'}|_i(y)$ and $|\psi_{d'}|_i(x) \in |F(\sigma_\ell)|_i$, $|\psi_{d'}|_i(y) \in |F(\sigma_r)|_i$ for some sufficiently large left and right fringe $F(\sigma_\ell)$, respectively $F(\sigma_r)$, in $F(\sigma_1)$. It follows that $x-y = |\phi_d|_i|\psi_{d'}|_i(x) - |\phi_d|_i|\psi_{d'}|_i(y)$ and if $a_{i+1}-a_i$ is sufficiently large, then one must have $x = |\phi_d|_i|\psi_{d'}|_i(x)$. Thus $\gamma_i(x) = 0$ and $C_i = P_i\oplus Q_i$ as desired.

If $\phi\colon F(\sigma) \longrightarrow F(\sigma')$ is an isomorphism, 4.9 gives, for all i with $a_{i+1}-a_i$ sufficiently large, an element $[P_i] \in \tilde{K}_0(\mathbb{Z}\pi_i)$. For fixed d, σ_1, and σ_1', if one alters σ_ℓ' or σ_r', then P_i is unchanged at least for

$a_{i+1}-a_i$ sufficiently large. For fixed d and σ_1, if one alters σ_1' by adding some generators, then P_i and Q_i will be stabilized by adding free $\mathbb{Z}\pi_i$ modules. For fixed d, if σ_1 is made smaller, then C_i will be stabilized by adding a free module. However, each generator in this free module will be situated either near a_i or near a_{i+1} (because the new $F(\sigma_1)$ is still full except near $\{a_i\}$). Hence, P_i will experience nothing but a stabilization. Finally, the bound d enters into the discussion only by deciding how small $F(\sigma_1)$ and how large $F(\sigma_\ell')$ and $F(\sigma_r')$ have to be. Altogether, we have shown the first part of the following proposition:

Proposition 4.10 *Let $\phi: F(\sigma) \longrightarrow F(\sigma')$ be an isomorphism of bfg based $\mathbb{Z}\mathcal{G}\hat{G}$ modules. Then taking left cokernels gives a well defined invariant*

$$\ell c(\phi) = \text{class}(\{[P_i]\}_i) \in U(\tilde{K}_0\mathbb{Z}\pi_i, a_{i+1}-a_i).$$

If $\tilde{\phi}: F(\tilde{\sigma}) \longrightarrow F(\tilde{\sigma}')$ and $\phi': F(\sigma') \longrightarrow F(\sigma'')$ are also such isomorphisms, then

(i) $\qquad \ell c(\phi'\phi) = \ell c(\phi') + \ell c(\phi)$; and

(ii) $\qquad \ell c(\phi \oplus \tilde{\phi}) = \ell c(\phi) + \ell c(\tilde{\phi})$

Corollary 4.11 *The function ℓc induces a homomorphism*

$$\ell c: \mathrm{Wh}\mathbb{Z}\mathcal{G}\hat{G} \longrightarrow U(\tilde{K}_0\mathbb{Z}\pi_i, a_{i+1}-a_i).$$

Proofs of 4.10(i) and (ii): One can easily choose essential submodules $F(\sigma_1)$ of $F(\sigma)$, $F(\sigma_1')$ of $F(\sigma')$, and $F(\sigma_1'')$ of $F(\sigma'')$ such that one has the required split monomorphisms $|\phi_e'|$ and $|\phi_d|$ and such that the morphism $\psi = |\phi_e'|_i \circ |\phi_d|_i$ is also a split monomorphism. Now consider the commutative diagram (where the superscript $\hat{}$ refers to the morphism $\phi'\phi$)

$$\begin{array}{ccccccccc}
 & & 0 & & 0 & & 0 & & \\
 & & \downarrow & & \downarrow & & \downarrow & & \\
0 & \longrightarrow & |F(\sigma_1)|_i & \xrightarrow{|\phi_d|_i} & |F(\sigma_1')|_i & \xrightarrow{\gamma_i} & C_i & \longrightarrow & 0 \\
 & & {\scriptstyle 1}\downarrow & & {\scriptstyle |\phi_e'|_i}\downarrow & & {\scriptstyle a}\downarrow & & \\
0 & \longrightarrow & |F(\sigma_1)|_i & \xrightarrow{\psi} & |F(\sigma_1'')|_i & \xrightarrow{\hat{\gamma}_i} & \hat{C}_i & \longrightarrow & 0 \\
 & & \downarrow & & {\scriptstyle \gamma_i'}\downarrow & & {\scriptstyle b}\downarrow & & \\
0 & \longrightarrow & 0 & \longrightarrow & C_i' & \xrightarrow{1} & C_i' & \longrightarrow & 0 \\
 & & \downarrow & & \downarrow & & \downarrow & & \\
 & & 0 & & 0 & & 0 & &
\end{array}$$

All the rows and the first two columns are exact. It follows that the last column is also exact. Moreover, since $|\phi_e'|_i$ must map $|F(\sigma_\ell')|_i$ into $|F(\sigma_\ell'')|_i$ when σ_ℓ'' is large enough (and $a_{i+1}-a_i$ is large), the map a takes the form $a_1 \oplus a_2\colon P_i \oplus Q_i \longrightarrow \hat{P}_i \oplus \hat{Q}_i$. Also $|F(\sigma_\ell'')|_i$ is mapped onto $\hat{P}_i$ by $\hat{\gamma}_i$ and onto P_i' by γ_i'. Hence $b = b_1 \oplus b_2\colon \hat{P}_i \oplus \hat{Q}_i \longrightarrow P_i' \oplus Q_i'$. Altogether,

$$0 \longrightarrow P_i \xrightarrow{a_1} \hat{P}_i \xrightarrow{b_1} P_i' \longrightarrow 0$$

is exact for all i with sufficiently large $a_{i+1}-a_i$. Hence $[\hat{P}_i] = [P_i] + [P_i']$ for such i. This finishes the proof of 4.7(i).

The proof of 4.7(ii) is a simple exercise in handling direct sums.

Proof of 4.11: Since every element of $\mathrm{Wh}\mathbb{Z}\mathscr{P}\hat{G}$ is of the form $[F(\sigma),\phi]$ for some bfg basis (S,σ) and some automorphism ϕ of $F(\sigma)$, it suffices to check that ℓc respects the relations IV.3.(1-3) and IV.3.7. The identities in 4.10 are easily seen to imply that ℓc respects IV.3.(1-3). If $\phi\colon F(\sigma) \longrightarrow F(\sigma)$ is of the form $-1_{F(\sigma)}$ or of the form $F(\alpha,\upsilon)$ for some automorphism (α,υ) of the basis (S,σ), then it is easily seen that essential submodules $F(\sigma_1)$ and $F(\sigma_1')$, which are suitable for computing $\ell c(\phi)$, can be chosen in such a way that each $|\phi_d|_i\colon |F(\sigma_1)|_i \longrightarrow |F(\sigma_1')|$ is an isomorphism. Hence $\ell c(\phi) = 0$, and the proof of 4.11 is complete.

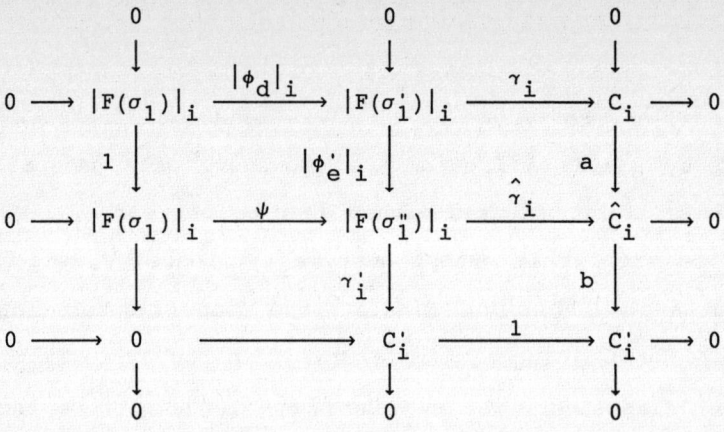

Proof of 4.5: To complete the proof of 4.5, we need only show

(4.12) ℓc is onto; and

(4.13) ℓc is one to one.

For (4.12), we assume given an integer n and for each i with $a_{i+1}-a_i \geq n$, a finitely generated projective $\mathbb{Z}\pi_i$ module P_i. Thus $\text{class}(\{[P_i]\}_i)$ is the typical element of the range of ℓc. We pick a complement Q_i so that $P_i \oplus Q_i = G_i = (\mathbb{Z}\pi_i)^{n_i}$, a based $\mathbb{Z}\pi_i$ module. For all i with $a_{i+1}-a_i < n$, let $P_i = Q_i = G_i = 0$. Also write $p_i: G_i \longrightarrow G_i$, respectively, $q_i: G_i \longrightarrow G_i$, for the projections onto P_i, respectively Q_i. Finally, let $Q_i \otimes_{\mathbb{Z}\pi_i} \mathbb{Z} = \mathbb{Z}^{m_i}$ and put $H_i = (\mathbb{Z}\pi_i)^{m_i}$.

For each $j \in (a_i, a_{i+1}) \cap \mathbb{Z}$, let S_j be a set with n_j elements and T_j a set with m_j elements. If $j = a_i$ for some i, let $S_j = T_j = \emptyset$. Put $S = \coprod_j S_j$, $T = \coprod_j T_j$ and define $\sigma: S \longrightarrow |\mathcal{P}\hat{G}|$, $\rho: T \longrightarrow |\mathcal{P}\hat{G}|$ by letting $\sigma(s) = \rho(t) = \{j\}$ for all $s \in S_j$ and all $t \in T_j$. Now $F(\sigma)$ and $F(\rho)$ are bfg free $\mathbb{Z}\mathcal{P}\hat{G}$ modules and we can define homomorphisms $\xi: F(\sigma) \longrightarrow F(\sigma)$, $\eta: F(\rho) \longrightarrow F(\rho)$ by the following diagram. In it we list every non zero component of (representatives of) ξ and η over an interval $[a_i, a_{i+1}]$. Also listed are components of the inclusions of the kernels, and the projections onto the cokernels, of ξ and η.

	a_i	a_i+1	a_i+2	$\cdots$	$a_{i+1}-2$	$a_{i+1}-1$	a_{i+1}
Ker ξ	0	Q_i	0	$\cdots$	0	0	0
		$\downarrow$					
$F(\sigma)$	0	G_i	G_i		G_i	G_i	0
		$p_i \downarrow$	$\swarrow q_i$; $p_i \downarrow$	$\swarrow q_i$ $\cdots$	$\swarrow q_i$; $p_i \downarrow$	$\swarrow q_i$; $p_i \downarrow$	
$F(\sigma)$	0	G_i	G_i		G_i	G_i	0
						$\downarrow$	
Cok ξ	0	0	0	$\cdots$	0	Q_i	0
Ker η	0	0	0	$\cdots$	0	H_i	0
						$\downarrow 1$	
$F(\rho)$	0	H_i	H_i	$\cdots$	H_i	H_i	0
		$\searrow 1$	$\searrow 1$	$\searrow 1$	$\searrow 1$		
$F(\rho)$	0	H_i	H_i	$\cdots$	H_i	H_i	0
		$\downarrow 1$					
Cok η	0	H_i	0	$\cdots$	0	0	0

The key observation now is that $\mathrm{Ker}\eta \oplus \mathrm{Ker}\xi \cong \mathrm{Cok}\eta \oplus \mathrm{Cok}\xi$ in $\mathbb{Z}\mathcal{G}\hat{G}$-*mod* (because these modules are supported over $\cup_i[a_i-1,a_i+1]$, $\hat{G}([a_i-1,a_i+1]) = 1$, and $Q_i \otimes_{\mathbb{Z}\pi_i} \mathbb{Z} \cong H_i \otimes_{\mathbb{Z}\pi_i} \mathbb{Z}$). Moreover, $\mathrm{Cok}\eta \oplus \mathrm{Cok}\xi \cong F(\rho_1)$ where the basis (T_1,ρ_1) has $T_1 = \{t \in T \mid \rho(t) = a_i \pm 1,\ i \in \mathbb{Z}\}$ and $\rho_1 = \rho|T_1$. There results an exact sequence

$$0 \longrightarrow F(\rho_1) \longrightarrow F(\sigma)\oplus F(\rho) \xrightarrow{\xi\oplus\eta} F(\sigma)\oplus F(\rho) \longrightarrow F(\rho_1) \longrightarrow 0.$$

for which we can choose a splitting to get an isomorphism of the form

$$\begin{bmatrix} \xi & 0 & ? \\ 0 & \eta & ? \\ ? & ? & 0 \end{bmatrix} = \Phi\colon F(\sigma)\oplus F(\rho)\oplus F(\rho_1) \longrightarrow F(\sigma)\oplus F(\rho)\oplus F(\rho_1),$$

where the maps labelled ? are unimportant since $F(\rho_1)$ is supported over $\cup_i[a_i-1,a_i+1]$. One can choose the essential submodules needed to compute $\ell c(\Phi)$ to be submodules of $F(\sigma) \oplus F(\rho)$. For example, one can take the following picture:

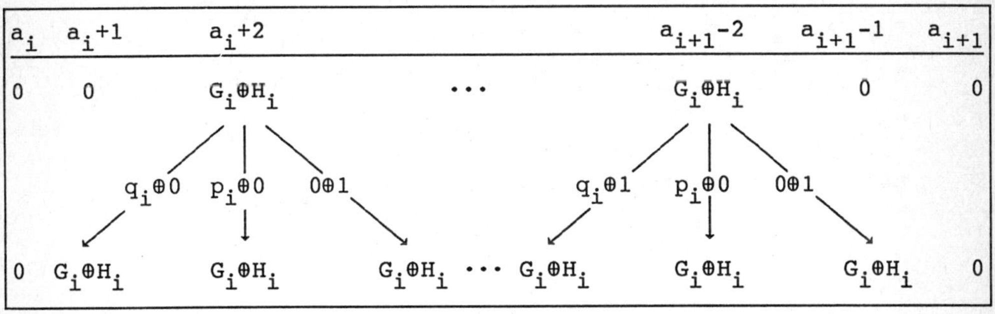

where each $G_i \oplus H_i$ is mapped the same way into three summands. The left cokernel then is $P_i \oplus H_i \oplus H_i$. Since $[P_i \oplus H_i \oplus H_i] = [P_i]$ in $\tilde{K}_0(\mathbb{Z}\pi_i)$, the proof of (4.12) is complete.

To prove (4.13), the following lemma is needed:

Lemma 4.14 *Let* $\phi\colon F(\sigma) \longrightarrow F(\sigma')$ *be an isomorphism with* $\ell c(\phi) = 0$. *Then there exist a bfg basis* $\rho\colon T \longrightarrow \mathcal{P}\hat{G}$, *2 elementary automorphisms* $\varepsilon_i\colon F(\sigma' \amalg \rho) \longrightarrow F(\sigma' \amalg \rho)$, $(i=1,2)$, *and essential submodules* $F(\rho_1)$ *in* $F(\sigma \amalg \rho)$, *respectively*, $F(\rho_1')$ *in* $F(\sigma' \amalg \rho)$, *such that* $\Phi = \varepsilon_2 \varepsilon_1(\phi \oplus 1)$ *maps* $F(\rho_1)$ *isomorphically onto* $F(\rho_1')$. *Moreover, for each* $i \in \mathbb{Z}$, $|\Phi|_i\colon |F(\rho_1)|_i \longrightarrow |F(\rho_1')|_i$ *is an isomorphism of* $\mathbb{Z}\pi_i$ *modules.*

Proof: Since $\ell c(\phi) = 0$, there is exists an integer n such that P_i is stably free whenever $a_{i+1}-a_i \geq n$. Clearly we can change the essential submodules $F(\sigma_1)$ and $F(\sigma_1')$ involved in the definition of P_i to make $|F(\sigma_1')|_i = |F(\sigma_1)|_i = 0$ whenever $a_{i+1}-a_i < n$. For those i, one then has $P_i = Q_i = 0$. We can stabilize the left cokernels by adding a morphism of the form $1\colon F(\sigma_0) \longrightarrow F(\sigma_0)$, where $F(\sigma_0)$ is supported on $U_i[a_i-2, a_i-1] \cup U_i[a_i+1, a_i+2]$. Thereby we can make each P_i and each Q_i free over $\mathbb{Z}\pi_i$. We assume that such stabilization has already been done, and define $F\colon \mathcal{P}\hat{G} \longrightarrow \mathbb{Z}\text{-}mod$ by letting

$$F(K) = \oplus\{P_i \mid a_i+1 \in K\} \oplus \oplus\{Q_i \mid a_{i+1}-1 \in K\}.$$

Then F is bfg free with a basis (T,ρ) where each $\rho(t)$ is either a_i+1 or $a_{i+1}-1$ for some i. Since P_i and Q_i are projective, there exist splittings $p_i\colon P_i \longrightarrow |F(\sigma'_\ell)|_i$ and $q_i\colon Q_i \longrightarrow |F(\sigma'_r)|_i$, and when we let i vary, these $\mathbb{Z}\pi_i$ homomorphisms define a homomorphism $j\colon F(\rho) \longrightarrow F(\sigma'_c)$, where $\sigma'_c = \sigma'_\ell \amalg \sigma'_r$. One therefore has a split exact sequence of $\mathbb{Z}\mathcal{P}\hat{G}$ modules

$$F(\sigma_1) \xrightarrow{\phi_{11}} F(\sigma'_1) \underset{j}{\overset{\gamma}{\rightleftarrows}} F(\rho)\ ,$$

where ϕ_{11} is the restriction of ϕ. Moreover, all of these modules are essential (in themselves) and the sequence

$$|F(\sigma_1)|_i \xrightarrow{|\phi|_i} |F(\sigma'_1)|_i \underset{|j|_i}{\overset{|\gamma|_i}{\rightleftarrows}} |F(\rho)|_i$$

of $\mathbb{Z}\pi_i$ modules is split exact for each i.

Now let $\sigma_2 = \sigma|S-S_1$, $\sigma'_2 = \sigma|S-S'_1$ and consider the decomposition

$$\phi\oplus 1 = \begin{bmatrix} \phi_{11} & \phi_{12} & 0 \\ 0 & \phi_{22} & 0 \\ 0 & 0 & 1 \end{bmatrix}\colon F(\sigma_1)\oplus F(\sigma_2)\oplus F(\rho) \longrightarrow F(\sigma'_1)\oplus F(\sigma'_2)\oplus F(\rho).$$

To finish the proof, let $\rho_1 = \sigma_1 \amalg \rho$, $\rho'_1 = \sigma'_1$, take

$$\varepsilon_1 = \begin{bmatrix} 1 & 0 & j \\ 0 & 1 & 0 \\ 0 & 0 & 1 \end{bmatrix}, \quad \varepsilon_2 = \begin{bmatrix} 1 & 0 & 0 \\ 0 & 1 & 0 \\ -\gamma & 0 & 1 \end{bmatrix},$$

and notice that

$$\varepsilon_2\varepsilon_1(\phi\oplus 1) = \begin{bmatrix} \phi_{11} & \phi_{12} & j \\ 0 & \phi_{22} & 0 \\ 0 & -\gamma\phi_{12} & 0 \end{bmatrix},$$

where $(\phi_{11},j)\colon F(\rho_1) \longrightarrow F(\rho'_1)$ is an isomorphism.

Proof of (4.13): Let $x = [F(\sigma),\phi] \in Wh\mathbb{Z}\mathcal{F}\hat{G}$ have $\ell c(x) = 0$. By Lemma 4.14, we can assume that there are essential submodules $F(\sigma_1)$ and $F(\sigma_1')$ of $F(\sigma)$ such that ϕ has the form

$$\phi = \begin{bmatrix} \phi_1 & \gamma \\ 0 & \phi_2 \end{bmatrix}: F(\sigma) = F(\sigma_1)\oplus F(\sigma_2) \longrightarrow F(\sigma_1')\oplus F(\sigma_2') = F(\sigma)$$

where ϕ_1, and hence also ϕ_2, is an isomorphism. By IV.2.6, there are isomorphisms of bases (α_i,ν_i): $(S_i',\sigma_i') \longrightarrow (S_i,\sigma_i)$, $(i = 1,2)$. We let (α,ν): $(S,\sigma) \longrightarrow (S,\sigma)$ be the disjoint union of (α_1,ν_1) and (α_2,ν_2). Then $x = [F(\sigma),\psi]$, where $\psi = F(\alpha,\nu)\phi$: $F(\sigma) \longrightarrow F(\sigma)$ has the form

$$\psi = \begin{bmatrix} \psi_1 & \gamma' \\ 0 & \psi_2 \end{bmatrix}: F(\sigma) = F(\sigma_1)\oplus F(\sigma_2) \longrightarrow F(\sigma_1)\oplus F(\sigma_2) = F(\sigma).$$

Hence $x = [F(\sigma_1),\psi_1] + [F(\sigma_2),\psi_2]$, and it suffices to show that each of the two terms vanishes.

For the second term, we note that $F(\sigma_2)$ is supported over $U_i[a_i-d,a_i+d]$ for some $d\geq 0$, and that $\hat{G}([a_i-d,a_i+d])$ vanishes. Therefore $[F(\sigma_2),\psi_2] \in \operatorname{Im} i_*$ where i: $\mathcal{F} \longrightarrow \mathcal{F}\hat{G}$ has $i(K\subseteq L) = (1,\text{incl})$: $K \longrightarrow L$. But $Wh\mathbb{Z}\mathcal{F} \cong Wh^c(\mathbb{R},1_{\mathbb{R}}) \cong \tilde{K}_0(\mathbb{Z}) = 0$ by VI.6.1 and 1.3.

To see that $[F(\sigma_1),\psi_1]$ vanishes, we use an infinite repetition trick adapted from Pedersen [Pe1] and dating back probably to Eilenberg. We can think of $F(\sigma_1)$ as a sequence of $\mathbb{Z}$-graded $\mathbb{Z}\pi_i$ modules $M_i = |F(\sigma_1)|_i$ where M_i vanishes in degrees outside (a_i,a_{i+1}). Also ψ_1 can be thought of as a sequence of bounded homomorphisms

$$\mu_i = |\psi_1|_i: |F(\sigma_1)|_i \longrightarrow |F(\sigma_1)|_i$$

all of which admit the same bound, say d. Likewise, all μ_i^{-1} can be assumed to be bounded by d. For each i, let $\beta(i)$ be the largest integer with $(M_i)_{\beta(i)} \neq 0$. We denote by $s(M_i)$ the $\mathbb{Z}$-graded $\mathbb{Z}\pi_i$ module having $(s(M_i))_\alpha = (M_i)_\alpha$ if $\alpha < \beta(i)-1$, $(s(M_i))_{\beta(i)-1} = (M_i)_{\beta(i)-1}\oplus(M_i)_{\beta(i)}$, and $s(M_i)_{\beta(i)} = 0$. Note that there is an obvious isomorphism s: $M_i \longrightarrow s(M_i)$ such that s and s^{-1} are bounded by 1. Moreover, if we let

$s(\mu_i) = s\mu_i s^{-1}$: $s(M_i) \longrightarrow s(M_i)$, then $s(\mu_i)$ and $s(\mu_i)^{-1}$ are bounded by d. We can iterate the construction s to get $s^j(M_i)$ for $j=1,2,\ldots,$ $\beta(i)-a_i-1 = \delta(i)$. Then the module $s^{\delta(i)}(M_i)$ is concentrated in degree a_i+1. We now stabilize $(F(\sigma_1),\psi_1)$ by means of $(F(\rho),\psi)$ constructed as follows. For each i we let $|F(\rho)|_i$ be the $\mathbb{Z}$-graded $\mathbb{Z}\pi_i$ module

$$\oplus_{j=1}^{\delta(i)}[s^j(M_i)\oplus s^j(M_i)]$$

and we let $|\psi|_i$: $|F(\rho)|_i \longrightarrow |F(\rho)|_i$ be given on $s^j(M_i)\oplus s^j(M_i)$ as $s^j(\mu_i^{-1})\oplus s^j(\mu_i)$. Since ψ has the form $\mu \oplus \mu^{-1}$, it follows easily from IV.3.2-3 that $[F(\rho),\psi] = 0$. On the other hand, we can also think of $|F(\sigma_1)\oplus F(\rho)|_i$ as

$$\oplus_{j=1}^{\delta(i)}[s^{j-1}(M_i)\oplus s^j(M_i)] \oplus s^{\delta(i)}(M_i).$$

Then $|\psi_1\oplus\psi|_i$ is given on $s^{j-1}(M_i)\oplus s^j(M_i)$ as $s^{j-1}(\mu_i)\oplus s^j(\mu_i^{-1})$. Since $s^{j-1}(\mu_i)$ is conjugate to $s^j(\mu_i)$, the above argument shows that $[F(\sigma_1)\oplus F(\rho),\psi_1\oplus\psi] = [F(\rho'),\psi']$, where $|F(\rho')|_i = s^{\delta(i)}(M_i)$ and $|\psi'|_i = s^{\delta(i)}(\mu_i)$. But $F(\rho')$ is supported on $\cup_i[a_i-1,a_i+1]$ so $[F(\rho'),\psi'] = 0$ by the argument in the preceding paragraph. This completes the proof of (4.13) and hence that of 4.5.

BIBLIOGRAPHY

[An] D.R. Anderson, *Geometric applications of lower algebraic K-theory* (Preliminary Report), Abstracts of the AMS, **4** (1983), 373.

[AH1] D.R. Anderson and W.C. Hsiang, *Extending Combinatorial PL Structures on Stratified Spaces*, Invent. Math. **32** (1976), 179-204.

[AH2] D.R. Anderson and W.C. Hsiang, *The Functors K_{-i} and Pseudo-isotopies of Polyhedra*, Ann. of Math. (2) **105** (1977), 201-223.

[AH3] D.R. Anderson and W.C. Hsiang, *Extending Combinatorial Piecewise Linear Structures on Stratified Spaces*, II, Trans. Amer. Math. Soc. **260** (1980), 223-253.

[AM1] D.R. Anderson and H.J. Munkholm, *The Algebraic Topology of Controlled Spaces*, Preprint, Odense University 1984.

[AM2] D.R. Anderson and H.J. Munkholm, *An Introduction to Boundedly Controlled Simple Homotopy Theory*, in Geometry and Topology (C. McCrory and T. Shifrin, eds.), Marcel Dekker, New York and Basel, 1987, 27-42.

[AM3] D.R. Anderson and H.J. Munkholm, *A Geometric Construction of the Boundedly Controlled Whitehead Group*, in Geometry and Topology, loc.cit., 13-26.

[Ba] H. Bass, *Algebraic K-theory*, W.A. Benjamin, New York 1968.

[BHS] H. Bass, A. Heller, and R.G. Swan, *The Whitehead group of a Polynomial Extension*, Publ. Math. IHES 22 (1964), 61-80.

[Br] R.A. Brualdi, *Transversal Theory and Graphs*, Studies in Graph Theory, Part 1 (D.R. Fulkerson, ed.), Mathematical Association of America, Washington, DC, 1975, 23-88.

[Ch] T.A. Chapman, *Controlled Simple Homotopy Theory and Applications*, Lecture Notes in Mathematics **1009**, Springer-Verlag, Berlin, Heidelberg, New York, 1983.

[CH] E.H. Connell and J. Hollingsworth, *Geometric Groups and Whitehead Torsion*, Trans. Amer. Math. Soc. **140** (1969), 161-181.

[Co] M.M. Cohen, *An Introduction to Simple Homotopy Theory*, Springer-Verlag, New York, Heidelberg, Berlin, 1973.

[EK] R.D. Edwards and R.C. Kirby, *Deformations of Spaces of Imbeddings*, Ann. of Math. (2), **93** (1971), 63-88.

[Fe] S.Ferry, *Homotoping ε-maps to Homeomorphisms*, Amer J. Math. **101** (1979), 567-582.

[FJ1] F.T. Farrell and L.E. Jones, *H-cobordisms with Foliated Control*, Bull. Amer. Math. Soc. **15** (1986), 69-72.

[FJ2] F.T. Farrell and L.E. Jones, *K-theory and Dynamics, I*, Ann. of Math. (2), **124** (1986), 531-569.

[Gr] D. Grayson, *Higher Algebraic K-Theory II* (after D. Quillen), Algebraic K-theory, Lecture Notes in Mathematics **551**, Springer-Verlag, Berlin, Heidelberg, New York, 1976, 217-240.

[Hu] J.F.P. Hudson, *Piecewise Linear Topology*, W.A. Benjamin, New York, 1968.

[Ki] R.C. Kirby, *Lectures on Triangulation of Manifolds*, Mimeo, UCLA, 1969.

[Ma] S. MacLane, *Categories for the Working Mathematician*, Springer-Verlag, New York, Heidelberg, Berlin, 1971.

[Mi1] J. Milnor, *Lectures on the h-cobordism Theorem*, Princeton Math. Notes, Princeton Univ. Press, Princeton, N.J., 1965.

[Mi2] J.W. Milnor, *Whitehead Torsion*, Bull. A.M.S. **72** (1966), 358-426.

[MA] H.J. Munkholm and D.R. Anderson, *Lower Simple-homotopy Theory, a Classical Approach*, Abstracts of the AMS, 5 (1984), 102.

[Pe1] E.K. Pedersen, *On the K_{-i} Functors*, J. of Alg., **90** (1984), 461-475.

[Pe2] E.K. Pedersen, *K_{-i}-Invariants of Chain Complexes*, Topology Proceedings, Leningrad, 1982, Lecture Notes in Math. **1060**, Springer-Verlag, Berlin, Heidelberg, New York, 1984, 174-186.

[Pe3] E.K. Pedersen, *On the Bounded and Thin h-Cobordism Theorem Parametrized over* $\mathbb{R}^k$, Transformation Groups, Poznan 1985, Lecture Notes in Math. **1217**, Springer-Verlag, Berlin, Heidelberg, New York, 1986, 306-319.

[PW1] E.K. Pedersen and C. Weibel, *A Nonconnective Delooping of Algebraic K-theory*, Algebraic and Geometric Topology, Lecture Notes in Math. **1126**, Springer-Verlag, Berlin, Heidelberg, New York, Tokyo, 1984, 166-181.

[PW2] E.K. Pedersen and C. Weibel, *K-theory Homology of Spaces*, preprint.

[Po] N. Popescu, *Abelian Categories with Applications to Rings and Modules*, Academic Press, London and New York, 1973.

[Q1] D. Quillen, *Higher Algebraic K-Theory I*, Algebraic K-Theory I, Lecture Notes in Mathematics **341**, Springer-Verlag, Berlin, Heidelberg, New York 1973, 85-147.

[Qn1] F. Quinn, *Ends of Maps, I*, Ann. of Math. **110** (1979), 275-331.

[Qn2] F. Quinn, *Ends of Maps, II*, Inv. Math. **68** (1982), 353-424.

[RS] C.P. Rourke and B.J. Sanderson, *Introduction to Piecewise-Linear Topology*, Springer-Verlag, Berlin, Heidelberg, New York, 1972.

[Sc] H. Schubert, *Categories* (E. Gray, translator), Springer-Verlag, Berlin, Heidelberg, New York, 1972.

[Si] L.C. Siebenmann, *Infinite Simple Homotopy Types*, Indag. Math. **32** (1970), 479-495.

[Sp] E.H. Spanier, *Algebraic Topology*, McGraw-Hill, New York, 1966.

[SW1] M. Steinberger and J. West, *Equivariant h-cobordisms and Finiteness Obstructions*, Bull. Amer. Math. Soc. **12** (1985), 217-220.

[SW2] M. Steinberger and J. West, *Equivariant Handles in Finite Group Actions*, in Geometry and Topology (C. McCrory and T. Shifrin, eds.), Marcel Dekker, New York and Basel, 1987, 277-295.

[Wh] G. Whitehead, *Elements of Homotopy Theory*, Springer-Verlag, New York, Heidelberg, Berlin, 1978.

INDEX

Vol. 1173: H. Delfs, M. Knebusch, Locally Semialgebraic Spaces. XVI, 329 pages. 1985.

Vol. 1174: Categories in Continuum Physics, Buffalo 1982. Seminar. Edited by F.W. Lawvere and S.H. Schanuel. V, 126 pages. 1986.

Vol. 1175: K. Mathiak, Valuations of Skew Fields and Projective Hjelmslev Spaces. VII, 116 pages. 1986.

Vol. 1176: R.R. Bruner, J.P. May, J.E. McClure, M. Steinberger, H_∞ Ring Spectra and their Applications. VII, 388 pages. 1986.

Vol. 1177: Representation Theory I. Finite Dimensional Algebras. Proceedings, 1984. Edited by V. Dlab, P. Gabriel and G. Michler. XV, 340 pages. 1986.

Vol. 1178: Representation Theory II. Groups and Orders. Proceedings, 1984. Edited by V. Dlab, P. Gabriel and G. Michler. XV, 370 pages. 1986.

Vol. 1179: Shi J.-Y. The Kazhdan-Lusztig Cells in Certain Affine Weyl Groups. X, 307 pages. 1986.

Vol. 1180: R. Carmona, H. Kesten, J.B. Walsh, École d'Été de Probabilités de Saint-Flour XIV – 1984. Édité par P.L. Hennequin. X, 438 pages. 1986.

Vol. 1181: Buildings and the Geometry of Diagrams, Como 1984. Seminar. Edited by L. Rosati. VII, 277 pages. 1986.

Vol. 1182: S. Shelah, Around Classification Theory of Models. VII, 279 pages. 1986.

Vol. 1183: Algebra, Algebraic Topology and their Interactions. Proceedings, 1983. Edited by J.-E. Roos. XI, 396 pages. 1986.

Vol. 1184: W. Arendt, A. Grabosch, G. Greiner, U. Groh, H.P. Lotz, U. Moustakas, R. Nagel, F. Neubrander, U. Schlotterbeck, One-parameter Semigroups of Positive Operators. Edited by R. Nagel. X, 460 pages. 1986.

Vol. 1185: Group Theory, Beijing 1984. Proceedings. Edited by Tuan H.F. V, 403 pages. 1986.

Vol. 1186: Lyapunov Exponents. Proceedings, 1984. Edited by L. Arnold and V. Wihstutz. VI, 374 pages. 1986.

Vol. 1187: Y. Diers, Categories of Boolean Sheaves of Simple Algebras. VI, 168 pages. 1986.

Vol. 1188: Fonctions de Plusieurs Variables Complexes V. Séminaire, 1979–85. Edité par François Norguet. VI, 306 pages. 1986.

Vol. 1189: J. Lukeš, J. Malý, L. Zajíček, Fine Topology Methods in Real Analysis and Potential Theory. X, 472 pages. 1986.

Vol. 1190: Optimization and Related Fields. Proceedings, 1984. Edited by R. Conti, E. De Giorgi and F. Giannessi. VIII, 419 pages. 1986.

Vol. 1191: A.R. Its, V.Yu. Novokshenov, The Isomonodromic Deformation Method in the Theory of Painlevé Equations. IV, 313 pages. 1986.

Vol. 1192: Equadiff 6. Proceedings, 1985. Edited by J. Vosmansky and M. Zlámal. XXIII, 404 pages. 1986.

Vol. 1193: Geometrical and Statistical Aspects of Probability in Banach Spaces. Proceedings, 1985. Edited by X. Femique, B. Heinkel, M.B. Marcus and P.A. Meyer. IV, 128 pages. 1986.

Vol. 1194: Complex Analysis and Algebraic Geometry. Proceedings, 1985. Edited by H. Grauert. VI, 235 pages. 1986.

Vol. 1195: J.M. Barbosa, A.G. Colares, Minimal Surfaces in $\mathbb{R}^3$. X, 124 pages. 1986.

Vol. 1196: E. Casas-Alvero, S. Xambó-Descamps, The Enumerative Theory of Conics after Halphen. IX, 130 pages. 1986.

Vol. 1197: Ring Theory. Proceedings, 1985. Edited by F.M.J. van Oystaeyen. V, 231 pages. 1986.

Vol. 1198: Séminaire d'Analyse, P. Lelong – P. Dolbeault – H. Skoda. Seminar 1983/84. X, 260 pages. 1986.

Vol. 1199: Analytic Theory of Continued Fractions II. Proceedings, 1985. Edited by W.J. Thron. VI, 299 pages. 1986.

Vol. 1200: V.D. Milman, G. Schechtman, Asymptotic Theory of Finite Dimensional Normed Spaces. With an Appendix by M. Gromov. VIII, 156 pages. 1986.

Vol. 1201: Curvature and Topology of Riemannian Manifolds. Proceedings, 1985. Edited by K. Shiohama, T. Sakai and T. Sunada. VII, 336 pages. 1986.

Vol. 1202: A. Dür, Möbius Functions, Incidence Algebras and Power Series Representations. XI, 134 pages. 1986.

Vol. 1203: Stochastic Processes and Their Applications. Proceedings, 1985. Edited by K. Itô and T. Hida. VI, 222 pages. 1986.

Vol. 1204: Séminaire de Probabilités XX, 1984/85. Proceedings. Edité par J. Azéma et M. Yor. V, 639 pages. 1986.

Vol. 1205: B.Z. Moroz, Analytic Arithmetic in Algebraic Number Fields. VII, 177 pages. 1986.

Vol. 1206: Probability and Analysis, Varenna (Como) 1985. Seminar. Edited by G. Letta and M. Pratelli. VIII, 280 pages. 1986.

Vol. 1207: P.H. Bérard, Spectral Geometry: Direct and Inverse Problems. With an Appendix by G. Besson. XIII, 272 pages. 1986.

Vol. 1208: S. Kaijser, J.W. Pelletier, Interpolation Functors and Duality. IV, 167 pages. 1986.

Vol. 1209: Differential Geometry, Peñíscola 1985. Proceedings. Edited by A.M. Naveira, A. Ferrández and F. Mascaró. VIII, 306 pages. 1986.

Vol. 1210: Probability Measures on Groups VIII. Proceedings, 1985. Edited by H. Heyer. X, 386 pages. 1986.

Vol. 1211: M.B. Sevryuk, Reversible Systems. V, 319 pages. 1986.

Vol. 1212: Stochastic Spatial Processes. Proceedings, 1984. Edited by P. Tautu. VIII, 311 pages. 1986.

Vol. 1213: L.G. Lewis, Jr., J.P. May, M. Steinberger, Equivariant Stable Homotopy Theory. IX, 538 pages. 1986.

Vol. 1214: Global Analysis – Studies and Applications II. Edited by Yu.G. Borisovich and Yu.E. Gliklikh. V, 275 pages. 1986.

Vol. 1215: Lectures in Probability and Statistics. Edited by G. del Pino and R. Rebolledo. V, 491 pages. 1986.

Vol. 1216: J. Kogan, Bifurcation of Extremals in Optimal Control. VIII, 106 pages. 1986.

Vol. 1217: Transformation Groups. Proceedings, 1985. Edited by S. Jackowski and K. Pawalowski. X, 396 pages. 1986.

Vol. 1218: Schrödinger Operators, Aarhus 1985. Seminar. Edited by E. Balslev. V, 222 pages. 1986.

Vol. 1219: R. Weissauer, Stabile Modulformen und Eisensteinreihen. III, 147 Seiten. 1986.

Vol. 1220: Séminaire d'Algèbre Paul Dubreil et Marie-Paule Malliavin. Proceedings, 1985. Edité par M.-P. Malliavin. IV, 200 pages. 1986.

Vol. 1221: Probability and Banach Spaces. Proceedings, 1985. Edited by J. Bastero and M. San Miguel. XI, 222 pages. 1986.

Vol. 1222: A. Katok, J.-M. Strelcyn, with the collaboration of F. Ledrappier and F. Przytycki, Invariant Manifolds, Entropy and Billiards; Smooth Maps with Singularities. VIII, 283 pages. 1986.

Vol. 1223: Differential Equations in Banach Spaces. Proceedings, 1985. Edited by A. Favini and E. Obrecht. VIII, 299 pages. 1986.

Vol. 1224: Nonlinear Diffusion Problems, Montecatini Terme 1985. Seminar. Edited by A. Fasano and M. Primicerio. VIII, 188 pages. 1986.

Vol. 1225: Inverse Problems, Montecatini Terme 1986. Seminar. Edited by G. Talenti. VIII, 204 pages. 1986.

Vol. 1226: A. Buium, Differential Function Fields and Moduli of Algebraic Varieties. IX, 146 pages. 1986.

Vol. 1227: H. Helson, The Spectral Theorem. VI, 104 pages. 1986.

Vol. 1228: Multigrid Methods II. Proceedings, 1985. Edited by W. Hackbusch and U. Trottenberg. VI, 336 pages. 1986.

Vol. 1229: O. Bratteli, Derivations, Dissipations and Group Actions on C*-algebras. IV, 277 pages. 1986.

Vol. 1230: Numerical Analysis. Proceedings, 1984. Edited by J.-P. Hennart. X, 234 pages. 1986.

Vol. 1231: E.-U. Gekeler, Drinfeld Modular Curves. XIV, 107 pages. 1986.

Vol. 1232: P.C. Schuur, Asymptotic Analysis of Soliton Problems. VIII, 180 pages. 1986.

Vol. 1233: Stability Problems for Stochastic Models. Proceedings, 1985. Edited by V.V. Kalashnikov, B. Penkov and V.M. Zolotarev. VI, 223 pages. 1986.

Vol. 1234: Combinatoire énumérative. Proceedings, 1985. Edité par G. Labelle et P. Leroux. XIV, 387 pages. 1986.

Vol. 1235: Séminaire de Théorie du Potentiel, Paris, No. 8. Directeurs: M. Brelot, G. Choquet et J. Deny. Rédacteurs: F. Hirsch et G. Mokobodzki. III, 209 pages. 1987.

Vol. 1236: Stochastic Partial Differential Equations and Applications. Proceedings, 1985. Edited by G. Da Prato and L. Tubaro. V, 257 pages. 1987.

Vol. 1237: Rational Approximation and its Applications in Mathematics and Physics. Proceedings, 1985. Edited by J. Gilewicz, M. Pindor and W. Siemaszko. XII, 350 pages. 1987.

Vol. 1238: M. Holz, K.-P. Podewski and K. Steffens, Injective Choice Functions. VI, 183 pages. 1987.

Vol. 1239: P. Vojta, Diophantine Approximations and Value Distribution Theory. X, 132 pages. 1987.

Vol. 1240: Number Theory, New York 1984–85. Seminar. Edited by D.V. Chudnovsky, G.V. Chudnovsky, H. Cohn and M.B. Nathanson. V, 324 pages. 1987.

Vol. 1241: L. Gårding, Singularities in Linear Wave Propagation. III, 125 pages. 1987.

Vol. 1242: Functional Analysis II, with Contributions by J. Hoffmann-Jørgensen et al. Edited by S. Kurepa, H. Kraljević and D. Butković. VII, 432 pages. 1987.

Vol. 1243: Non Commutative Harmonic Analysis and Lie Groups. Proceedings, 1985. Edited by J. Carmona, P. Delorme and M. Vergne. V, 309 pages. 1987.

Vol. 1244: W. Müller, Manifolds with Cusps of Rank One. XI, 158 pages. 1987.

Vol. 1245: S. Rallis, L-Functions and the Oscillator Representation. XVI, 239 pages. 1987.

Vol. 1246: Hodge Theory. Proceedings, 1985. Edited by E. Cattani, F. Guillén, A. Kaplan and F. Puerta. VII, 175 pages. 1987.

Vol. 1247: Séminaire de Probabilités XXI. Proceedings. Edité par J. Azéma, P.A. Meyer et M. Yor. IV, 579 pages. 1987.

Vol. 1248: Nonlinear Semigroups, Partial Differential Equations and Attractors. Proceedings, 1985. Edited by T.L. Gill and W.W. Zachary. IX, 185 pages. 1987.

Vol. 1249: I. van den Berg, Nonstandard Asymptotic Analysis. IX, 187 pages. 1987.

Vol. 1250: Stochastic Processes – Mathematics and Physics II. Proceedings 1985. Edited by S. Albeverio, Ph. Blanchard and L. Streit. VI, 359 pages. 1987.

Vol. 1251: Differential Geometric Methods in Mathematical Physics. Proceedings, 1985. Edited by P.L. García and A. Pérez-Rendón. VII, 300 pages. 1987.

Vol. 1252: T. Kaise, Représentations de Weil et GL_2 Algèbres de division et GL_n. VII, 203 pages. 1987.

Vol. 1253: J. Fischer, An Approach to the Selberg Trace Formula via the Selberg Zeta-Function. III, 184 pages. 1987.

Vol. 1254: S. Gelbart, I. Piatetski-Shapiro, S. Rallis. Explicit Constructions of Automorphic L-Functions. VI, 152 pages. 1987.

Vol. 1255: Differential Geometry and Differential Equations. Proceedings, 1985. Edited by C. Gu, M. Berger and R.L. Bryant. XII, 243 pages. 1987.

Vol. 1256: Pseudo-Differential Operators. Proceedings, 1986. Edited by H.O. Cordes, B. Gramsch and H. Widom. X, 479 pages. 1987.

Vol. 1257: X. Wang, On the C*-Algebras of Foliations in the Plane. V, 165 pages. 1987.

Vol. 1258: J. Weidmann, Spectral Theory of Ordinary Differential Operators. VI, 303 pages. 1987.

Vol. 1259: F. Cano Torres, Desingularization Strategies for Three-Dimensional Vector Fields. IX, 189 pages. 1987.

Vol. 1260: N.H. Pavel, Nonlinear Evolution Operators and Semigroups. VI, 285 pages. 1987.

Vol. 1261: H. Abels, Finite Presentability of S-Arithmetic Groups. Compact Presentability of Solvable Groups. VI, 178 pages. 1987.

Vol. 1262: E. Hlawka (Hrsg.), Zahlentheoretische Analysis II. Seminar, 1984–86. V, 158 Seiten. 1987.

Vol. 1263: V.L. Hansen (Ed.), Differential Geometry. Proceedings, 1985. XI, 288 pages. 1987.

Vol. 1264: Wu Wen-tsün, Rational Homotopy Type. VIII, 219 pages. 1987.

Vol. 1265: W. Van Assche, Asymptotics for Orthogonal Polynomials. VI, 201 pages. 1987.

Vol. 1266: F. Ghione, C. Peskine, E. Sernesi (Eds.), Space Curves. Proceedings, 1985. VI, 272 pages. 1987.

Vol. 1267: J. Lindenstrauss, V.D. Milman (Eds.), Geometrical Aspects of Functional Analysis. Seminar. VII, 212 pages. 1987.

Vol. 1268: S.G. Krantz (Ed.), Complex Analysis. Seminar, 1986. VII, 195 pages. 1987.

Vol. 1269: M. Shiota, Nash Manifolds. VI, 223 pages. 1987.

Vol. 1270: C. Carasso, P.-A. Raviart, D. Serre (Eds.), Nonlinear Hyperbolic Problems. Proceedings, 1986. XV, 341 pages. 1987.

Vol. 1271: A.M. Cohen, W.H. Hesselink, W.L.J. van der Kallen, J.R. Strooker (Eds.), Algebraic Groups Utrecht 1986. Proceedings. XII, 284 pages. 1987.

Vol. 1272: M.S. Livšic, L.L. Waksman, Commuting Nonselfadjoint Operators in Hilbert Space. III, 115 pages. 1987.

Vol. 1273: G.-M. Greuel, G. Trautmann (Eds.), Singularities, Representation of Algebras, and Vector Bundles. Proceedings, 1985. XIV, 383 pages. 1987.

Vol. 1274: N. C. Phillips, Equivariant K-Theory and Freeness of Group Actions on C*-Algebras. VIII, 371 pages. 1987.

Vol. 1275: C.A. Berenstein (Ed.), Complex Analysis I. Proceedings, 1985–86. XV, 331 pages. 1987.

Vol. 1276: C.A. Berenstein (Ed.), Complex Analysis II. Proceedings, 1985–86. IX, 320 pages. 1987.

Vol. 1277: C.A. Berenstein (Ed.), Complex Analysis III. Proceedings, 1985–86. X, 350 pages. 1987.

Vol. 1278: S.S. Koh (Ed.), Invariant Theory. Proceedings, 1985. V, 102 pages. 1987.

Vol. 1279: D. Ieşan, Saint-Venant's Problem. VIII, 162 Seiten. 1987.

Vol. 1280: E. Neher, Jordan Triple Systems by the Grid Approach. XII, 193 pages. 1987.

Vol. 1281: O.H. Kegel, F. Menegazzo, G. Zacher (Eds.), Group Theory. Proceedings, 1986. VII, 179 pages. 1987.

Vol. 1282: D.E. Handelman, Positive Polynomials, Convex Integral Polytopes, and a Random Walk Problem. XI, 136 pages. 1987.

Vol. 1283: S. Mardešić, J. Segal (Eds.), Geometric Topology and Shape Theory. Proceedings, 1986. V, 261 pages. 1987.

Vol. 1284: B.H. Matzat, Konstruktive Galoistheorie. X, 286 pages. 1987.

Vol. 1285: I.W. Knowles, Y. Saitō (Eds.), Differential Equations and Mathematical Physics. Proceedings, 1986. XVI, 499 pages. 1987.

Vol. 1286: H.R. Miller, D.C. Ravenel (Eds.), Algebraic Topology. Proceedings, 1986. VII, 341 pages. 1987.

Vol. 1287: E.B. Saff (Ed.), Approximation Theory, Tampa. Proceedings, 1985–1986. V, 228 pages. 1987.

Vol. 1288: Yu. L. Rodin, Generalized Analytic Functions on Riemann Surfaces. V, 128 pages, 1987.

Vol. 1289: Yu. I. Manin (Ed.), K-Theory, Arithmetic and Geometry. Seminar, 1984–1986. V, 399 pages. 1987.